Olympic Games, Mega-Events and Civil Societies

Global Culture and Sport

Series Editors: **Stephen Wagg** and **David Andrews**

Titles include:

Mahfoud Amara
SPORT, POLITICS AND SOCIETY IN THE ARAB WORLD

John Harris
RUGBY UNION AND GLOBALIZATION
An Odd-Shaped World

Graeme Hayes and John Karamichas (*editors*)
OLYMPIC GAMES, MEGA-EVENTS AND CIVIL SOCIETIES
Globalization, Environment, Resistance

Roger Levermore and Aaron Beacom (*editors*)
SPORT AND INTERNATIONAL DEVELOPMENT

Jonathan Long and Karl Spracklen (*editors*)
SPORT AND CHALLENGES TO RACISM

Pirkko Markula (*editor*)
OLYMPIC WOMEN AND THE MEDIA
International Perspectives

Peter Millward
THE GLOBAL FOOTBALL LEAGUE
Transnational Networks, Social Movements and Sport in the New Media Age

Global Culture and Sport
Series Standing Order ISBN 978–0–230–57818–0 hardback
   978–0–230–57819–7 paperback
(*outside North America only*)

You can receive future titles in this series as they are published by placing a standing order. Please contact your bookseller or, in case of difficulty, write to us at the address below with your name and address, the title of the series and one of the ISBNs quoted above.

Customer Services Department, Macmillan Distribution Ltd, Houndmills, Basingstoke, Hampshire RG21 6XS, England

Olympic Games, Mega-Events and Civil Societies

Globalization, Environment, Resistance

Edited By

Graeme Hayes
Aston University, UK

and

John Karamichas
Queen's University Belfast, UK

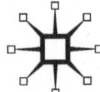

First published 2012 by
PALGRAVE MACMILLAN

Palgrave Macmillan in the UK is an imprint of Macmillan Publishers Limited,
registered in England, company number 785998, of Houndmills, Basingstoke,
Hampshire RG21 6XS.

Palgrave Macmillan in the US is a division of St Martin's Press LLC,
175 Fifth Avenue, New York, NY 10010.

Palgrave Macmillan is the global academic imprint of the above companies
and has companies and representatives throughout the world.

Palgrave® and Macmillan® are registered trademarks in the United States,
the United Kingdom, Europe and other countries.

ISBN 978-1-349-31881-0   ISBN 978-0-230-35918-5 (eBook)
DOI 10.1057/9780230359185

This book is printed on paper suitable for recycling and made from fully
managed and sustained forest sources. Logging, pulping and manufacturing
processes are expected to conform to the environmental regulations of the
country of origin.

A catalogue record for this book is available from the British Library.

A catalog record for this book is available from the Library of Congress.

10 9 8 7 6 5 4 3 2 1
21 20 19 18 17 16 15 14 13 12

Contents

Tables

Figures

Abbreviations

ALP	Australian Labour Party
BC	British Columbia
BIE	Bureau International des Expositions
BOCOG	Beijing Organizing Committee for the Olympic Games
CCP	Chinese Communist Party
CNOSF	Comité National Olympique et Sportif Français
CO_2e	carbon dioxide equivalent, including CO_2 and other greenhouse gases
COJO	Comité d'Organisation des Jeux Olympiques
EIA	Environmental Impact Assessment
EIS	Environmental Impact Statement
EM	ecological modernization
EMS	Environmental Management System
FIFA	Fédération Internationale de Football Association
FINA	Fédération Internationale de Natation
GHG	greenhouse gases
ICC	International Cricket Council
IOC	International Olympic Committee
IRB	International Rugby Board
LOCOG	London Organizing Committee for the Olympic Games
MCC	manual for candidate cities
NGO	non-governmental organization
NO_2	nitrogen dioxide
OAP	Olympic Action Plan
OCA	Olympic Coordination Authority
PASOK	Panhellenic Socialist Movement
SEA	Strategic Environmental Assessment
SEPC	Shanghai Environmental Protection Committee
SOCOG	Sydney Organizing Committee for the Olympic Games
TOROC	Turin Organizing Committee for the Olympic Games
UEFA	Union of European Football Associations
UNDP	United Nations Development Programme
UNEP	United Nations Environmental Programme
VANOC	Vancouver Organizing Committee for the 2010 Olympic Games
WWF	Worldwide Fund for Nature

Acknowledgements

The genesis of this book can be traced back to the identification of a shared interest in mega-event politics between the two editors, through their interaction in the ECPR Green Politics Standing Group online message board. For us, sports mega-events are a sphere where our main research interests, civil contestation (protest and social movements) and environmental sociology and politics, coalesce.

In 2007 we directed a session on 'Mega/Hallmark Events and the Environment' for the Environment and Society Research Network of the European Sociological Association at the 8th ESA Conference in Glasgow, from 3–6 September. That experience reinforced our commitment in further pursuing and developing our interests by bringing together, under the same roof, scholars from various disciplinary specializations who have produced work in those areas. The end result was a two-day workshop, in June 2008, at Queen Mary, University of London. The papers presented at that workshop form the basis of the chapters that make up the greater part of this volume. The financial sponsorship provided by the School of Sociology, Social Policy and Social Work at Queen's University Belfast and the invaluable hospitality and support offered to us by Queen Mary's Department of Politics in hosting that event are gratefully acknowledged.

We also wish to thank Philippa Grand and Olivia Middleton at Palgrave, and Adam Fagan at Queen Mary, in particular, for their invaluable help and encouragement at various phases of this project.

Graeme Hayes
John Karamichas

Contributors

Anne-Marie Broudehoux is Associate Professor in the School of Design, Université du Québec à Montréal (UQAM), Canada. She is the author of *The Making and Selling of Post-Mao Beijing* (2004), which received the International Planning History Society book prize in 2006.

Pietro Caratti works for the Presidency of the Lombardy Region, Italy, and was Senior Research Fellow at FEEM (Fondazione Eni Enrico Mattei). He is the author of numerous publications in the areas of environment and sustainability, including *Analysing Strategic Environmental Assessment* (2004) and *Development and Environmental Protection in Mega Events: The Case of the Turin 2006 Olympics* (2006).

Egidio Dansero is Associate Professor in the Faculty of Political Science at the University of Turin, Italy, where he teaches political and economic geography, and environmental policy. He is a founder member of the University of Turin's OMERO (Olympics and Mega Events Research Observatory) group. His international publications include articles in *Leisure Studies, Journal of Environmental Planning and Management* and *Geojournal*.

Hugh Dauncey is Senior Lecturer in French Studies at Newcastle University, UK, where he works on popular culture in France. He is co-editor of *France and the 1998 World Cup: The National Impact of a World Sporting Event* (1999) and *The Tour de France 1903–2003: A Century of Sporting Structures, Meanings and Values* (2003, both with Geoff Hare), and *Stereo: Comparative Perspectives on the Sociological Study of Popular Music in France and Britain* (2011, with Ph. Le Guern). He is currently finishing a monograph on the social history of French cycling.

Barbara Del Corpo is a graduate in International Relations of the Faculty of Political Science at the University of Turin, Italy, and is a collaborator with the university's OMERO (Olympics and Mega Events Research Observatory) group. She published an article on the Turin Games in *Bollettino della Società Geografica Italiana* (2006).

Ludovico Ferraguto is Research Fellow at the Rome-based I-COM, where his research focuses on risk management and public consensus for mega-projects in the field of energy and environment. He is a contributor to Italian specialized magazines such as *Staffetta Quotidiana* and *Lavoce.info*.

John Horne is Professor of Sport and Sociology in the School of Sport, Tourism and the Outdoors at the University of Central Lancashire, UK, where he is Director of the International Research Institute for Sport Studies

(IRISS). He is currently Managing Editor in Chief of the journal *Leisure Studies* and a member of the editorial boards of the *International Review for the Sociology of Sport* and *Sport in Society*. His publications include, as author, *Sport in Consumer Culture* (2006); co-author, *Understanding Sport* (1999, with Alan Tomlinson); and co-editor, *Sports Mega-Events* (2006) and *Japan, Korea and the 2002 World Cup* (2002, both with Wolfram Manzenreiter).

Graeme Hayes is Senior Lecturer in the School of Languages and Social Sciences at Aston University, UK, and Marie Curie fellow, CNRS, in the Centre de Recherches sur l'Action Politique en Europe (CRAPE) at the Institut d'Etudes Politiques in Rennes, France. He is Editor in Chief of *Social Movement Studies*, author of *Environmental Protest and the State in France* (2002) and co-editor of *Cinéma et engagement* (2005, with Martin O'Shaughnessy). He is currently working with Sylvie Ollitrault on a book titled *La Désobéissance Civile* (forthcoming).

John Karamichas is Lecturer in Sociology at Queen's University, Belfast, UK. He has recently published articles on environmental politics and policy in the *Journal of Modern Greek Studies, South European Society and Politics, European Environment, Human Ecology Review* and *Environmental Politics*.

Alfredo Mela is Professor in the Faculty of Architecture at the Politecnico di Torino, Italy, where he teaches urban sociology and environmental sociology. He is founder member of the University of Turin's OMERO (Olympics and Mega Events Research Observatory) group. He is the author of *Sociologia delle Città* (2006).

Arthur P. J. Mol is Chair and Professor of Environmental Policy at Wageningen University, the Netherlands, and Professor of Environmental Policy at Renmin University, China. His research interests are in social theory and the environment, globalization, informational governance, greening production and consumption, and environment and development. His latest books are *Environmental Reform in the Information Age* (2008) and *The Ecological Modernisation Reader* (2009, co-edited with Gert Spaargaren and David Sonnenfeld).

Jean-François Polo is Associate Professor in the Institut d'Etudes Politiques, Rennes, France, and member of the Centre de Recherches sur l'Action Politique en Europe. He has contributed a chapter to an edited volume titled *L'Europe telle qu'elle se fait: Européanisation et Sociétés Politiques Nationales* (2007).

Xavier Renou is an activist and researcher. He is the founder of *Les Désobéissants*, an informal network of activists, based in France, and a trainer in techniques of non-violent direct action. He is the author of *La Privatisation de la Violence* (2006), *Petit Manuel de Désobéissance Civile* (2009) and is series editor of *Désobéir*, short handbooks on civil disobedience.

Irene Ropolo is a freelance sociologist working on local development and participatory planning projects and processes at local and supra-local levels. She is a collaborator with the University of Turin's OMERO (Olympics and Mega Events Research Observatory) group.

David Whitson is Professor in the Department of Political Science at the University of Alberta, Edmonton, Canada. He is co-author of *Hockey Night in Canada: Sport, Identities & Cultural Politics* (1993, with Rick Gruneau) and of *The Game Planners: Transforming Canada's Sport System* (1990, with Don Macintosh). He has also written articles on the hosting of major sport events in *Third World Quarterly* and *Sociology of Sport Journal*.

Lei Zhang is Assistant Professor in the environmental policy research group at Wageningen University, The Netherlands. Her recent publications include articles on environmental policy in China in *Environmental Politics* and the *Journal of Contemporary China*.

1
Introduction: Sports Mega-Events, Sustainable Development and Civil Societies

Graeme Hayes and John Karamichas

On 24 March 2008, at Olympia, the site of the ancient Olympic Games in Greece, a ceremony was held to mark the quadrennial ceremonial lighting of the Olympic torch. The event itself promised added value as a media spectacle for reasons beyond the symbolism of Olympic pageantry. The Olympic host nation, China, was subject to widespread criticism for its human rights record in general, and its violent repression of protest in Tibet in particular; the global media were expecting to capture the possible hijacking of the day's events by pro-Tibet campaigners. A year earlier, Greece had experienced the most devastating forest fires in its modern history; the last-minute salvation of the world heritage site of Olympia, as the flames had already entered the site, provided an opportunity for the Greek government to demonstrate the country's symbolic survival. Notwithstanding the extensive security operation that was mounted on the day by the authorities, three members of the Paris-based media advocacy group Reporters sans Frontières (Reporters Without Borders) managed to evade security and disrupt the speech of BOCOG (Beijing Organizing Committee for the Olympic Games) president (and Beijing Communist Party Secretary) Liu Qi and were arrested as they were about to unfurl a banner representing the Olympic Rings as handcuffs. Greek and Chinese state media acted promptly: live television coverage suddenly cut away to carefully selected footage of the ancient landscape. No footage of the incident was shown in the live transmission by either the Greek or Chinese state-run TV channels. 'If the Olympic flame is sacred, human rights are even more so', Reporters sans Frontières said in a statement. 'We cannot let the Chinese government seize the Olympic flame, a symbol of peace, without denouncing the dramatic situation of human rights in the country.' 'It's always sad when there are protests. But they were not violent and I think that's the important thing', countered IOC (International Olympic Committee) president Jacques Rogge (both quoted in *The Guardian* 24 March 2008).

1

This volume is about sports mega-events; their social, political and cultural characters; the value systems that they inscribe and draw on; the claims they make on us and the claims the organizers make for them; the spatial and ethical relationships they create; and the responses of civil societies to them. Our premise is that sports mega-events – in Maurice Roche's now familiar formulation, 'large-scale [...] events, which have a dramatic character, mass popular appeal and international significance' (2000: 1), and which generally encompass Olympic and Commonwealth Games, FIFA (Fédération Internationale de Football Association) World Cups and UEFA (Union of European Football Associations) Championships (but also perhaps, in some observers' eyes, rugby and cricket world cups, and other recurrent multi-national sports or multi-sports tournaments) – are not simply sporting or cultural phenomena. They are also political and economic events, characterized by the generation and projection of symbolic meanings – most obviously over the nature of statehood, economic power and collective cultural identity – and by social conflict, especially over land use, and over the extent and contours of public spending commitments. Because of their peculiar spatial and temporal organization, they raise questions about the relationships between global cultural and economic flows and particular local and national spaces; because of what Hiller terms the 'phases in their evolution' (2000: 192), or not simply the time of the event itself but those of pre-event bid construction, tournament implementation and post-event 'legacy', they ask us to consider the effects of the event on the long-term direction, implementation and consequences of public policy. Most fundamentally, sports mega-events interest us because of their capacity to reveal the orientation of national and global political systems and processes, and the ideological assumptions and operations that underpin them.

This volume is designed to fill a major lacuna in the literature on sports mega-events. Given the size of their global audience and their political, economic and cultural importance, it is perhaps unsurprising that mega-events have begun to generate an impressive academic literature, particularly in the fields of tourism and leisure studies, business studies, urban studies, media and communication studies and sports sociology. Yet despite (as we shall see) the dominant trends and seemingly inherently controversial nature of the conditions under which sports mega-events are staged, they have given rise to relatively little in the way of analysis addressing the importance of globalization, environmental performance, claims to sustainable development, and social and civic responses from either sociology or political science. Such analysis and discussion are particularly important given the nature of pre-event claims made by metropolitan political, business and media ('booster') elites. Typically, these claims are designed to convince domestic publics and international regulatory bodies, such as UEFA, FIFA and the IOC, of the necessity of hosting sports mega-events for the delivery of desirable urban policy goals, conceived in terms of the creation of post-event legacies. Increasingly, these projected legacies take the form of large-scale, top-down, transformative urban projects: new housing; leisure, corporate and retail construction projects; the development of transport infrastructures; the implementation of sustainable development and environmental best practice programmes; and so on.

This book therefore focuses on a series of specific characteristics of these events, characteristics which appear to us to be increasingly central to their staging and design, and of our understanding of their function. These are the questions of globalization, be it political, economic or cultural, and particularly in its the neo-liberal guise, and the effects of mega-events on urban infrastructural development; of the increasingly corporate nature of sports mega-events, and their consequent social impacts; of the role of mega-events in showcasing and promoting sustainable development programmes, but also the impacts of mega-events on the physical environment; of their elite nature, and of the relationships between political elites and publics; and finally, especially given their promotion as popular cultural celebrations, of the nature of democratic participation in their design, and the subsequent responses of civil societies to mega-events. The contributors to this volume come from different academic disciplines – from sociology and from political science most obviously, but also from architecture and design, from management and urban studies, not to mention from social movements themselves – and we anticipate that the readers of this volume will likewise have backgrounds in different disciplines and sub-disciplines – comparative politics, environmental politics, political sociology, sports history, sports sociology, cultural studies and so on. We also hope this book will appeal to the general lay reader. Given the heterogeneous nature of the volume's audience, the aim of this introduction is to give a flavour of the nature of the different contributions the book brings together, and to establish the main developments and debates which the subsequent chapters will discuss. The introduction has therefore been divided into three sections: mega-events and globalization; mega-events, sustainable development and environmental politics; the corporate character and social impacts of mega-events, and civic responses and resistances to them. The volume as a whole replicates this structure.

Mega-events and globalization

For many observers, sports mega-events seem to crystallize, or reveal, the processes at the heart of contemporary globalization. Anthony Giddens, for example, in his foundational *Sociology* textbook, introduces his discussion of globalization by pointing to the FIFA World Cup as a key example of the globalizing effects of information and communication technologies (2006: 50), whilst Richard Giulianotti, introducing his selection of key sociological writings on sport, underlines that modern sport 'illustrates *par excellence* the globalization of cultural practices and social relations' (2005: xvii). That sports mega-events are global events barely needs substantiation. This is central to the legitimacy of such events and is most obviously cast as a function of two factors. First, the geographical representativity of the organizations which administer them: FIFA boasts 208 national member Football Associations; the IOC, whose most distinctive symbol represents the union of athletes from five continents, federates 205 National Olympic Committees (NOCs). In contrast, the United Nations can claim only 192 member states. And second, the size of the global television audience which tunes into them. Each of the last Summer Olympics have been broadcast to 220 national territories. The IOC's sales of broadcast rights generated a revenue of USD 2570 million for the four-year period up to and including the Beijing Games (2005–8) – double the revenue generated by such sales for the four years to Atlanta (1993–6), and three times the size of the revenue generated by its lucrative commercial sponsorship programme (see below) (IOC 2010: 6, 24). Figures for the FIFA World Cup Finals reveal similar patterns: the 2006 Finals, held in Germany, were broadcast to 214 national territories, clocking up a total broadcasting time of over 73,000 hours – five times the figure for Italia 1990 (FIFA 2007a). It is important to stress that without global real-time television broadcasting there would be no sports mega-event.

The relationships between broadcasters, sponsors, organizers and public authorities are a pre-condition for the discussion on globalization which this volume develops. By globalization, we primarily mean epochal transformations in the contours and strength of political, cultural and economic integration at the international level, where 'the extensive reach of global relations and networks is matched by their relative high intensity, high velocity and high impact across many facets of social life' (Held & McGrew 2002: 2). Our contention is that sports mega-events are intimately linked to the globalizing processes of neo-liberalism, defined by David Harvey as 'a theory of political economic practices that proposes that human well-being can best be advanced by liberating individual entrepreneurial freedoms and skills within an institutional framework characterized by strong private property rights, free markets and free trade', and whose hallmarks are privatization, deregulation and the withdrawal of the state from its established welfare functions (2005: 2–3). The role of the state in this model is to free private actors, who in turn are 'the prime movers behind globalization, integrating markets and societies, breaking the constraints of space and time, and erasing local variations' (Lake 1999: 43). Beyond their capacity to showcase high-level international sport to global audiences, sports mega-events are promoted by corporate, media and political elites concerned with the development and extension of consumption-driven market opportunities. This is visible through the regulatory and organizational structures of their staging as an event, through the decisional processes by which the event is

awarded and defined, and through the patterns of infrastructural development apparently demanded by their staging. It signals, also, the shift of
sports mega-events from inter-state politics to inter-urban politics. A defining (though, as a number of contributors to this volume argue, heavily
nuanced) feature of Olympic Games over the past two decades (but also,
though to a much lesser extent, of World Cups) has been a shift in the locus
of the sports mega-event from the national to the metropolitan, with cities
instrumentalizing mega-events within their renewal and re-imaging strategies, competing for capital, prestige, reputation and visibility in the global
inter-city marketplace (Shoval 2002: 584–5, Greene 2003: 166). In the terms
set out by Saskia Sassen, host cities are increasingly *global cities*, 'a space of
power that contains the capabilities needed for the global operation of firms
and markets' (2007: 23–4), vying with each other for competitive leverage
through the visibility of event hosting.

Beyond the internationalization of sport, and the geographical coverage
of the international sporting bodies which host the competitions, we are
thus interested in the relationships between sports mega-events and three
features of what we can roughly call late modern capitalist societies. These
are the development of global governance regimes by non-state actors, and
their attendant capacity for the diffusion of norms and regulations affecting
public policy goals, particularly those articulated around notions of sustainable development; the relationships between sports mega-events and
the promotion of both cultural standardization and universalizing value
systems; and the relationship between sports mega-events and the development of new markets (and consolidation of existing ones) by major national
and, especially, transnational corporations. Our premise here is that sports
mega-events are regulated by multi-level, multi-actor governance regimes,
in which authority to define the event and its terms and conditions is negotiated between public and private non-state actors, such as FIFA, UEFA and
the IOC, and their national emanations. In the conceptual terms staked
out by James Rosenau (1990: 36), these latter international organizations
are 'sovereignty-free actors' who engage in contractual relationships with
other sovereignty-free actors (transnational corporations, sub-national governments, bureaucratic agencies, non-governmental organizations (NGOs)
etc.) and with 'sovereignty-bound' state actors. Their increased political
saliency reflects the diffusion of political and economic authority in the
multi-centric post-Cold War context, and of the effects of rapid technological development and ideological entrenchment on the political authority
and democratic capacity of nation-states (Held 2004: 73–88). The point here
is not that sovereignty-free actors such as FIFA and the IOC act as regulatory bodies for the conduct or the registration of the athletes or the athletics
organizations that they federate: this is, after all, their raison d'être (and,
again, key to the standardization of sport which has permitted its internationalization in specific forms). Rather, the point is that sports mega-events

impinge increasingly on the definition of public policies (urban planning, transport infrastructure, environment, social welfare, health, etc.) and the allocation of the increasingly scarce public resources available to achieve these (or other, competing) collective goals. And that they do so, even in liberal democratic regimes, by circumventing normative, established, collective decision-making structures and processes. We shall return to these points below.

Sports mega-events thus pose problems of resource allocation and of collective accountability. But just as acutely they raise the question of cultural standardization, or rather the projection of a Western, liberal model of social relations on local host cultures. Of course, there may be much about this which is laudable, at least from the vantage point of a Western, liberal worldview. There are two broad aspects to this. The first is the association of sports mega-events, and the Olympic Games in particular, with a universalizing rights-driven discourse. As Hoberman (2004) points out, de Coubertin's original vision for the revived Olympics was the promotion of international peace in an age of rampant nationalism; the Olympic Charter's first 'Fundamental Principle of Olympism' dedicates the Olympic Movement's mission to be the creation of 'a way of life based on the joy of effort, the educational value of good example and respect for universal fundamental ethical principles' (IOC 2007: 11). Perhaps the most cherished of Olympic symbols is the ceremonial lighting of the torch in Olympia, the relay that follows, and the idea of the Olympic Truce (the ceasing of all warfare during the time of the event itself) that it marks. These are carefully orchestrated myths which, through repetition, have become established systems of meaning, though it was of course Goebbels who concocted the Olympic torch ritual for the 1936 Berlin Games (Krüger 2004: 45–6, 2005), whilst the Truce – though it has undisputable ancient origins – was revived only in the 1990s (see Briggs et al. 2004). This is not to reject the constructive potential of such symbolism; we might ask whether it really matters that the origins of the Olympic Torch can be traced back to the Nazi regime, especially from a vantage point where it has become an aspiring 'toolkit' for world peace. In global public consciousness, the Olympic flame and the related torch relay – as exemplified by the action of Reporters sans Frontières – are important symbolic acts for not only global peace, but also the promotion of human rights and liberal democracy.

In effect, therefore, the Olympics are constructed as promoters of historically specific 'universal world views' (Roche 2000: 198). Compared to the Olympics, the emphasis on the promotion of universal values and liberal social programmes has come late to sport's other mega-events. However, in 2005, FIFA established a corporate social responsibility department, and added a third pillar, 'Build a Better Future', to its existing commitments to 'Develop the Game' (improve and promote the sport) and 'Touch the World' (extend its geographical reach). The core of this third pillar is the commitment to 'share best practice, systematize successful experiences and develop new and innovative solutions, ready to be translated into investment opportunities or to be an integral part of regional or global development strategies' (FIFA n.d.). In other words, in conjunction with its corporate partners, member associations and civil society organizations, FIFA provides logistical, technical and financial support through its Football for Hope programme; the idea is that football help achieve the UN's Millennium Development Goals, using sport as a lever for health promotion, social integration, children's rights and education, environmental amelioration and so on. FIFA stresses football's social responsibility in a world 'where many are still deprived of their basic rights' (2007b: 5). It currently commits 0.7 per cent of its annual revenue to such corporate social responsibility programmes.

The second aspect of this process which interests us here is the relationship between sports mega-events and the erosion of cultural particularity. To be sure, we are not suggesting that sports mega-events are alone responsible for such processes, that the process is even and inevitable, or that cultural standardization is a unidirectional imposition; Theodore Levitt, in his seminal article announcing the globalization of markets and the dominance of the global corporation, was already provocatively announcing nearly 30 years ago that the 'world's needs and desires have been irrevocably homogenized' (1983: 4). Rather, we are interested in what Ien Ang identifies as the 'checkered process of systemic desegregation in which local cultures lose their autonomous and separate existence and become thoroughly interdependent and interconnected' (1996: 153). As John Tomlinson underlines, the heightened connectivity which defines contemporary globalization is threatening to our conception of culture and its association with territorial boundedness; globalization matters for culture, he argues, as 'it brings the negotiation of cultural experience into the centre of strategies for intervention in the other realms of connectivity: the political, the environmental, the economic' (1999: 30–1).

To put it another way: the tight association of sports mega-events with transnational capital, Western broadcast media, and neo-liberal urban governance regimes places them at the heart of two dynamics. The first is the transformation of urban landscapes. The evolution of sports mega-events and infrastructures over the past 20 years has been spurred by the desire of local entrepreneurial elites to connect their cities with the competitive global economy (Burbank et al. 2001); culturally, the conditions of mega-event staging are dependent on the norms and standards set by their regulatory organizations, and are synonymous with the deterritorialization and general decentring of recreation throughout the North, as industrial and manufacturing spaces have given way to landscapes of consumption or 'fantasy cities' (Hannigan 2002). Our collection therefore seeks answers to the processes of urban transformation associated with mega-events: who benefits, politically, socially, culturally? What type of transformation is on offer, how is it planned and managed, what are the values that underpin it? The second dynamic focuses on the explicit Westernization of cultural mores and values in non-Western host cities. This is neither necessarily inevitable nor the only view: Susan Brownell (2008), for instance, writes persuasively of the Beijing Games as a moment for cultural renewal in the flow of ideas between East and West, and of the capacity of Olympism to reinscribe history as global interconnection, rather than Western domination, through sport. What values, therefore, are articulated by mega-event hosting coalitions and hosting bids? What worldviews are promoted? How, in particular, do non-Western hosts react and respond culturally and civically to hosting sports mega-events?

Mega-events, sustainable development and environmental performance

The second principal concern of this volume is the relationship between sports mega-events and the environment. Prior to the 1990s, the IOC had shown little interest in the development of an environmental policy, despite the physical impacts of Games, the increasing awareness of environmental problems amongst Western publics, and the public concern which led Denver to decline the IOC's offer to host the 1976 Winter Games. The integration of environmental concerns into the hosting of and bidding for Olympic Games dates from the mid-1990s, with 'the environment' declared the 'third dimension' of Olympism, alongside sport and culture, by the then IOC president Juan Antonio Samaranch in 1995, following the December 1994 Centennial Olympic Congress. In 1996, a paragraph on environmental protection was added to the Olympic Charter, defining the IOC's role with respect to the environment such that

> the IOC sees that the Olympic Games are held in conditions which demonstrate a responsible concern for environmental issues and encourages the Olympic Movement to demonstrate a responsible concern for environmental issues, takes measures to reflect such concern in its activities and educates all those connected with the Olympic Movement as to the importance of sustainable development. (IOC 2007: x)

As Cantelon and Letters point out (2000: 299–303), despite its initial slowness, the IOC's development of an environmental policy was rapid, achieved within a precise and potentially ruinous context for the movement. They also imply that it was, at base, opportunistic: just as environmental protection moved to the top of the international policy agenda, the 1992 Winter Games, held in Albertville and the Savoy region of France, was revealed to have caused 'massive environmental damage'. Faced with a crisis of legitimacy and further loss of reputation following performance-enhancing drug use and bribery and corruption scandals, the IOC sought to protect its global mandate – and thus its ability to secure highly lucrative sponsorship and television revenues – by first denying the charges of poor environmental management, then by promoting itself as an environmental custodian. The IOC's subsequent conversion was enabled by the staging two years after Albertville of the Lillehammer 'Green' Games, whose environmentally sensitive character was particularly influenced by the Norwegian state environment ministry's will to showcase its commitment to the sustainable development principles set out in *Our Common Future*, the 1987 report of the UN's World Commission on Environment and Development, which had been chaired by the country's prime minister, Gro Harlem Brundtland (Lesjø 2000: 290).

The IOC has since developed substantial internal organizational and procedural capacity for maintaining this emphasis: the conclusion in 1994 of a Cooperative Agreement with the UN Environment Programme (UNEP), the creation in 1995 of a Sport and Environment Commission, the adoption in 1999 of an Olympic Agenda 21, the appointment for each Games of an environmental advisor to ensure host cities comply with IOC stipulations and so on (IOC 2007: 1–2). Following the 2000 Sydney Games, awarded in 1993 on the basis of the host city's integration of environmental considerations into planning and staging, and subsequently widely promoted by the IOC as environmentally conscious, Rogge has sought to consolidate this orientation whilst attempting to find ways of managing the seemingly ever-increasing scale of the Games. On succeeding Juan Antonio Samaranch as IOC president in 2001, Rogge set up a Study Commission to establish a blueprint for 'good governance', which reported to the 2003 Prague IOC meeting. Chaired by Dick Pound, the commission produced 117 recommendations to manage the size, cost and complexity of future Games. These recommendations were designed to create a philosophy for hosting decisions: China, yes, to bring the Games to the world's most populous nation – and expand the Olympic market's reach – but thereafter the watchwords would be the end of gigantism, the avoidance of over-spending, the importance of ethics and the nature of the legacy for future generations. The commission's recommendations include the development of venues in clusters, the preference for temporary installations over permanent ones where post-Games use does not justify the latter, and the encouragement of the use of public transportation and car sharing where possible (Pound 2003: 23, 35).

Despite the widely perceived environmental failures of the 2004 Athens Games, the Olympics now generally operates as an international showcase for the development and dissemination of environmental and sustainability best practice. Of course, professional sports tournaments require a basic level of environmental quality both for high performance competition to take place and for the health of the competitors to be protected. Los Angeles

(1984), Seoul (1988) and Athens (2004) notoriously suffer from high levels of atmospheric pollution; the Athens Organizing Committee particularly sought to address this issue through massive investment in public transport infrastructure (Tian & Brimblecombe 2008). But the decision to hold the 2008 Games in Beijing brought specific environmental remediation measures (as well as human rights) sharply into focus.

What of FIFA? Sustainable development and environmental management programmes are now increasingly important for individual World Cup tournaments. The German Football Association (DFB) bid dossier set out an environmental management programme for the 2006 World Cup, leading to the definition in 2003 of the Green Goal scheme, targeting significant cuts in energy and water consumption, improved waste management (through the use, for example, of reusable and returnable drinking cups in the stadiums) and promotion of public transport for spectator to stadium journeys (Öko-Institut 2003, 2007). The 2006 World Cup Finals were thus the first FIFA Finals to claim a 'green legacy'. Yet this scheme, drawn up in 2003 by NGOs WWF (Worldwide Fund for Nature) Germany and Öko-Institut at the request of the organizing committee, though successful, was voluntary. FIFA (and its corporate partners) were not actively involved until around six months before the event itself (Öko-Institut 2007: 16–21). FIFA has, unlike the IOC, not (yet) made environmental commitments binding for event hosting; there was little in South Africa's bid for either the 2006 or 2010 World Cup Finals which centred on the environment, with only the proximity of host cities to national parks and what we might call 'tourist wildlife' mentioned in the bid dossier, and transport addressed only in terms of logistics. However, the South Africa NOC was delegated by FIFA to draw up local environmental action plans with host cities, in conjunction with national and local government, such as the 'Host City Cape Town Green Goal Programme'. The aim of the programme – much like the German Green Goal scheme on which it is based – was to minimize the environmental impacts of the tournament, raise long-term standards and promote sustainable development programmes (City of Cape Town 2009).

As will be apparent, environmental and sustainable development programmes are dependent on partnerships between public authorities, private regulatory bodies, corporate partners and civil society actors, particularly NGOs. The positioning of the latter with respect to event staging is conditioned by a number of factors, from domestic institutional and cultural arrangements to public responses to sustainable development agendas. Sports mega-events increasingly provide a platform for economic growth oriented approaches to environmental protection and amelioration. In this scenario, the IOC (and FIFA) function as regulatory authorities for the development and dissemination of environmental best practice and sustainable technologies, facilitating the creation and growth of new markets. One sense in which we should see mega-events as globalizing events therefore is through their capacity to act as powerful agents of technology transfer and technical norm diffusion, such as the generalization of international certification systems (Leadership in Energy & Environmental Design (LEED), Forest Stewardship Council (FSC), ISO 14001 etc.) and the introduction or extension of processes such as monitoring and benchmarking. Zbicz argues with respect to Beijing that one function of the Games is to consequently raise expectations – here, of a cleaner environment – amongst the country's middle-class (2009: 40).

The Games (and perhaps other sports mega-events) accordingly operate as showcases for the internalization of environmental values and norms, with a wide mimetic potential in both geographic and public policy sector terms. In other words, high environmental standards are no longer seen as antagonistic to the development of economic growth regimes, becoming instead a key source of market innovation and future economic growth (Barry 2005: 303–4). There are, of course, problems with this apparently rosy and consensual outlook. The first is conceptual: what exactly do we mean by sustainable development? There are multiple definitions of sustainable development, different emphases and contrasting visions, neatly summarized and discussed by Mawhinney (2002: 1–24). Perhaps the most famous definition is set out by the Brundtland report, as 'development that meets the needs of the present without compromising the ability of future generations to meet their own needs' (WCED 1987: 43), a definition broadly adopted by the IOC, which also sets out its goals such that

> the starting point of sustainable development is the idea that the long-term preservation of our environment, our habitat as well as its biodiversity and natural resources and the environment will only be possible if combined simultaneously with economic, social and political development particularly geared to the benefit of the poorest members of society. (IOC 1999: 17)

For the London 2012 Organizing Committee, 'sustainability is fundamentally about people and how we live; it is not simply a technical discipline' (LOCOG (London Organizing Committee for the Olympic Games) 2009: 9). Such definitions of sustainable development thus generally place economic growth in tension with social and environmental justice. Yet one might ask a series of questions of the operationalization of claims to sustainable development. What priority should we give to different collective public goals, for example? Through what policy mechanisms and instruments should sustainable development strategies be implemented? What is the trade-off between effectiveness and accountability? What prioritization should there be for improving environmental performance and for collective, participatory decision-making? How should 'people' be included – in the project definition or just in its implementation? What priority should be given to the reduction of social and economic inequalities? What level of economic growth is attainable, or desired? Which sectors of the population should bear the cost? It is relatively easy to see, for example, how an Olympic politics of sustainable development becomes immediately problematic when the Games are awarded to an authoritarian regime like China, notwithstanding the remedial, positivist discourse in which it is cloaked. But the definitional problem is of course not restricted to non-democratic regimes: the problematic trade-off between growth, social justice, environmental protection and civic accountability is also acute in liberal democratic regimes.

In the context of sports mega-events, it is thus our task as observers to investigate the trade-offs and tensions, functions and definitions, processes and structures, actions and programmes that constitute sustainable development, in their time and place specific event iterations, and in their global governance formulations. But there is a second problem here. Sports mega-events, because of their global audience, bring a spotlight to obvious environmental problems. But they do not just create a persuasive space for knowledge transfer and environmental problem-solving; they also carry a carbon cost and create material environmental impacts, through the production of waste, the consumption of energy, the development of transport, communications and leisure infrastructures, through national and international travel, and not least through the symbolic promotion of individual mobility and consumption-based lifestyles. Local land-use conflicts are common.

Mega-event organizers are clearly mindful of the reputational costs both of negative publicity and of environmental damage. Beyond specific local planning impacts, they typically seek not only to minimize secondary negative environmental impacts, but also to negotiate them against carbon remediation schemes such as offsetting. Carbon offsetting was introduced for the first time in the FIFA World Cup in Germany in 2006, as part of the organizing committee's Green Goal campaign (see above). The scheme achieved 'climate neutrality' by offsetting 92,000 tonnes of CO_2e (Carbon dioxide equivalent, including CO_2 and other greenhouse gases (GHG)), meeting the WWF's 'Gold Standard'. Since the 2002 Winter Olympics in Salt Lake City, NOCs have used some form of carbon offsetting scheme to mitigate their carbon emissions. The Beijing Organizing Committee developed a campaign to offset CO_2 emissions that the athletes caused by flying to the Games, in partnership with Greenpeace China; Vancouver was the first NOC to appoint an Official Supplier of Carbon Offsets (Offsetters Green Technology Inc.), parlayed against sponsorship rights for the Games and for the Canadian Olympic team, with offsets meeting the regional government's relatively stringent standards. It is evident that national and transnational environmental NGOs have been instrumental in establishing and legitimizing such schemes, and in measuring carbon footprints.

But the methodology of 'carbon neutral' offsetting actions for sports mega-events has so far tended to address only the most superficial of carbon expenditures. The German 2006 Green Goal campaign, for instance, included only event-time emissions within Germany itself, excluding the carbon emissions from international travel. Similarly, Vancouver's much heralded offsetting scheme covered only the 'direct' emissions of Games staging such as athlete travel, venue construction and facility heating (which it calculated to be some 118,000 tonnes of CO_2 in total), but not the estimated 130,000 tonnes of CO_2 generated by spectator or sponsor travel to the Games (VANOC 2009). To cover this, Vancouver asked Games visitors to offset their own travel voluntarily. In practice, therefore, carbon neutrality – even in the terms defined by the organizers – is highly unlikely to be achieved.

In contrast, London 2012's carbon footprint methodology seeks to calculate emissions 'when they happen', with the projected reference footprint covering the period from the point of the bid win to the closing Games ceremony, assuming development as set out in the bid dossier. Unsurprisingly, this footprint, at 3.4 million tonnes of CO_2e (or around 0.5 per cent of annual UK emissions), dwarfs those calculated for previous Games or WC Finals; unsurprisingly also, especially given the extent of infrastructure development for the 2012 Games, more than half the reference footprint is produced by infrastructure and venue construction in the Olympic Park (CSL 2009: 17). In contrast, BOCOG calculated a carbon footprint of 1.2 million tonnes of CO_2e for the Beijing Olympics, of which only 2 per cent was ascribed to venue construction (UNEP 2009a: 103–7). The advanced picture from London 2012 is one of progressively robust methodology. But this developing methodology also reveals, on the global level of event iteration, incoherence and confusion. Sochi 2014, for example, is to use the 'HECTOR' (HEritage Climate TORino) methodology of the 2006 Winter Games, which counted only Games-time energy use and transport emissions. It is evident that neither the IOC nor FIFA plays a regulatory role in terms of defining standards or methodologies for such schemes (and thus in holding NOCs to public account, diffusing best practice or requiring collective transparency), or indeed in funding them. It is not currently possible to compare successive mega-events with one another given the absence of reliable, transferable, reporting and auditing systems. This is particularly important given the proliferation of schemes and objectives, from 'low-carbon' to 'zero net emissions' to 'carbon neutral': in practice, it is extremely difficult for publics to understand the significance, extent and operation of carbon mitigation schemes, or the details and differences between competing claims, pledges or actions.

More fundamentally, the offsetting schemes on which mega-event claims to climate neutrality are inevitably based are themselves highly contentious, to say the least. For all the apparent practical attractiveness of offsetting, additionality is hypothetical and difficult to measure (would the carbon reduction schemes have been implemented anyway? how are they vetted?); piecemeal top-down measures run the risk of causing the simple displacement, or 'leakage', of harmful activities; the permanence of offsetting projects is impossible to verify (in marked contrast to our knowledge of the permanence of carbon emissions in the atmosphere); schemes are frequently confusing, lack credibility and are open to fraud; offsetting typically places the burden of action on countries in the global South, consequently reducing the scope of developing countries to reduce their own emissions or to raise basic living standards; offsetting schemes are essentially an alibi for the failure of public and private actors to address systemic carbon emissions in developed countries (hence the notion of 'carbon indulgences', 'sub-prime carbon' and even 'carbon laundering'); offsetting does little to address, and in many cases exacerbates, global inequalities and environmental degradation; offsetting is simply another form of neo-liberal market expansion, commodification and capitalization (see inter alia Muhovic-Dorsner 2005: 239–40, Gössling et al. 2007, Smith 2007, Böhm & Dabhi 2009, Clifton 2009: 25–30, Densham et al. 2009, Nerlich & Koteyko 2009, Hari 2010).

A rather poignant example of one aspect of the problem is provided by the Green Goal campaign. The greater part of the carbon offsetting from the 2006 FIFA World Cup Finals in Germany was achieved by investing in two greenhouse gas reduction projects in South Africa: a sawdust-fired boiler for a citrus fruit farm in Letaba in the north of the country, near the Kruger National Park, and a project to generate electricity from sewage gas in the Sebokeng Township outside Johannesburg (Öko-Institut 2007: 90–3). But South Africa had great difficulty securing the funding for offsetting the much larger domestic carbon footprint of its own Finals, and as a consequence host cities such as Cape Town scrapped the carbon neutral goal and preferred to aim for a 'low-carbon' event (City of Cape Town 2009: 17). Developed countries can offset in the South; Southern countries cannot do the reverse. Here, they were unable to fund offsetting at all; indeed, the 11 teams that offset their travel did so as part of a UNEP programme, with sportswear multi-national Puma covering the costs for the seven teams it sponsored (including four African teams, Algeria, Cameroon, Côte d'Ivoire and Ghana). In late 2009, LOCOG quietly dropped offsetting for London 2012.

The social impacts of mega-events and civic responses to them

The third concern of this volume is the corporate character and social impacts of sports mega-events. Against the opportunities presented for policy and technology development, and for the substantive and procedural benefits that hosting might bring in its wake, mega-events generate wider concerns. The suicide bombings on the London transport system on 7 July 2005, the day after London was awarded the 2012 Olympics, killing 56 people (including the four bombers) and injuring around 700, focused public and state attention on sports mega-events as an opportunity for terrorist attack. Whilst sports mega-events present exceptional risk profiles, this is likely to be of a different order for football tournaments (where venues are dispersed, there is likely to be public disorder, and conflict is likely to be generated by national/territorial rivalry) and Olympic Games (where events are concentrated, and conflicts are likely to be geopolitical) (Jennings & Lodge 2009). The Olympics have already been the victim of terrorism, of course, in Munich, in 1972. It is unsurprising therefore that security is a main consideration in mega-event bidding and hosting. But if we underline the stakes for states and publics, we should also stress the opportunities for markets. Kimberley Schimmel (2006: 167–70), discussing the 2002 Salt Lake Winter Olympics, argues that major sports structures and events are being used to intensify and accelerate the transformation of urban spaces into militarized environments characterized by increased surveillance and control, whose goal is not simply to protect public safety but to secure corporate contracts. Taking this logic further, Naomi Klein argues that the discourse of Olympic safety enabled the Chinese authorities to ramp up their repression and surveillance apparatus, creating a 'Police State 2.0, an entirely for-profit affair that is the latest frontier for the global Disaster Capitalism Complex' (Klein 2008).

Concerns about mega-event staging and civic rights are thus also important for liberal democratic regimes. Such concerns include the close association between the staging of mega-events and the channelling of public resources to national and transnational corporate interests, the attendant privatization of public space, the suspension and loss of civil liberties, the reduction of democratic accountability, the downgrading of social policy priorities, and the entrenchment of social polarization. Another of the basic premises of this volume therefore is that we cannot address the environmental capacities and consequences of mega-events without addressing their social impacts.

To take a few examples: as noted above, the Sydney Games were notable for their environmentally sensitive character. But they were also notable for the restrictions placed on democratic accountability and freedom of assembly. The New South Wales state government introduced fast-track planning procedures in 1995 denying Sydney residents the right to initiate court appeals against Olympic construction projects, and its Olympic Coordination Authority (OCA) Act suspended all projects linked with the Games from the usual requirements of Environmental Impact Statements (Searle & Bounds 1999: 171, Hall 2001: 172–3). Cunneen argues that the series of legislative acts passed in the run-up to the Games – the Homebush Bay Operations Act, the Security Industry (Olympic and Paralympic Games)

Act, the Sydney Harbour Foreshore Authority Regulation and the Olympic Arrangements Act – individually and cumulatively imposed draconian controls on basic civic rights. The legislation

> significantly infringes basic democratic rights of public protest, and freedom of movement and assembly; it dramatically increases the power of police and security guards in many of the most important public places in Sydney; it introduces wide ranging public order offences without safeguards or accountability against the abuse of power; and it has a legislative life-span well beyond the period in which the Olympics and Paralympics are scheduled to take place. (Cunneen 2000: 26)

The 2004 Athens Games reveal a similar picture. Here, faced with the IOC's insistence that all construction work comply with the operative standards set by national and international planning and impact assessment legislation, the Greek government opted to revise the national constitution to downgrade forest protection, effectively curtailing the power of environmental and citizens' initiative groups to oppose potentially damaging projects. John Karamichas (2005: 139) concludes that 'the Greek authorities not only violated their own constitutional framework, but they essentially failed to engage the public even in the most rudimentary form of debate and consultation'. Ahead of London 2012, one of the key concerns raised by George Monbiot (2007), a British media commentator and environmental activist, was that even where formal planning requirements were being respected by the Olympic Delivery Authority, the structure of the consultation procedure, the sheer volume of documentation and the imbalance in resources between developers on the one side and community groups on the other effectively meant that civic participation in Olympic Park land-use planning decisions was precluded.

The restriction on basic civic freedoms is also typically apparent in the overt commercialization of sporting mega-events and the enforcement of a comprehensive security and control doctrine. The 1996 Atlanta Games were notorious for their corporate nature, seen as distasteful even within the Olympic Movement. But the corporate nature of mega-events has become structural. In his analysis of the 1998 Nagano Winter Olympics, for example, Atsushi Tajima stresses the imposition of 'draconian commercial control' on local businesses throughout the host city, to the benefit of the market-dominant transnational corporations who had negotiated lucrative category-exclusive world-wide commercial sponsorship rights with the IOC as part of the organization's ongoing TOP (The Olympic Partner) programme (2004: 245–7, 250–3). As should be apparent, organizations such as the IOC, FIFA and UEFA have clearly defined private enterprise goals; these are not public bodies, but private transnational actors. FIFA steadily increased its annual revenue from USD 575 million in 2003 to USD 1291 million in 2010; its equity, from USD 76 million to USD 1280 million over the same period. Its primary source of income is broadcasting and marketing rights for World Cup Finals (FIFA made a profit of USD 249 million in 2006, by some distance the highest in its history, and USD 202 million in 2010; television rights for the 2010 World Cup Finals generated a revenue of USD 2408 million) (FIFA 2011: 14–17). The Olympic Movement generates revenue through a combination of IOC (broadcasting rights, TOP, official suppliers and licensing) and OCOG managed programmes (domestic sponsorship, host country ticketing and licensing). For the 2005–8 period, the total revenue was approximately USD 5.45 billion – in excess of double the amount generated over the 1993–6 four-year cycle (IOC 2010: 6).

Domestic and global sponsorship agreements are of course now fundamental to the business model of mega-event staging. FIFA has negotiated a series of global advertising, promotion and marketing contracts with 'partners' – Adidas, Coca-Cola, Emirates, Hyundai, Sony and Visa, and for the 2009 Confederations Cup and the 2010 World Cup Finals, both in South Africa, two further commercial rights packages, of World Cup Sponsors (McDonald's, MTN, Budweiser, Castrol, Continental, Satyam Computer Services) and National Supporters (First National Bank, Telkom South Africa). These agreements confer on the sponsors the exclusive right to display their products and marketing materials within venues and with all associated FIFA and/or tournament marketing; the sale of these marketing rights for the 2010 World Cup Finals generated revenue of USD 1072 million (FIFA 2011: 17). The London 2012 Olympics reveals a similar logic and list: nine world-wide partners, from Coca-Cola to Visa, seven official partners (Adidas to Lloyds TSB), six official supporters (Adecco to UPS) and 12 official suppliers and providers (Airwave to Trident).

The reasons why corporations should wish to conclude such deals is relatively obvious: the brand is associated with the positive values of support, competition and success, and the event is seen as a key opportunity to bolster the strategic presence of a brand in both global markets and existing or new national or continental markets. For civic populations, on the other hand, the interest is rather less clear, for two reasons. First, we may be sceptical about the corporate hijacking of collective social goals and agendas. For instance, what should we make of the 'first-ever Environmental Torchbearer Program for an Olympic Torch Relay', established in the lead-in to the Vancouver Games by WWF, the David Suzuki Foundation and Coca-Cola? The scheme encouraged Canadians to 'pledge small lifestyle changes for the environment in exchange for the opportunity to carry the Olympic Flame'. The first torch bearer was Gerald Butts, president and chief executive officer of WWF-Canada; ordinary Canadians, on the other hand, could sign up by visiting www.icoke.ca. One can legitimately wonder what the collective benefits of such scheme are; the corporate benefits appear far more obvious, both in terms of sales marketing and global reputational enhancement

(Coca-Cola's bottling operations in India and South America have been the site of a series of environmental and social conflicts over the last decade, including allegations of pesticide contamination and groundwater depletion in Kerala, and of anti-union intimidation and major human rights abuses in Colombia). It is probably to be assumed that the 'small lifestyle changes' advocated by the scheme do not include drinking less coke or joining organized labour.

Second, as Tajima's argument underlines, exclusive commercial agreements imply the exclusion of plurality, whether corporate or cultural. Event organizers pressurize national governments to introduce stringent anti-ambush marketing laws so as to protect their contractual sponsorship agreements. In Portugal, the government made it a criminal offence for a non-endorsed brand to gain promotional advantage through association with the 2004 UEFA Football Championships. But this sort of legislation doesn't simply affect brands. At the 2006 World Cup Finals held in Germany, FIFA negotiated a series of deals with alcohol sponsors Anheuser-Busch (AB), brewer of (American) Budweiser beer. Under the terms of the deal, AB gained exclusivity over selling and marketing of beer in the event venues. However, in Holland, brewer Bavaria offered a commercial promotion of orange lederhosen or *Leeuwenhose* (an ironic as well as patriotic statement) to Dutch football fans backing the national team; the brewer reportedly sold a quarter of a million pairs, which bore its logo and a lion tail. Heineken, the official sponsor of the Dutch football association, tried and failed in a legal action against the lederhosen. At the Finals themselves, FIFA took action to protect itself against the trousers, which it considered to be ambush marketing: hilariously, over a thousand Dutch fans attending the Holland versus Ivory Coast match in Stuttgart were forced to take off their lederhosen and watch the game in their underpants (*The Guardian* 19 June 2006, *BBC News* 21 June 2006).

Hilariously? Sure. The Dutch won; there were a few grumbles; the anecdote adds colour, especially when repeated – as a corporate stunt this time – at the 2010 World Cup Finals. Let's take another example from South Africa 2010. For the Finals, FIFA forced the nine host cities to comply with a bylaw protecting the event from ambush marketing in event zones (Cornelissen 2010). The bylaw controls access to such zones and also covers traffic, parking and street trading in contiguous areas to event zones. Quite apart from how we might feel about this democratically, what are its cultural implications, for a Finals predicated on its supposed 'pan-African' nature? According to Desai and Vahed (2010), the corporatization and securitization of the Finals mean that though the tournament is held in Africa, the exclusion of traders and food sellers from event zones enforces a Western/Northern corporate cultural standardization on the Finals, creating a disjuncture between the global world of FIFA corporate governance and the local social, political, economic and cultural contexts of its application.

As hosting sports mega-events has, since the staging of the Barcelona Olympics in particular, been increasingly incorporated into the strategic planning of cities seeking enhanced global economic competitivity, so the potential social impacts of sports mega-event staging have correspondingly increased. The negative effects of staging the Games can be expected to have the greatest impact on the poorest and most socially deprived parts of the urban population. These populations suffer in two ways: first, because event staging is typically concentrated in poor areas earmarked for regeneration, they are forced to relocate from their residential areas by pre-Games construction projects and post-Games gentrification, as the establishment of new infrastructures and the influx of new populations are targeted by developers; second, because mega-events require the underwriting and expenditure of extraordinary amounts of investment, the city's financial and policy focus on mega-event projects crowds out its ability to maintain and develop its key welfare functions (Olds 1998, Preuss 2004: 24–5). Thus sports mega-events may simply misdirect scarce public resources away from pressing social and housing needs. For instance, at the insistence of FIFA, which refused the upgrading of the city's existing rugby stadium for the 2010 World Cup, Durban is now home to the magnificent 70,000-seat Moses Mabhida Stadium. It is 'an asset from which every resident and ratepayer will benefit', according to the City Council (http://www.durban.gov. za/durban/government/spu/moses). Yet one might legitimately wonder at the 2.5 billion rand (USD 330 million) spent on elite-driven infrastructure in the city when 180,000 of Durban's 3 million residents are shack dwellers, or at the estimated 20 billion rand (USD 2.6 billion) budgeted for stadium construction and refurbishment in South Africa for the four years to the Finals – equal, according to the economist Stephen Gelb, to the cost of 90,000 low cost housing units per year over the same period (Ngonyama 2010, Desai & Vahed 2010).

In her March 2010 report to the UN's Human Rights Council, the UN's Special Rapporteur Raquel Rolnik argues that, though staging mega-events can have positive impacts on housing – through the provision of new infrastructures, environmental improvement, improved public service provision and the renovation of the housing stock itself – experience has shown that this has rarely been the case. According to Rolnik,

Numerous past experiences have shown that redevelopment projects adopted in preparation for the games often result in extensive human rights violations, particularly of the right to adequate housing. Allegations of mass forced evictions and displacement for infrastructural development and city renewal, reduced affordability of housing as a result of gentrification, sweeping operations against the homeless, and criminalization and discrimination of marginalized groups are frequent features in cities staging the events. The impact of these practices is mainly endured by the most disadvantaged and vulnerable sectors of society, such as low-income populations, ethnic minorities, migrants, the elderly, persons with disabilities, and marginalized groups (such as street vendors and sex workers). (2009: 4)

Sports mega-events are particularly critical for the displacement of poor and homeless populations. This is because, as discussed, they appear now to be structurally entwined with major, transformative urban renewal programmes – commonly referred to as 'mega-projects' – which are dependent on market-driven profitability. Of course, the contours of mega-projects can be expected to vary as a function of their political and cultural contexts. The extent to which these developments are counter-balanced by socially just policies will vary: a key argument for London 2012, for example, is that the Olympic Park will provide a link opening out the city to the east, enabling policy solutions to the acute and endemic social and economic problems of the five Olympic Boroughs. But as Susan Fainstain argues, the dependency of mega-projects in post-industrial cities on private sector capitalization means that commitments to socially just housing policy are a secondary concern. The primary orientation is towards profitability and competitivity, which inevitably means three types of development: luxury residences and hotels, large-footprint office towers and shopping malls (Fainstain 2009: 783). But this is also a function of the two key structural properties of sports mega-events that we have outlined above: time compression in the run-up to the event which necessitates rapid solutions, combined with the mass mediatization of the event itself which gives the city a global visibility which is real as well as symbolic. The problem here therefore is that urban beautification and renewal does not typically feature long-term systemic solutions to poor housing or homelessness, but rather event-urbanization displacement measures (Olds 1998, Greene 2003, COHRE 2007).

Gentrification is of course the democratic, liberal, market economy mechanism of social displacement. Games staging in non-democratic regimes has been characterized by its authoritarian version. In December 2007, COHRE (Centre on Housing Rights and Evictions), a Geneva-based advocacy NGO, jointly named the Beijing Municipality and BOCOG as one of its three Housing Rights Violators for widespread evictions and displacements ahead of the Olympics. In all, COHRE estimates that 1.5 million people were displaced in the eight years to the Olympics, to make way for the principal venues and other Olympic-related facilities and infrastructure, to undertake 'beautification' measures in expectation of increased tourism, and for improvements to the city's general infrastructure. In a July 2008 report, COHRE wrote that 'despite hosting the Olympic Games, China has been undeterred in undertaking massive housing rights violations and promises to continue along this path once the Games are over' (2008: 5).

There is, in other words, a well-established pattern here, spanning mega-events, continents and regime types. The pattern is one where corporate profit and effective delivery are valued more highly in event hosting than the values of participatory democracy or social justice. It is perhaps little wonder then that, for observers such as Horne and Manzenreiter, undemocratic organizations such as the IOC and the staging of sports mega-events are part of the 'ideological assault on citizenship that has occurred since the 1980s, which prefer global consumers to local publics' (2006: 18). Given this situation, what then of civic responses to sports mega-events? Given the social impacts of event staging outlined above, one might expect regular civil society campaigns and social mobilizations either seeking to correct the negative impacts, to reform the ideological or organizational apparata of events or to prevent the staging of these events. Along with most standard definitions, we conceive civil society to be the conscious set of social networks, associations and relationships existing in a tripartite structure alongside those regulated by the state on one side, and the market on the other (Cohen & Arato 1992: vii–xi). Broadly, civil society organizations include social movements, activist networks, advocacy NGOs, civic associations and church organizations.

Two broad contrasting potential sets of relationships concern us here. In the first set of potential relationships, civil society organizations (primarily NGOs) are integrated into event governance structures. Again speaking broadly, we can identify this as the neo-liberal model of civil society, where civil society (or third sector) organizations take on functions traditionally provided by the state, such as service delivery, but also some advocacy. For both NGOs and event organizers, the potential benefits of such integration are clear: NGOs provide *scientific expertise* which may generate constructive solutions to difficult design and implementation problems, but they also have the capacity to confer *moral authority* through their ideational systems and *democratic legitimacy* through their popular representativity, especially where they are membership based (see Uhlin 2009: 7). Because of these attributes, environmental NGOs are thus increasingly regarded as key stakeholders toward the initiation and implementation of environmental programmes in sports mega-event planning, whilst FIFA's social justice programmes make increasing reference to civil society partnerships (though again we must differentiate between regime types and specific local contexts whilst making clearly broad assertions).

This type of development is in evidence in the relationship between NGOs and events such as Games staging. For the IOC as for host cities in the Western industrialized world, environmental NGOs are representative civil society interlocutors, able to provide both expertise and participatory democratic legitimization. Since Sydney, national and transnational NGOs have been able to use their strategic advantages – scientific expertise (over the use of PVC in construction, or the climate change effects of HFCs in cooling systems), media responsiveness and the potential mobilization of consumer activism – to raise expectations. Pre- and post-Games reporting (such as by WWF and Greenpeace) and consultation during the delivery stage are effective tools in the adoption of best current standards by organizing committees intent on avoiding negative publicity, particularly given the now familiar promises to stage the 'greenest', 'most sustainable' Games ever. In Sydney, for example, Greenpeace adopted a constructive critical position, working on the design for the Olympic village and praising the organizing committee's environmental rehabilitation and sustainable technology advances, but also adopting a watchdog position on the committee's failure to adhere to environmental guidelines, and pressuring institutional and corporate actors through public action campaigns and the initiation of legal action (Lenskyj 1998: 349–50, 2002: 156–62). Greenpeace subsequently drew up a set of outline environmental guidelines for future event staging (Greenpeace 2000), and it has worked on relationships with national organizing committees, Olympics sponsors, and other civil society groups.

Yet, as Hiller points out (2000: 193), mega-event planning is top-down planning, in which the 'idea of citizen participation is [...] primarily merely responding to a plan conceived by others'. Perhaps at bid stage, where everything remains hypothetical, citizen participation is bound to be minimal (though a more participatory approach has been successfully implemented on occasion, such as for Toronto's and Calgary's respective (failed) Olympic bids; see Lenskyj 2000, Gursoy & Kendall 2006: 604). For Andranovich et al. (2001, see also Burbank et al. 2001), citizen participation and democratic accountability in decision-making for the respective Los Angeles, Atlanta and Salt Lake Games were notoriously absent. More widely civil society 'does not have the same say in this arena of public life as it does in others; citizens are typically kept at a substantial distance from megaproject decision making' (Flyvberg et al. 2003: 5). As Hiller (2000: 198) points out, mega-projects, like mega-events, 'have fixed completion dates that must follow a tight schedule which, on the one hand, ensures results rather than unending deliberations but, on the other hand may produce autocracy against which opposition may arise'. The integration of civil society actors into mega-event decision-making is thus limited. But the integration of NGOs into mega-event governance regimes can help organizers achieve desirable outcomes (legitimacy, efficiency and perhaps pluralism), and can also potentially help avoid one of the key costs of civil society organization: unpredictability, the imposition of delays and the generation of negative publicity. Alongside the neo-liberal civil society model therefore sits one based on active citizenship and collective social mobilization.

Two of the defining structural attributes of sports mega-events – their mass mediatization, including in real-time, and the fixed and often very tight timeframes for their organization and staging – can therefore be expected to make them potentially vulnerable particularly to social mobilizations.

In Vancouver, the anti-Olympic Convergence coalition organized a series of actions immediately before and during the Games, twice forcing the re-routing of the Olympic torch relay on the final day of the relay, and disrupting transport to the opening ceremony. These protests were the culmination of a long campaign led in particular by the Olympic Resistance Network and No 2010 Olympics on Stolen Native Land, arguing inter alia that the Olympics are environmentally destructive, force evictions and increase homelessness, criminalize the poor, increase violence against women, increase public debt, divert public funding from social programmes to repressions and surveillance on the one hand and corporate profit on the other (No 2010, 2007a).

About this volume

Of course, the specific symbolic meanings, sociological significance, policy lessons and collective actions that characterize each event will depend on the configuration of political and social contexts within which it takes place. Therefore mega-events pose a fascinating series of questions about the relationships between the global and local worlds of the event, and about how we as analysts and observers might attempt to systematize their multiple meanings. This volume therefore presents a series of comparative and nation-specific analyses, exploring the social and environmental impacts, responses, consequences and opportunities provided by the staging of sports mega-events. As underlined in the opening pages of this introduction, this book has been organized into three parts: Part I deals with Sports Mega-Events, Citizenship and Civil Society; Part II with Environmental Impacts and Sustainable Development; Part III with Constructing Civic Resistances to Mega-Events. The focus is predominantly on Summer Olympic Games, though not exclusively so: two chapters discuss World Expos, two Winter Games, and all chapters are careful to place event hosting within the wider contextual frameworks of mega-events as a whole.

Part I, with contributions from Horne, Broudehoux, Polo, and Dauncey, focuses on the globalization of mega-events, and the (changing) roles of and relationships between institutional, civil and economic actors within these globalization processes. For John Horne, mega-event hosting is akin to the 'disaster capitalism' identified by Naomi Klein within the operation of neo-liberalism: the time-specific demands of mega-event hosting create a crisis within deliberative and representative democratic structures, enabling the imposition of material and processual structural reform on civic populations mobilized by the prospect of collective cultural manifestation. Mega-events, Horne argues, generate their own forms of shock and awe.

Anne-Marie Broudehoux's chapter on Beijing 2008 spells out the extent of the possibilities of this type of systemic shock within China's authoritarian state capitalist regime. Here, hosting the Games entailed large-scale urban transformation and the radical improvement of environmental standards (in Beijing at least), but was used as an opportunity to engineer structural behavioural change in the populace at large – all in the name of projecting the image of China as a more attractive destination for Western tourist and trade flows. She examines how the 2008 Olympic Games offered the Beijing leadership the opportunity of a powerful tool of social reform, used to transform and control both the body and the mind of the Beijing population and to rewrite terms of belonging to Chinese society. Through an investigation of some of the main approaches used in this process, ranging from ideological means of reform and embodied civilizing practices to more coercive measures of control and discipline, her chapter raises questions about the role of mega-events in the conscious and planned management of human activity, the regulation and normalization of human behaviour and the organization of social stratification and domination.

Jean-François Polo's chapter identifies, in different cultural and regime contexts, similar undercurrents in repeated Turkish bids to host mega-events: Games and Finals hosting is widely understood as a sign of global legitimacy, and thus integrated within Turkey's candidacy for membership of the European Union. Moreover, if the stakes of these bids is acceptance within the global club of developed economies, then paramount also are the cultural ramifications: behavioural change, the idea of 'being civilized', is as integral to the Turkish bids as it is to Chinese strategies identified by Broudehoux. Polo argues that hosting international sports events is a means for the Turkish state to demonstrate its modernity, but this creates numerous contradictions: between the multi-cultural, diverse image projected by Istanbul's Games bids and the human and minority rights issues which dog Turkey's EU candidacy; between the open and positive image of Turkey that the authorities seek to promote to the outside world, and the wide currency of nationalist discourses in Turkish political and, especially, sporting culture.

Finally in this section, Hugh Dauncey takes the long view of French bids to host Summer and Winter Olympic Games. The failure of Paris to host the 2012 Summer Olympic Games has focused attention on France's relationship with the IOC and on the internal functionings of her Olympic Movement; by reflecting on previous French bids, particularly Lyon's failed bid to host the 1968 Summer Olympics, Dauncey underlines how Gaullist governments of the 1960s aimed to harness the developmental impetus of Olympic mega-events to their project for France, and how French cities and regions hoped to instrumentalize the Olympics in their own 'regional' agendas. His analysis reveals the constantly shifting dynamics between scales of public action and projection, nuancing the relationship between nation-state power and global urban competitivity in hosting motivations. For Dauncey, from a French perspective, the primary motivation remains geopolitical, with neoliberal competitivity a significant but secondary impulse.

Part II, featuring analyses from Caratti and Ferraguto, Mol and Zhang, Karamichas and Hayes, interrogates the development of environmental and sustainable development programmes and discourses within mega-event hosting requirements. The chapter by Pietro Caratti and Ludovico Ferraguto focuses on the role of environmental planning and management tools (such as SEAs (Strategic Environmental Assessments), EIAs (Environmental Impact Assessments) and EMSs (Environmental Management Systems)) in addressing environmental issues and considering stakeholder issues. Through the analysis of the experience accumulated thus far in the management of five political, economic, religious and sports mega-events, they evaluate how environmental planning and management tools have been devised and applied throughout the whole cycle of an event. They argue that, though uneven, the rational use of these tools can provide a number of environmental benefits, particularly in site location and in improved coordination and consultation procedures.

Arthur Mol and Lei Zhang, in their chapter on Beijing 2008 and the Shanghai World Expo of 2010, also offer a tempered optimistic view of the operation of mega-event hosting: in their analysis, major, highly mediatized events such as the Olympic Games both reveal and further the development of sustainability as a global norm. They argue that mega-events are increasingly perceived as sustainability attractors, which restructure physical objects, institutions, socio-material networks and mobile flows, and that, given the status of these events as high-profile and very visible happenings that attract world-wide attention, common norms on the environment, democracy, transparency and equality can hardly be ignored in the route towards an event. Mol and Zhang therefore develop a perspective for understanding and investigating the short- and long-term impact of mega-events on the environment, using ecological modernization theory and the sociology of networks and flows, and apply this framework to investigate current developments in China – asking finally whether we should celebrate these mega-events for their contribution to enduring sustainability, or whether these contributions will melt into thin air once the events are past.

Interrogating the data for the pre- and post-event periods of the first two Summer Games to have taken place within the era of Olympic environmentalism, John Karamichas offers a sceptical view on the claims of hosting coalitions that mega-events can have positive long-term effects on key environmental indicators. His chapter evaluates the environmental legacies left by the Sydney 2000 and Athens 2004 Olympiads by delineating the most important parameters of Ecological Modernization (EM), and developing an analytical framework that facilitates the evaluation of the potential of the Summer Olympics to act as an impetus for the host nation to improve or develop its capacity for environmental sustainability. Karamichas concludes by suggesting that earlier procedural barriers are so entrenched that the hosting of the Games is unlikely to have a lasting change, on its own, on existing negative practices; event-delivery gains must be divorced from long-term processes, which are much more directly affected by the prevalent political culture, economic structure and institutional arrangements of hosts.

Finally in this section, Graeme Hayes's analysis looks at the urban infrastructural promises associated with the French bid to host 2012. The promotion of the putative Games as 'ecological', with notably a series of public transport developments, ensured that the Greens (institutional coalition partners of the ruling Socialist Party within Paris city council) supported the bid. The lesson from this bid, Hayes suggests, is that hosting relatively small-scale Olympics, in urban transformative terms, is possible, but that, despite successive adaptations to IOC demands, this was also less attractive to the Olympic Movement than the type of vast redevelopment project on offer in London, Beijing and Rio. Moreover, subsequent development shows, from a sustainability perspective, that the Games offer relatively little in additionality – and that, in purely instrumental terms, *not* hosting the Games may in fact be the most attractive outcome of Games bidding, as they may accelerate the mainstreaming of sustainability practices across public policy, whilst avoiding urban mega-projects. For new infrastructures, the removal of the event-driven hard completion timetable may create potential space for (perhaps) effective counter-mobilization and resistance.

Part III, which features contributions from Dansero, Del Corpo, Mela and Ropolo, Whitson and Renou, analyses civic responses to the Games, through the prism of the participation of civil society organizations in shaping and contesting the event. Egidio Dansero, Barbara Del Corpo, Alfredo Mela and Irene Ropolo evaluate the local consultation procedures put in place for the Turin 2006 Winter Games. They argue that despite the Olympic discourse of peace, consensuality and celebration through competition, the staging of the Games is invariably characterized by social conflict; for the authors, mega-events are inherently 'disputed places'. Torino 2006 is a particularly interesting case study as its bipolar spatial organization produces confrontation between development strategies and actors and because the regional population had already proved itself to be highly reactive when faced with top-down projects. The authors consequently discuss the attempts of a wide variety of local actors, especially drawn from the environmental and counter-globalization movements, to provide an effective counter-weight to the staging of the Games. They argue that neither constructive criticism from within the Olympic tent nor opposition from without was able to produce significant positive outcomes – in part, indeed, because of the strategic divergences between groups.

Two accounts of high-profile conflicts surrounding event staging close the collection. David Whitson focuses on the land-use conflict of Eagleridge Bluffs, a four-lane road transport highway connecting Whistler to Vancouver, deemed necessary for the 2010 Vancouver Winter Games, despite its dubious value-for-money, and severe environmental impacts on protected habitats and species in the context of a Games promised to be the 'greenest Olympics ever'. Yet here again, activists were unable to form effective coalitions. Whitson's analysis focuses primarily on a consideration of class politics: in particular, how the NIMBY interests of affluent home-owners whose million-dollar views stood to be spoiled by the new highway may have hindered their cause. His analysis asks why the struggle to save Eagleridge Bluffs briefly became headline news and then disappeared, and whether environmental opposition to mega-events that is led by privileged protestors can be effectively undercut by this very fact.

Finally, activist and organizer Xavier Renou gives an illuminating account of the successful staging of action by human rights and pro-Tibet activists against the Olympic torch as it passed through Paris en route to Beijing in April 2008. Olympic rituals and events are important for civic actors not simply as a target for social mobilization, but also as protest event opportunities in themselves; aside from the emphasis on organization, resources and differential political opportunities for and costs of action over mega-events in different regimes, Renou's account is not just a reminder of the contradictions that lie at the heart of the Games, but also of the possibilities of collective action in formulating a public challenge to these contradictions.

The concluding chapter ties each of these analyses together. Here, Hayes and Karamichas expand on the key ideas put forward across the volume, arguing that we should see the staging of mega-events in the three overlapping frameworks. They are, first, critical moments in the development of host territories, as hosting coalitions use the mega-event to create long-lasting social and cultural change. Second, they are characterized by a series of systemic contradictions, between their legitimizing discourses on the one hand and processes and impacts on the other. Finally, they are disputed places, moments and loci of social conflict.

Part I

Sports Mega-Events, Citizenship, and Civil Society

2
The Four 'Cs' of Sports Mega-Events: Capitalism, Connections, Citizenship and Contradictions

John Horne

The purpose of this chapter is to outline some theoretical frameworks and conceptual tensions around the study of sports mega-events, and thus provide a non-prescriptive structure for the discussions and debates that appear elsewhere in this volume. Borrowing the language of Perry Anderson (2007) writing in the *New Left Review*, these should be regarded as 'Jottings more than theses' and as a result 'they stand to be altered or crossed out' as time progresses. I will attempt to identify, in journalistic fashion, what's the story? What is worth telling about sports mega-events? What things catch the interest of researchers and students and what does not? In this respect I will connect an earlier paper of mine on the 'four "knowns" of sports mega-events' to this one (Horne 2007).

I have noted elsewhere how it is surprising that the sociological and social scientific study of sport – ritualized, rationalized, commercial spectacles and bodily practices that create opportunities for expressive performances, disruptions of the everyday world and affirmations of social status and belonging – was still seen as unserious by mainstream sociologists until recently (Horne & Manzenreiter 2006: 1). Yet there is no obvious reason why social aspects of sport cannot be considered from most classical, modern and post-modern sociological theoretical perspectives, even if the 'founding fathers' did not have much explicitly to say about them. Organized sport and sports competitions feature ritualized, civic, events and ceremonies (Durkheim); rationalized, bureaucratically organized, science-driven behaviour (Weber); commercial, global spectacles (Marx); expressivity and the everyday (Simmel and post-modernism); and opportunities for predominantly male cultural displays and the building of cultural centres (feminism). These are just a few of the issues that have concerned sociological theorists and inform the sociological analysis of sport. It was Pierre Bourdieu however who recognized the difficulty in taking sport seriously from a sociological point of view: 'the sociology of sport: it is disdained by sociologists, and despised by sportspeople' (1990: 156). The reasons for this, and whether Pierre Bourdieu was being unduly pessimistic about sociology ever taking sport seriously or sports people taking sociology seriously, we will leave aside in this chapter.

This chapter, and the collection overall, will hopefully confirm that just as modern competitive sport and large-scale sport events were developed in line with the logic of capitalist modernity, sports mega-events and global sport culture are central to late modern capitalist societies. As media events, the Summer Olympic Games and the FIFA World Cup provide cultural resources for reflecting upon identity and enacting agency.[1] More generally they provide resources for the construction of 'a meaningful social life in relation to a changing societal environment that has the potential to desta-bilize and threaten these things' (Roche 2000: 225). Sports 'mega-events' are important elements in the orientation of nations to international or global society. Hence sport, here in its mega-event form, comes to be an increas-ingly central, rather than peripheral, element of urban (post-)modernity.

The chapter has four sections. First, it will briefly discuss the scope and growth of sports mega-events in the past 25 years and review debates about their attractions and impacts, or legacies. Second, it will briefly discuss the four 'Cs' in the title – capitalism, connections, citizenship and contradic-tions – and their significance for understanding sports mega-events. Third, it offers reflections on the specific relationship between the Olympic Games and civil societies – national and global. Finally the conclusion suggests why sociologists and other social scientists will remain interested in analysing sports mega-events.

What makes a mega-event 'mega', and how did they get that way?

When he was asked to clarify the meaning of the phrase 'Actor-Network Theory' (ANT), French sociologist Bruno Latour (1999) replied that there were four things wrong with it: the words 'actor', 'network', 'theory' and the hyphen. Sometimes it feels the same with the phrase 'sports mega-event'. First, with the obvious exception of the Olympics, there are actually very few sports involved in mega-events. After the Olympics and the FIFA World Cup there is a real struggle to find an agreement about which is the third larg-est sports mega. The UEFA Euro (or as some would prefer, Euro™) Football Championship, the Rugby Union World Cup and even the Ryder Cup golf competition between Europe and the USA, all claim to be the third big-gest according to media audience. The ICC (International Cricket Council) Cricket World Cup has a vast media audience in South Asia, and obvious interest in the other cricket-playing countries, but as with the American Football Superbowl Final it has a narrow fan-base, regionally speaking, that makes it difficult to call it 'mega'. Second, the question is how big does an event have to be to be described as 'mega'? There are different 'orders' of mega-event according to size, scope and reach of the sports involved, their geographical location and their appeal (Black & Van der Westhuizen 2004, Cornelissen 2004, Gratton et al. 2006). The most obvious measure is to refer to broadcast audience size. Yet if the size of an event is primarily related to the overall television audience, this is an estimated figure for much of the world. The difference between TV audience numbers claimed and those actually verifiable can be enormous. Hence whilst media audience size is a key driver of the definition, related promotional opportunities for hosts and corporate sponsors and the potential for the transformation of a location's infrastructure also play a part in defining particular sports events as megas. Third, when it comes to talking about 'events' we find that for many the actual live experience of the mediated spectacle is a lot more mundane. The televised show is spectacular; but from the stands the small figures performing are only recognizable because of TV screens in the stadium. Fourth, do we need the hyphen? Arguably in this case it is useful to link mega and event because otherwise the three words have a habit of moving about – sports mega-event can become mega-sporting event. Rather than using the word 'mega' to simply express that something is big or important, as it has come to be used in everyday speech, I suggest that research interest in sports mega-events should be understood as a specific social scientific focus on a particular genre of mega-event, in Maurice Roche's (2000) terms.

Roche's definition provides the best way into understanding mega-events – as 'large-scale cultural (including commercial and sporting) events, which have a dramatic character, mass popular appeal and international significance' (2000: 1). Two defining features of contemporary sports mega-events are, first, that they are deemed to have highly significant social, political, economic and ideological consequences for the host city, region or nation in which they occur, and second, that they will attract considerable media coverage. By this definition, therefore, an unmediated mega-event would be a contradiction in terms, and for this reason the globally mediated sports genre of mega-event has tended to supplant other forms of mega such as World's Fairs or Expos, although these latter do continue to be hosted and attract substantial numbers of visitors.

Sports 'megas' are thus defined by the ability to transmit promotional messages to billions of people via television and other developments in telecommunications and in doing so have attracted an increasingly international audience and composition. Whilst there remain concerns about the methods of collection and hence accuracy of the television viewing figures announced by organizers after events, the figures act as a proxy for anything more accurate. Illustrations from the Olympics and football's leading tournaments will illustrate this.

The estimated television audience for the 2004 Athens Olympic Games was 3.9 billion people, and the cumulative TV audience estimate was 40 billion; 35,000 hours were dedicated to its media coverage – an increase of 27 per cent over the Summer Olympics held in Sydney in 2000. The most recent 2008 Beijing Summer Olympics drew an estimated cumulative global television audience of 4.7 billion over the 17 days of competition, according to market research firm Nielsen. The 2008 Beijing Olympics was also the most-viewed event in US television history – according to Nielsen, 211 million viewers watched the first 16 days of Olympic coverage on US network NBC. With respect to the FIFA World Cup, staged in Italy in 1990, sales of television rights were estimated to amount to USD 65.7 million (41 per cent), sales of tickets to USD 54.8 million (34 per cent) and sales of advertising rights to USD 40.2 million (25 per cent). Twelve years later the world TV rights (this time excluding the US) were sold for USD 1.97 billion for the 2002 and 2006 Football World Cup Finals. This included a six-fold increase on the USD 310 million paid by the European Broadcasting Union (EBU) for the three tournaments held in the 1990s. The 2002 FIFA World Cup, staged in Japan and South Korea, provided an estimated 41,000 hours of programming in 213 countries and produced a cumulative audience of 28.8 billion viewers (Madrigal et al. 2005: 182).

In the official programme for the UEFA EURO 2008™ competition, the president of the Austrian Football Federation, Friedrich Stickler, claimed that the cumulative audience for the Finals would be 8 billion. By comparison, the UEFA EURO 2004™ Finals in Portugal attracted an estimated total of 7.9 billion TV viewers, and more than 150 million live television viewers for each of the 31 matches. On a global scale, EURO 2004™ was one of the most-viewed sports events ever on TV (Marivoet 2006): 279 million TV viewers watched the final between Portugal and Greece, and the match between Portugal and England attracted the largest ever TV audience for a sports event in the UK. As many as 120 million of the Finals' viewers were outside Europe. This was an increase of 157 per cent on corresponding figures for UEFA EURO 2000™ held in Belgium and the Netherlands. Each match in 2004 drew an average 74 million viewers from outside Europe. After the first round of matches, organizers, host cities, TV executives and not least corporate sponsors anticipated that the UEFA EURO 2008™ would be even more popular among TV viewers world-wide (Carlsberg Media Presskit).

In the past three decades of growth, it is possible to detect four interlinked trends affecting sports mega-events: increased frequency, expansion, proliferation and hierarchical consolidation. The rhythm of the four-year cycle of sports mega-events is indicative of the order and ranking of them. 'Lower order' megas, in terms of audience size, reach and impact, have been moved to accommodate to the rhythm of 'higher order' megas – hence the Winter Olympic Games were held in 1992 and 1994, and the Asian Football Confederation Championship was staged in 2004 and 2007 to adjust to a new cycle, avoid congestion with the higher order megas and to maximize revenue and their potential audience reach. Since 1992, when the Summer and Winter Olympic Games took place in the same year for the last time, there has effectively been a two-year cycle of higher order sports mega-events. The Summer Olympic Games occupies the same year as the UEFA European Football Championship, whilst the Winter Olympics now shares its year with the FIFA World Cup Finals and the Commonwealth Games. The Winter Olympic Games is roughly one-quarter the size of the Summer Games in terms of athletes and events and so some argue that it is not a true 'mega-event' (Matheson & Baade 2003). What counts as a mega-event can differ from one society to another. In those countries that do well in winter sports, the Winter Olympics may be considered as more important than the summer edition. Likewise, where cricket is valued over football, the importance and hence television audience response will be different. In terms of global reach though the Winter Olympics certainly qualifies as a 'second order' mega-event. The UEFA European Football Championship is in a similar category, and is referred to as the second major sports event in the world on the UEFA website.[2] Despite the decision of the Olympic Programme Commission of the IOC (International Olympic Committee) in July 2005 to reduce the number of sports from 28 to 26 from the 2012 Olympics by removing baseball and softball after Beijing (though the number will be reinstated to 28 after London, with rugby sevens and golf added for Rio in 2016), it is evident that the size of these events, as well as the enthusiasm to host and participate in sports mega-events like the Euro, the Olympic Games and the FIFA World Cup, has grown in the past 20 years.

Between 1980 and 2000 seven new sports and 79 events were added to the programme of the Summer Olympics; 28 sports have featured in the Summer Olympics since 2000. After 1998, the FIFA World Cup Finals expanded from 24 to 32 football teams. The Euro Finals expanded to 16 teams in 1996 and will expand to 20 teams from 2016, thus jeopardizing the chances of smaller countries from acting as hosts in the future. Possibly as a response to this growth in size, co-hosting of the Euro by relatively small European nations was first permitted in 2000 (Belgium/Netherlands) and, after 2008 (Austria/Switzerland), will also be the format for the 2012 edition in Poland and Ukraine. This expansion and growing attraction of sports mega-events have been for three fairly well-known reasons.

First, as we have already noted, new developments in the technologies of mass communication, especially the development of satellite television, have created the basis for global audiences for sports mega-events. Since the 1960s, US broadcasting networks have substantially competed to 'buy' the Olympic Games. Next in order of magnitude of rights payments is the consortium representing the interests and financial power of Europe's public broadcasters, the EBU, that buy the rights to transmission in Europe. In addition media rights fees are paid by the Asian broadcasters (including Japan and South Korea) and national media organizations, such as CBC and CTV in Canada. So in addition to the USD 300 million paid by the

US corporation NBC to the International Olympic Committee in 1988 (for the Seoul Summer Olympics), the EBU paid just over USD 30 million and Canada paid just over USD 4 million for media broadcasting rights. For the Beijing Summer Olympics in 2008, NBC paid USD 894 million, the EBU paid over USD 443 million and Canadian broadcasters paid USD 45 million just for the rights to transmit pictures of the action (Coakley & Donnelly 2004: 382, Westerbeek & Smith 2003: 91).

Similar to the Olympic Games, since the 1980s the FIFA World Cup has attracted substantial media interest and commercial partners and has thus become a huge media event. The resources made available for the communications systems, the enormous media centres and the amounts paid by national broadcasting systems to televise the event provide ample evidence for this. Not surprisingly therefore representatives of the media now easily outnumber the athletes at sports mega-events – in Sydney in 2000 there were 16,033 (press and broadcasting) reporters and during the Winter Olympic Games in Salt Lake City in 2002 there were 8730 reporters covering the performances of 2399 athletes (Malfas et al. 2004: 211).

The second reason for the expansion of mega-events is the formation of a sport-media-business alliance that transformed professional sport generally in the late twentieth century. Through the idea of packaging, via the tripartite model of sponsorship rights, exclusive broadcasting rights and merchandising, sponsors of the Olympics and the two biggest international football events have been attracted by the association with the sports and the vast global audience exposure that the events achieve.

The two largest sports mega-events led the way in the 1980s in developing the transnational sport-media-business alliances worth considerable millions of dollars, but UEFA football competitions (at club and national levels) have closely followed the same pattern. The idea of selling exclusivity of marketing rights to a limited number of sponsoring partners began in Britain in the 1970s with Patrick Nally and his associate Peter West at the media agency WestNally. In the early 1980s, the idea was taken up by Horst Dassler, son of the founder of Adidas and at the time chief executive of the company. With the blessing of the then FIFA president João Havelange, Dassler established the agency ISL Marketing in 1982. Later in the 1980s, ISL linked up with the IOC, presided over by Juan Antonio Samaranch. It was ISL that established TOP, or 'The Olympic Programme', in which a few select corporations were able to claim official Olympic world-wide partner status. Whilst the TOP programme supports the Olympic Movement internationally, sponsorship agreements by Olympic host cities create even further opportunities for making money. Hence for the 2008 Olympics in Beijing, organizers created three additional tiers of support at the national level (Beijing 2008 Partner, Sponsor and [exclusive] Supplier). In light of the enormous attraction of the Chinese market it is not surprising that revenues from national sponsorship arrangements were likely to be considerably more than those gained from the TOP programme. For example there were three companies – Anheuser-Busch ('Budweiser'), Qingdao-based 'Tsingtao' and the Beijing-based 'Yanjing' – acting as beer sponsors at the national level in 2008. With respect to London 2012, despite being caught up in worsening global economic conditions, sponsorship has remained as important as ever. In November 2009 the London Games organizers could boast nine Worldwide partners (from the IOC's TOP Programme), seven official partners, six official supporters and 11 official suppliers and providers. As Sugden and Tomlinson (1998: 93) noted in relation to the World Cup, 'Fast foods and snacks, soft and alcoholic drinks, cars, batteries, photographic equipment and electronic media, credit sources – these are the items around which the global sponsorship of football has been based, with their classic evocation of a predominantly masculinist realm of consumption: drinking, snacking, shaving, driving.' The same applies to the sponsors and partners for the UEFA Euro competition and the Olympic Games.

The third reason why interest in hosting sports mega-events has grown is that they have come to be seen as valuable promotional opportunities for cities and regions. The aim is to generate increased tourism, stimulate inward investment and promote both the host venues and the nation of which they are a part to the wider world, as well as internally. An element of this is what John Hannigan (1998) calls the growth of 'urban entertainment destinations' (UEDs) as one of the most significant developments transforming cities throughout the developed world. The breath-taking 'Bird's Nest' stadium built for the 2008 Summer Olympics now offers a centrepiece of this kind for both foreign and domestic tourists travelling to Beijing. Hannigan argues that the ideal 'fantasy city' of the late twentieth and early twenty-first century has been formed by the convergence of three trends. First, through the application of the four principles of efficiency, calculability, predictability and control (or 'McDonaldization' as Ritzer (1993) described it), there has been a rationalization of the operation of the entertainment industries. Second, theming, as exemplified by the Disney Corporation (or 'Disneyization', see Bryman (2004)), produces new opportunities for commercial and property developers in urban areas. Third, accompanying synergies between previously discrete activities, such as shopping, dining out, entertainment and education, lead to 'de-differentiation' – what some analysts regard as a feature of 'post-modernization' and others regard as key experiential commodities in the growth of consumer capitalism.

The four Cs

The next section turns attention to the four 'Cs' in the title – capitalism, connections, citizenship and contradictions – and their significance for understanding sports mega-events. The developments relating to the staging of sports mega-events in the past 30 or so years outlined in the previous section, alongside the pursuit of enhanced, or even 'world class', status by politicians and businesses, raise questions for some analysts about the social distribution of the supposed benefits of urban development initiatives. Which social groups actually benefit, which are excluded and what scope exists for contestation of these developments are three important questions that are often ignored (Lowes 2002). In the build-up to bidding for mega-events, Gruneau (2002: ix–x) argues that local politicians and media often focus on the interests and enthusiasms of the developers, property owners and middle-class consumers as 'synonymous with the well-being of the city'. As a result, sectional interests are treated as *the* general interest, and ongoing 'class and community divisions regarding the support and enjoyment of spectacular urban entertainments' are downplayed, if not ignored altogether (Gruneau 2002: ix–x).

This downplaying is what I have called, after Slavoj Žižek's retort to the former US Defence Secretary Donald Rumsfeld, one of the 'known unknowns' of sports mega-events (Horne 2007). Rumsfeld, engaging in speculation about the situation in Iraq in March 2003, had stated:

> There are known knowns. These are things we know that we know. There are known unknowns. That is to say, there are things that we know that we don't know. But there are also unknown unknowns. There are things we don't know we don't know.

Žižek (2005) felt that Rumsfeld had forgotten to add a crucial fourth term, '"unknown knowns", things we don't know that we know'. I agree with Žižek and consider that it is an academic's duty to look critically and self-critically at the assumptions, beliefs and sometimes obscene practices undertaken by those involved with sports mega-events that are often suppressed, or perhaps more accurately, repressed – the 'unknown knowns'.

Capitalism

When it comes to sports mega-events, politicians, senior administrators of sport, corporate leaders and even some academic researchers encourage the pretence that we do not know about many of the most significant things that actually form the background to them. This is the case because, just as with other aspects of urban planning, sports mega-events are highly political affairs, surrounded by sports, urban and corporate interests. Even the language used is highly nuanced – as Hiller amongst others has pointed out, the aftermath or repercussions of sports megas are often discussed now in terms of their 'legacies', rather than their 'impacts'. Yet legacy is a warm word, sounding positive, whereas if we consider the word 'outcomes' it is a more neutral word, permitting the discovery of both negative and positive outcomes. Whilst outcomes can be tangible and material or intangible and symbolic – and economists and urban planners have tended to focus their research attention on the former (Gratton et al. 2006), whereas sociologists, political scientists and social geographers have often been more interested in the intangible, symbolic and representational outcomes (Manzenreiter 2006, McNeill 2004) – the relationship of sports megas to developments in contemporary capitalism is evident.

Contemporary capitalist development has been underpinned by 'the shock doctrine' according to media commentator and author Naomi Klein (2007a). She argues that the use of shock, or violence, is a technique or tool in order to impose an ideology – what she calls the free market fundamentalist ideas underpinning neo-liberal economic thought and policy. The shock doctrine is also a philosophy of how political change can happen and be brought about. Charting the rise of free market fundamentalism over the past 35 years reveals that when ideas are unpopular advocates of free market neo-liberalism have exploited shocks to help push through their policies without popular democratic consent. The product is what Klein calls 'disaster capitalism' – a form of capitalism that uses large-scale disasters in order to push through radical neo-liberal capitalist policies and its related privatization agenda for (formerly) public services. In addition disaster capitalism also creates disasters and responses to them in what Klein refers to as its 'post-modern' form (Klein 2007b).

Klein in her investigation explores the roots of disaster capitalism back to General Pinochet's military coup on 11 September 1973, and monetarist economist Milton Friedman's prescriptions for Chile afterwards. In addition to the use of torture, and the imprisonment and murder of dissenters in Latin America, Klein views Thatcher's Falklands/Malvinas War in the 1980s as an equivalent shock that enabled privatization to be implemented in the UK. More recently environmental disasters, terrorist attacks (such as the other September 11th, in 2001, when the World Trade Center in Manhattan was destroyed) and economic crises have been used to bring about free market reforms. Hence another of Klein's arguments is that the equation of the free market with democracy and freedom is misplaced – privatization and the spread of capitalist market relations more generally have often been accompanied by violence, terror and crises. The idea that capitalism advances on the back of disasters, or violent circumstances, is not a new one. Marx (1867/1973) wrote at length in *Capital* about the extra-economic coercion required to bring about primitive accumulation in the eighteenth century in Western Europe, for example. But Klein's book is a valuable insight into recent history.

What relationship does sport and specifically sports mega-events have with disaster capitalism? Within the 35-year period that Klein refers to sports megas have become more prominent as we have seen. Anderson (2007: 5) writes that, since the 1980s, America's 'two great protagonists of the Cold War period, China and Russia', have been 'integrated into the festivities of the global spectacle (St Petersburg summit, Beijing

Olympics, etc.)'– and we can now add Sochi, selected as host of the 2014 Winter Olympic Games to this list. So I want to suggest that these sporting spectaculars can be viewed as the twin of disaster capitalism's shock therapy, involving their own shocks and generating their own forms of awe. Winning a bid to host a mega-event, putting the fantasy financial figures of the bid document into operation, dealing with the proposed location before and dealing with it after the event has taken place are just some of the moments where shock and awe is generated by sports mega-events.

Connections

Appadurai (1996) recognized that there are economic, technological, financial, ethnic and image mobilities that help construct the contemporary globalizing social 'scape'. Sport can be seen as both a metric and a motor of this globalization process (Giulianotti & Robertson 2007). Sport – in its professional-, commercial- and consumer-oriented forms – is inevitably part of the expansion of capitalist social relations on a global basis. Sport has become increasingly commodified. Sport reflects this in many ways, including in terms of the unequal distribution of involvement and participation in sport and the growth of the global sports market. Sports mega-events have perhaps their greatest contemporary allure as an element in these globalizing processes.

For athletes, sports fans and many citizens the appeal of hosting a 'once in a lifetime' experience on home soil is palpable. The ability to make connections with global flows, possibly as a new hub in the networks of financial, media or tourist flows, is a primary motivation for the involvement of city, regional and national governments in the competitions to host sports megas. Sports megas have promotional leverage, and enable (transnational and national) corporations to leverage business opportunities out of them. But in doing so the risky nature of sports mega-events also comes into focus.

Citizenship

Along with Rick Gruneau previously mentioned, many other writers, including Whitson and Horne (2006), have raised concerns about the distribution of the benefits and the costs of sports mega-events. As Michael Hall (2005) points out, the selling of a city in order to host a sports mega-event or develop it in ways to attract inward investment can lead to some local citizens being sold short. Modern Western cities are based in large part on activities of 'repair and maintenance' according to Nigel Thrift (2005: 135). Citizens there are 'surrounded by the hum of continuous repair and maintenance' (Thrift 2005: 136). Cities in the Global South may be in a continuous state of emergency; they operate repair and maintenance on the basis of social networks, based on kin and friendships (ibid.: 138). A key feature of repair and maintenance is the idea of 'regeneration', and this takes material and representational forms, just like legacies or outcomes promised from mega-events. The promotional value of sports mega-events for cities in the Global North, and some in the South, relates to the search for international esteem and 'world class' ranking, via image generation, and both external and internal promotion. For example, the use of the phrase 'Expect Emotions' as the slogan for UEFA EURO 2008™ reminds us that, as Thrift suggests, 'the systematic engineering of affect has become central to the political life of Euro-American cities' (Thrift 2004: 57). Cleverly the phrase 'EMotions' also enabled the co-organizers of Euro 2008 to allude to the abbreviation of the competition understood by their German-speaking populations and the German team which also qualified for the Finals – 'EM' = 'Europameisterschaft 2008'. As Bennett (1991), with respect to Australian cities, and more recently Whitson (2004), with respect to Canada, have suggested, staging (sports mega-) events is as much about engineering the emotions of the local populations as welcoming foreign visitors.

Contradictions

Mega-events are short-life events with longer-life pre- and post-event social dimensions, not least because of their scale, their occupation and maintenance of a time cycle, and their impacts (whether conceived of as positive or negative). As sports mega-events have become global media events they have taken priority over World's Fairs or Expos as a result. I have focused much attention on the structural impacts of sports mega-events. Following improvements in global mediation, corporations use sports mega-events to leverage business opportunities more than ever before, and this neatly sums up one of the major concerns with them. To add to these we might mention the following four significant contradictions: the potential for patriotic promotional discourse to sow the seeds for heightened xenophobia, the imbalance between local democratic control and autocratic (international) sports and other organizational demands, the growing imbalance between dependency on global media rights on the part of organizers and Internet streaming of content, and the way that the biggest multi-sport mega-event (the Olympic Games) relies upon its anti-commercial ideology as its major commercial asset. In addition to the overestimation of their benefits and underestimation of their costs, the related uneven internal development as the host location benefits from the 'lightning rod' effect of the mega-event on public infrastructural spending decisions, makes the hosting of sports mega-events one of the most fundamentally political acts of the current age. Sport is thus now fully interconnected economically, ideologically and politically into society through its mega-events.

Sports mega-events and civil societies

Roche (2000: 41) outlines four ways in which megas 'provided opportunities and arenas for the display and exercise of "civil society" in addition to "the state" both at national and international levels'. Focusing on the modern Olympic Games and the IOC established in 1892, he argues that the Games were connected to the following aspects of global citizenship: universal citizenship, and the associated discourse on human rights; mediatized citizenship, and the rights to participate in the Olympics as a media event; movement citizenship, and the rights to participate in the Games as a sports organization and a movement; and finally corporate citizenship, or the position of the IOC as a collective actor in global civil society. His assessment of these aspects is briefly considered before we outline some elements of civil society that can be produced by mega-events.

Regarding the first, universal citizenship claims, Roche states, 'arguably the negatives outweigh the positives in the Olympic record' (2000: 203). That within 27 years of the cessation of hostilities in 1945 all three axis powers had hosted at least one Olympic Games might suggest otherwise, but Roche argues the IOC has not tended to take 'a consistent and strong line on the human rights record' of the host nations (ibid.). The likelihood that the Olympics as a media event will become fully available to all people in the world, via the Internet, is another of those arguments about new media technologies that is based as much on hope as experience. It is difficult to imagine that the IOC can allow Internet coverage of the Games to compromise the major element in its funding – exclusive broadcasting rights revenue. Hence the media coverage of the mega-event has tended to be both commercialized and nationalized – insofar as the sports covered (the 'feed') tends to be determined by national TV companies' choices, in line with the involvement of its athletes and the anticipated tastes of its viewers. It is in this way that international mega-events can be transformed into fora for national(ist) introspection.

As a 'movement' the IOC claims to be quite different from other sports mega-events. The problem is that even if the Olympics were seen, by progressives and reactionaries alike, as a positive cultural innovation 100 years ago, this was essentially a dream built upon a particular set of values and relationships (embodied in the ideology of nineteenth-century amateurism, and based on Western, masculine and (upper) social class-based moral conceptions) that simply no longer apply. Members of the IOC today, and many sports people and physical educators, still believe that sport has a higher social and moral purpose, but elite sport has become part of consumer culture, and mega-events its commercial spectacles. The Olympics uses its difference as a 'movement' with an ideology (Olympism) from other world cups and commercial events (such as the FIFA and IRB (International Rugby Board) World Cups) to provide it with its own distinctive 'brand': anti-commercialism can thus enhance its commercial value! However the IOC faces two main challenges, around democracy and fair play. Under siege since the 1990s about its undemocratic procedures, it remains 'a self-recruiting and secretive elite international club, directly accountable and accessible to nobody but itself' (Roche 2000: 207). Despite the establishment of an ethics commission and various other sub-committees attempting to bring about Olympic reform, since then the IOC has had difficulty, outside of those people ready to accept its ideology, in convincing others that it is operating according to the highest standards of democratic governance. This situation has not been aided by the increase in revelations of cheating in sport, and especially the use of performance-enhancing drugs, in a context where the chasm between the rewards from success and the anonymity derived from losing has widened considerably.

Finally, as a collective actor in global civil society, the IOC has had to deal with another two issues concerning its integrity: the process of bidding to act as host and the development of the idea of an 'Olympic truce'. Regarding the first issue, Olympic city bidding corruption and the role of agents in helping to win a bid were the focus of investigative journalism for much of the 1990s, and especially after 1998, and the revelations surrounding the bribes that enabled Salt Lake City to obtain the 2002 Winter Olympic Games. The Olympic truce idea, in conjunction with the United Nations, is an Olympic contribution to international civil society in so far as it seeks the preservation of human life and peaceful coexistence. Yet through this the UN risks 'being associated with an association which is committed to commercialism, global capitalism and consumer culture' (Roche 2000: 214–15).

In addition to their structural properties, Roche also draws attention to the phenomenological impacts of sports mega-events. In particular he looks at the role of 'megas' in providing time-structuring resources – both interpersonal and public – and suggests that the 'once in a lifetime' opportunity is one of the main reasons for their popularity, at least amongst those who live in the cities and places that host them. He argues elsewhere (Roche 2003) that mega-events are socially memorable and culturally popular precisely because they mark time between generations, and thus provide a link between the everyday life world (micro social sphere) with the meso and macro social spheres. They are 'a special kind of time-structuring institution in modernity' (ibid.: 102). Hughes (1999) too notes that underpinning the economic strategies captured by such notions as selling places, place marketing and the creative city, the idea that ludic space might be an economically valuable use of land has come to the fore. The ludic city however might also be seen as valuable for the growth of sociality and the consideration of alternative ways of relating to each other as human beings (see Latham 2003). Consumer identities and consumer spaces are produced by trademarked mega-events.

At the same time sport, culture and (pop) music events enable flows and mobilities of people and non-human entities. In the midst of these, new social identities and understandings – interlinked through social class, gender, ethnic, religious and national differences – may be produced, resisted and sustained. Cashman (2006: 21–2) suggests that memory regarding sports mega-events such as the Olympic Games can take three forms: individual or private memory, spontaneous collective memory and cultivated public memory. This begs a question however about who does the sustaining of memory – at the grassroots, citizens, the media or politicians – and for what ends? There can be a tendency, when recalling events, toward what Cashman refers to as 'Olympic reductionism' (ibid.: 25). Here memories are reduced to the highlights – 'a few events which are repeatedly mentioned in public discourse' – and usually only the official achievements. In the popular memory of sports mega-events, how it is possible to go beyond these official accounts is an important question that needs to be addressed.

Conclusions: what's the story?

In conclusion I want to reaffirm why sociologists and other social scientists should be interested in analysing sports mega-events and what the four 'Cs' I have identified can contribute to an understanding of a fifth – civil societies. There is now a well-established rhythm to the sports mega-event cycle or calendar. Every 'leap year' there is a Summer Olympic Games and a UEFA football championship. Two years later there will be a Winter Olympic Games, a Commonwealth Games and the FIFA World Cup. Outside of these even years there are decisions being made about the location of either a Summer or a Winter Olympics by the IOC, competitions leading to qualification for football's major events (the football World Cup or Euro) and many other lower order but socially significant sports mega-events (2011, for example, sees the IRU Rugby World Cup in New Zealand and the ICC Cricket World Cup in India, Sri Lanka and Bangladesh). The same issues and concerns sketched here in general apply in specific focus to these and other lower order sports megas.

Sports mega-events are obviously not essential to living, but the organization of contemporary social life would be different without them. As Naomi Klein suggests, if the development of the contemporary capitalist order has been underpinned by a 'shock doctrine', then the spectacular impact and impression that contemporary sports mega-events make on their host locations and global audiences can be seen as an ally of, rather than in conflict with, this doctrine. Creating competitions between cities, regions and nations to host events; exemplifying the ideal typical person in contemporary capitalism; offering promotional opportunities for both corporations and public agencies; promising, if successful, the generation and regeneration of significant parts of the urban infrastructure; and attracting large flows of capital, people and technological attention – sports mega-events entail many if not all of the dominant features of contemporary capitalism. Social scientists are attracted to study sports mega-events precisely because of this, and the dynamic processes involved and the political struggles they provoke are central to understanding the way contemporary social life is organized in all its facets. The contributions to this volume, and hopefully this chapter, also invite us to consider how an adequate politics of sports mega-events might be developed.

Notes

1. Although it has become conventional to refer to the 'Olympic and Paralympic Games', as hosts of both Summer and Winter editions of the Games are now expected to also host the Paralympic event three to four weeks after the 'main event', we do not address the specifics of this event here. For discussion of some of the issues associated with the Paralympics see Howe (2008) and Cashman & Darcy (2008).
2. 'In terms of viewing figures, the UEFA European Championship is the world's second major sports event, only topped by the FIFA World Cup', http://www1.uefa.com/history/background/development/index.html [accessed 30 November 2009].

3
Civilizing Beijing: Social Beautification, Civility and Citizenship at the 2008 Olympics

Anne-Marie Broudehoux

In recent years, mega-events have been increasingly studied in the urban lit-erature for their capacity to enhance the global visibility of a city and their role as catalysts for the modernization of urban infrastructure in host cities. Yet another important urban aspect of mega-events-led city marketing and urban image construction that is too often overlooked in the literature is the way these high-visibility global-scale events also foster state interventions to reform and control social behaviour. Hosting mega-events often pressures host cities to reinvent their image by transforming their human environ-ment through social beautification and disciplining programmes. Urban research on mega-events is generally concerned with the socio-spatial impli-cations of Olympic redevelopment, the impacts of mega-events on local pol-itics and public policy, the economic legacy of mega-events and their role in city marketing (De Lange 1998, Essex & Chalkley 1998, Burbank et al. 2001, Gold & Gold 2008). Only rarely are the important 'civilization campaigns' that accompany these initiatives examined (Choi 2004, Lenskyj 2002).

In reality, mega-events, and particularly the Olympic Games, are often harnessed by local authorities as a powerful tool of social engineering, to transform both the body and the mind of their population and produce a tame and obedient citizenry that fits modern norms of public behaviour and global expectations of civility. Far from being the monopoly of auto-cratic states, attempts to regulate human activity and normalize behaviour through social reform and disciplining programmes have been observed in a variety of settings across the political spectrum. However, authoritarian regimes do have an advantage in their ability to unleash great collective energy and to elicit or coerce self-sacrifice with limited popular resistance, allowing them to carry out social beautification campaigns on a more exten-sive scale than in more democratic states.

Over the past 20 years, with the opening of China to the world, the Chinese leadership has been increasingly involved in the staging of national and international-scale events, and has developed an expertise in showing the national capital and its residents in the best possible light, through a now familiar scenario that combines urban and social beautification initiatives with more coercive disciplinary measures. The 2008 Beijing Olympics provided the opportunity to refine this scenario.

Indeed, the picture perfect image of a 'harmonious society' seen by millions on televisions screens and in the streets of the Chinese capital in August 2008 is a testimony to Beijing's impressive capacity for propaganda and control. Beijing's Olympic image construction efforts were not limited to spectacular architecture projects and modern infrastructure construction, but they also included a series of social engineering programmes, which sought to show the world that China has not only developed economically, but socially and culturally as well. Beijing Olympic organizers and civic leaders perceived the Olympic Games as an 'opportunity for societal advancement' (*China Daily* 30 March 2007), a stage for popular acculturation into global cosmopolitan society, and an occasion to promote state-sanctioned ideals of Chineseness. The momentum and civic pride attached to hosting the Olympic Games helped hasten the pursuit of a civilization campaign, initiated in the 1990s, that sought to turn Beijing residents into well-disciplined representatives of twenty-first-century China.

This chapter examines how the 2008 Olympic Games offered the Beijing leadership the opportunity of a powerful tool of social reform, used to transform and control both the body and the mind of the Beijing population, and to rewrite the terms of belonging to Chinese society. Through an investigation of some of the main approaches used in this process, ranging from ideological means of reform and embodied civilizing practices to more coercive measures of control and discipline, this chapter raises questions about the role of mega-events in the conscious and planned management of human activity, the regulation and normalization of behaviour, and the organization of social stratification and domination.

The civilization process

Norbert Elias (2000) defines the historical normalization of behaviour in Europe as the 'civilization process'. He views the civilization process as an intrinsic part of the modernization and urbanization of society, justifying the development of social control mechanisms, the normalization, rationalization and regulation of social conduct, and the imposition of limits and contexts of obedience to the law.

Michel Foucault (1975) vastly documented how nineteenth-century social reformers – who conceived of the social as an object of regulation – developed different forms of social engineering and disciplinary procedures to reform bodies and produce efficient, well-behaved and productive individuals; using public education and other forms of training to help internalize norms of behaviour and rationality, to maintain a certain sense of order and to safeguard morality. The modern state would use similar disciplinary practices to facilitate the implementation of the universalizing norms of modern society, and subjected bodies to disciplinary regimes that sought to uniformize, pacify and control their behaviour (Foucault 1976).

The implementation of modern social reforms has not been limited to concerted, state-led actions, but was also carried out by members of civil society, motivated both by the idea that certain norms of behaviour are desirable for the benefit of all, and by a desire for social differentiation. For Foucault (1975), although questions of security, health and hygiene historically justified the enforcement of social norms, they generally concealed social motivations of status and a concern for appearances and distinction, since health justifications often appeared after the fact. Social distinction is fundamental in the development of 'civilized' social norms, and civilizing projects are closely linked to the emergence of modern social hierarchies. The civilization process is thus predicated upon social inequality and competition for status, and follows a logic of social exclusion; Elias (2000), indeed, identifies good manners and etiquette as resources historically used by upper classes to set themselves apart socially and dominate lower classes, regularly adding new behaviour restrictions to ensure the maintenance of a comfortable distance with the masses.

Civilizing projects are also defined by a fundamental inequality between civilizing agents and the people upon which they act. Notions of civility are therefore closely related to the right to participate in society and to the constitution of citizenship. It is through embodied practices that mark individuals and groups as appropriately or insufficiently civilized that eligibility for inclusion in or exclusion from an idealized body politics is established, and that citizenship is defined (Friedman 2004).

Mega-events and the civilizing process

Mega-events have long played a role in the spread of a civilizing ideal and the globalization of modern cultural norms. As mass celebrations of progress and modernity, early World's Fairs were driven by an implicit civilizing impulse and often acted as sites for the acculturation of the masses into Western industrialized culture (Rydell 1993). The modern Olympic Movement has also been marked by a drive to inculcate new morals, values and embodied practices upon their hosts, visitors and participants (Tohey & Veal 2004, Brownell 1995).

More recently, the intensified mediatization of mega-events, and their rising role in city marketing and national boosterism, have placed citizen behaviour at the forefront for image-conscious civic leaders and event organizers, and prompted the adoption of social beautification programmes. The symbolic power of mega-event 'brands', such as the Olympics or FIFA World Cups, has helped legitimize such social engineering initiatives, which would otherwise be perceived as oppressive and elicit popular resistance.

While the role of powerful international federations like the FIFA and the IOC in the management of social behaviour and the development of civilization programmes is generally indirect and diffuse, it is clear that their low tolerance for civic disturbances and their insistence upon media image, public order and security exert tremendous pressures on local organizing committees and host governments to take social behaviour seriously. Chapter 5 of the IOC's Olympic Charter explicitly states that 'No kind of demonstration or political, religious or racial propaganda is permitted in the Olympic areas' (IOC 2003: 92). The contracts signed between federations and host cities often contain a series of covert disciplinary protocols, and these unworded demands are generally internalized by local authorities, who are eager to comply with federation rules and perceived expectations to create the best possible social environment for the event (Bennett 1991, Choi 2004).

There are important social costs associated with the way mega-events influence the management of human behaviour. By urging local leaders and event organizers to provide secure urban spaces, safe from violence or political agitation, mega-events can foster the development of complex tactics of social disciplining, with a tightening of the social control apparatus, and the imposition of limits on civil liberties (Lenskyj 2002, Broudehoux 2004, 2007). As host cities and their residents are urged to play up to international standards, civilization programmes associated with the hosting of mega-events can also be experienced as a form of cultural imperialism, where imported global cultural norms are imposed upon local reality. In the run-up to the 2002 FIFA World Cup it co-hosted with Japan, for example, South Korea implemented image construction programmes to 'raise [the] standards to match that of an advanced nation'; measures initiated focused on global etiquette, cultural mannerism and the Anglicization of the urban landscape, and were criticized for Westernizing Korean society by effacing its own cultural standards with an American replacement (Choi 2004). Mega-events also promote social exclusion, especially when urban sanitization and social beautification campaigns arbitrarily target, criminalize or simply camouflage those deemed detrimental to the city's positive image or threatening to the smooth realization of the event. Recent examples include a 1984 pre-Olympic 'gang sweep' in Los Angeles, which sent hundreds of young black men to jail, and the confinement of the homeless and mentally ill and deportation of refugees and asylum seekers by the Greek government for the 2004 Athens Games; Lenskyj (2002, 2008) discusses similar examples in Sydney, Atlanta and Vancouver.

The civilizing process in China

In China, where the moral education of the people has long been viewed as a function of good government, the civilization process has gone hand in hand with modernization. Throughout the twentieth century, Chinese elites, intellectuals and officials have pursued a modern civilization ideal, both as an ideological mission and as a national project, advocating *wenming* (civilization) as a national strategy for radical social transformation (Anagnost 1997: 81–2, Duara 2001: 122). China's masses have been subjected to diverse civilizing campaigns, motivated by ideological discourses that had their roots both in neo-Confucian principles and in enlightenment ideals of evolutionary progress. But in spite of important differences in the ideological construction of civilization under different leaderships, civilizing projects were always defined by a fundamental inequality between the civilizing centre and the peripheral people subjected to civilization, and thus justified the domination of different population groups that were perceived as needing to be reformed; this is especially the case with ethnic minorities who were considered to need to be reformed and assimilated during the Mao years (Harrell 1995).

Early urban modernization efforts in pre-Socialist China were accompanied by civilizing initiatives. Already in the late 1920s, a series of public education campaigns, led by an elite upholding modern values of rationality, discipline and order, was launched throughout urban China to reform both the body and mind of the Chinese people. These campaigns were generally conducted in the public spaces of the city, which were turned into sites of acculturation into proper behaviour and as supports for public education messages on public behaviour and civic virtues. Madeleine Yue Dong (2000) describes how, in the late 1920s, China's new republican government strove to reach out to the masses through the creation of public parks where public behaviour could be reformed. While this acculturation process was fostered in part through emulation, as the less civilized masses could learn from the example set by educated elites, proper demeanour was also promoted through repetitive and ubiquitous propaganda. Early urban parks were filled with didactic messages and mottoes preaching the moral principles of modesty, tolerance, honesty, patriotism, diligence, self-reflection, critical thinking, healthy living and public mindedness (Dong 2000, Shi 1993).

In Communist China, strict norms and forms were also imposed upon the Chinese experience of modernity. Early Communist leaders saw their mission not simply as maintaining or improving society but as transforming and restructuring it as well. Consequently, they initiated important social programmes, which aimed at remoulding both the body and mind of their population, with intense public education and mass indoctrination campaigns.[1]

In the Mao years, civilizing discourses were mainly anti-feudal in nature and sought to promote modernization, socialism and revolutionary ideals such as class struggle and allegiance to the party. They also touched upon issues of civility, personal discipline and public behaviour, albeit with a more Socialist outlook. If traditional Chinese customs were very strict about manners and public demeanour, after the Communists took power in 1949, etiquette and refinement were actively rooted out, condemned as elitist and bourgeois. Under the principles of 'social egalitarism', it was thus considered revolutionary to defy conventions, especially Confucian norms of civility, and to endorse coarse and simple peasant manners.[2] The eradication of the upper class, followed by the crackdown on the intelligentsia during the Cultural Revolution (1966–70), similarly affected previous social norms and manners. Public security relied upon the positive cooperation of the citizenry through political indoctrination, public education and normative appeals to patriotism, on a more coercive policing system based on routine public discipline campaigns to apprehend unruly citizens and on periodic demonstrations of state repression to deter disruptive behaviour (Vogel 1971).

In the post-Mao period, especially under Jiang Zemin, civilization became one of the keywords of the new state ideology, reoriented to serve the ideals of market socialism. The new doctrine focused on the 'construction of a socialist spiritual civilization' *(shehui zhuyi jingshen wenming jianshe)*, which sought to instil a commitment to both state and market, while reinvigorating national identity. The *spiritual civilization* project aimed to build a new, modern set of values for Chinese society to match China's modern 'material civilization', or economic development and infrastructure. A central tenet of the spiritual civilization project, which was administered by the Office of Spiritual Civilization *(Jingshen wenming bangongshi)*, a sub-agency of the Central Propaganda Department, thus focused on the promotion of civilized behaviour, through various national campaigns and mass pedagogical programmes.

These campaigns were clearly marked by a self-conscious desire to upgrade China's world image, which had become a major concern in the face of growing contacts with the international community. For Ann Anagnost, the new civilizing project exposed the failure of the Chinese people to embody international standards of modernity, civility and discipline. It sought to raise the 'quality' of the people *(tigao renminde suzhi)* by refashioning them into the modern, disciplined and productive citizens who would help China assume its rightful place in a global community of civilized nations (Anagnost 1997).

In 2006, Hu Jintao's administration formally endorsed the principle of 'Harmonious society' *(hexie shehui)* as its new official doctrine, to emphasize social unity, cohesion and peaceful harmony, in China's relationship with other nations, between state and society, and with its environment. The concept of harmony *(hexie)* – a classic Confucian term that connotes humanism, decency and honourable behaviour – has been reinterpreted since imperial times by those in power to foster popular obedience and respect for authority. While its resurgence as the main keyword of the Hu government's ideology sought to announce the advent of a more humanistic leadership, it also follows similar interpretations of Confucian doctrine as practised by Asian states such as Singapore, where it is used to justify marrying authoritarian politics with capitalist prosperity. According to critics, in such selective rereading of the Confucian concept of harmony, compliant behaviour and submission to authority are emphasized, whereas notions of social justice, political dissent and the moral duty of citizens to criticize abusive or oppressive rulers are conveniently ignored.[3]

The harmonious society doctrine focuses on notions of national unity, shared values, social stability, public order and the rule of law, and is seen by many as an instrument of pacification and national unification in a period when growing disparities and differences are disrupting social peace. Civility has long been known to serve as a social lubricant meant to harmonize social relations, minimize social frictions and facilitate peaceful coexistence. As China prepared to host the world at the 2008 Beijing Olympics, the ideal of the harmonious society appeared as a timely response to the mounting popular discontent and worsening social tensions that increasingly threaten national stability. As we will see, the Olympic Games will act both as a means and an end in implementing the harmonious society ideal and play a key role in the official orchestration of the civilizing discourse.

Social reforms and the 2008 Olympics

Part of Beijing's Olympic preparation thus involved an important programme for reforming the demeanour of Beijing residents, as both hosts and potential members of the Olympic audience, and to transform them into well-mannered, modern Chinese citizens. With a budget of USD 2.5 million from the Beijing Organizing Committee of the Olympic Games, the Capital Ethics Development Committee, a subset of the Office of Spiritual Civilization of the Beijing city government, was charged with 'raising the quality and civilization level' of the city's 15 million inhabitants, to ensure they were on their best behaviour for the event.

This Olympic civilization programme was overseen by three main agencies with close links to the central government's propaganda apparatus: the Central Propaganda Department, the Central Office of Spiritual Civilization and BOCOG. Since 2005, BOCOG has been headed by Liu Peng, who had been the deputy director of the Central Propaganda Department from 1997 to 2002. Liu was also concurrently head of the State General Administration of Sports, which is under the guidance of the Central Propaganda Department and part of the Chinese propaganda system. BOCOG has its own propaganda bureau, led by officials who concurrently head the propaganda sections of the Beijing Party Committee and the State General Administration of Sports (Brady 2009).

The Olympic Games thus provided a unique opportunity to reform Beijingers, through the concerted action of diverse levels of government – at the national, local and neighbourhood level – with the support of Olympic authorities. Aspiring social elites and other members of civil society also played an important part in the implementation of this programme.

Such social reform and control interventions served several goals. Internally, they sought to create a more convivial, law-abiding, disciplined and orderly urban culture to help achieve social stability, in a period of rising volatility and insecurity. It also served a deeper political purpose, sustaining long-term attempts to link China's successful Olympics bid to ongoing efforts to maintain the political credibility of the Chinese Communist Party (CCP) and the legitimacy of the political system it represents (Brady 2009). Externally, these programmes also sought to improve China's image as a rising world power. Just as Beijing's shiny new buildings and modern infrastructure helped project the image of a forward looking, progressive modern society, well-mannered and disciplined citizens also strengthened the image of Beijing as a twenty-first-century world metropolis. This image was not only aimed at the thousands of foreign visitors who would tour the city during the Games, but it also targeted potential investors with the reassuring image of a content, docile and obedient workforce, and of a stable and well-managed society, that could sustain business confidence.

In order to facilitate its public outreach, the civilization campaign rested upon three main approaches: the first was mainly ideological, and consisted in the orchestration of an official discourse surrounding the Beijing Olympiads, through the repetitive and ubiquitous promotion of a cultural ideology that could foster popular compliance and facilitate reform. The second approach sought to promote social reform through embodied practices, using active participation and mimetic practices to teach people new behaviour and to reshape them into ideal citizens. The third approach was coercive, and focused on more traditional means of enforcing public order and social control, through the tightening of security, by limiting freedom of movement and by restricting popular accessibility. The following sections discuss each of these three approaches in more detail.

Discursive means of reform

One of the dominant approaches used to transform Beijingers ahead of the Olympic Games rested upon a series of discursive strategies which sought to manipulate popular consciousness in order to facilitate compliance. To maximize their effectiveness as both an image construction strategy and a means of social control, civilizing reforms had to be achieved through wilful participation and persuasion rather than through more constraining means. The Chinese authorities understood the importance of striking a delicate balance between their desire to host harmonious Olympic Games and their fear of presenting the image of a police state to the world. Authorities thus turned to propaganda, mass persuasion techniques and discursive strategies to engineer a consensus around the campaign, and to invite popular participation – using nationalism, patriotism, self-sacrifice and voluntarism as means of enticement.

In the two years leading up to the Olympics, the Beijing population was relentlessly bombarded with public interest messages, using the well-established propaganda tactic of media saturation as a tool of mass communication and indoctrination (Brady 2009). Olympic messages and slogans were seen and heard on billboards throughout the city, at bus stops, in classrooms, in all forms of public offices, in the print media, on television and on the Internet, which were infused with a mix of political and moral discourse, encouraging self-reform, civic pride and compliance with the harmonious society ideal.[4]

This strategy largely relied upon the framing of Olympic propaganda in accordance with the harmonious society and spiritual civilization discourses. In many ways, the Olympic ideals of brotherhood among people, solidarity, equality, peace, harmony, mutual respect and fair play resonate with several aspects of these two doctrines. Beijing's official Olympic propaganda fused these ideals with nationalist notions of a 'harmonious society' and remnants of the spiritual civilization dogma, allowing these three discourses to merge into a new, overarching rhetoric, which I call the Olympic civilization discourse. Examples of this discourse were found in Olympic propaganda posters bearing slogans such as 'Safe Olympics, Harmonious Beijing: Serving with Hearts, Mind and Spirit' or, similarly, 'Harmonious China, Courteous Beijing', which combine references to the Olympics, to Harmony and to 'civilized' social behaviour (safety, courtesy, service). These slogans were generally accompanied by depictions of individuals engaged in actions suggesting civic values such as diligence, altruism, solidarity and selflessness. Another message, commonly found on city walls, in taxi cabs and on subways platforms, was 'Welcome the Olympics, Be Civilized, Follow the New Trend' (*ying aoyun, jiang wenming, shu xinfeng*) which similarly associates notions of civilization and social change with the Olympic Games, weaving all three official discourses into an encompassing Olympic civilizing discourse.

The framing of the Olympic civilization programme in concordance with the harmonious society discourse allowed the Beijing Olympics to become a key umbrella concept, a kind of magic signifier that would serve as a powerful tool of reform and help promote different aspects of state ideology. The Olympic civilization discourse thus acted as a legitimating discourse that would justify efforts to remake both the human and physical face of the city through pragmatic, intrusive and at times punitive interventions. By drawing upon popular support for the Olympics, this ideological campaign would contribute to the creation of a consensus over the need for social reforms, facilitate the implementation of new modes of control and discipline, and provide the state with renewed legitimacy.

Such blurring of official discourses also helped concretize the role of the Olympic Games as an exercise in national image construction, by projecting the state-sanctioned vision of China as a friendly, modern, disciplined and united society. This ideal of peaceful unity with the nation and the rest of the world is clearly stated in the 'One World One Dream' official Olympic slogan. The framing of the Olympic torch as the 'sacred flame' of Chinese patriotism also emphasizes the patriotic recuperation of the Games intended to rally unflinching popular support for the ideals the event was made to embody.

A second discursive strategy devised to ensure popular compliance with the civilizing process and to facilitate its wide diffusion was to infuse the discourse emanating from Olympic promotional material with nationalist rhetoric, to the point that the nation and the Olympics became indissociable in the collective imagination, and that serving one meant serving the other. In Beijing, embracing the Olympic Games was not only a civic duty, but a patriotic gesture and a contribution to the advancement of the motherland. By extension, non-compliance to social reforms became antipatriotic. Citizens were exhorted to help China rise to greatness by changing their behaviour and improving their manners with complete devotion. A common slogan, 'Win Honour For Our Motherland And Glory To The Olympics', exemplifies the fusion of notions of patriotism and personal commitment with the ideal of a successful Olympics. Every Beijing classroom was also adorned with a list of the *Ten acts of civic virtue and decorum*, starting with 'Love the Motherland', which similarly links civic virtue with national allegiance.

The patriotic framing of the Olympic civilizing programme helped reinforce the symbolic importance of the event in the collective imagination. By constructing the Olympics as a paramount historical marker, a unique occasion to redress collective injuries and past humiliation, to restore national pride, and to regain China's deserving status in the world system, local leaders and Olympic organizers succeeded in creating a social consensus around both the event and the reform programmes that would ensure its success. Merging civilizing and patriotic ideals with the noble idea of the Olympic Games would not only ensure compliance with the programme and limit potential resistance, it also promoted unabated devotion to the great cause, to the point that any kind of sacrifice, no matter what the social, cultural, economic or ecological costs, became justified.

A last discursive strategy used in state propaganda to ensure the success of the Olympic civilization campaign was to appeal directly to the individual as the responsible bearer of the Olympic civilization ideal. A ubiquitous poster series depicting various individuals actively contributing to the Olympic spirit by serving the community or helping those in difficulty, carried the highly subjectifying slogan 'My Games, My Contribution, My Happiness' (*wo canyu, wo fengxian, wo kuaile*). This repetitive use of the first person, which directly calls out to individuals and makes them feel personally involved in the fate of the Olympics, helped fuse notions of personal well-being (*my happiness*) with the success of the Games, suggesting that one was conditional upon the other. The lyrics of the School Children's Olympic Song, taught throughout the Beijing school system, are also deeply infused by the responsibilizing rhetoric of the Olympic civilization discourse:

> The Olympics will be held in 2008,
> Our civic virtue environment must be great!
> Spitting everywhere is really terrible
> Littering trash is also unbearable
> To get a 'thumbs up' from foreign guests,
> Beijing's environment depends on us![5]

This responsibilization process was further expanded through the vast solicitation of individual participation in Olympic preparations. Early on in the Olympic planning process, public opinion was symbolically solicited on different issues, including the design of Olympic venues. People were also encouraged to participate in developing the Olympic civilization programme itself, by devising slogans, writing thematic songs or submitting photos and cartoons illustrating proper behaviour and Olympic ideals.

Such personal involvement helped instigate the notion that individuals played an important role in this collective image construction process, and made people feel personally responsible for the progress of the Socialist nation. It pressured citizens to comply to state-sanctioned notions of civility by suggesting that the failure to do so would not only embarrass the leadership, tarnish China's reputation, and shame the nation, but also compromise Beijing's costly Olympic image construction efforts.

By presenting the adoption of prescribed norms of behaviour as a personal commitment and as the self-imposed decision of a sovereign agent, this joint personalization and responsibilization process helped foster voluntary compliance, as people internalized the civilizing process, and minimized resistance against the reform movement. Peer pressure also played a part, as people felt compelled to conform and adapt to the behaviour of those around them and thus to subscribe to the dominant cultural norm.

This individualizing process not only fostered voluntary compliance with the civilizing programme but also contributed to its successful implementation, by transferring the civilizing burden from the state and its institutions onto members of civil society. It was highly successful in autonomizing the civilizing process and in devolving the role of enforcing the civilized norm into the hands of individuals, pushing some citizens to take it upon themselves to reform Beijing's population. In the months leading up to the Games, several members of China's civil society voluntarily began acting as civilizing agents, forming citizens groups, incorporated as NGOs (non-governmental organizations), to take on the task of reforming their peers. Examples include the Green Woodpeckers, an anti-spitting activist group whose volunteers offered tissues to people as an alternative to spitting on the ground, and tried to convince offenders to change their habits.[6]

Another volunteer group focused on civilizing Beijing's linguistic landscape. Acting as Beijing's English police, this group helped correct more than 50,000 signs across the city displaying unorthodox English translations – from monument plaques and museum captions to roadsigns and restaurant menus – by providing offenders with correct translations, free of charge (Yardley 2007). As one of the main interfaces between Chinese hosts and their international visitors, Beijing's English language signs were often ridiculed in the foreign media as testimonies to Chinese unwordliness and cultural opacity. Direct translations – often uncensored by polite, euphemistic language – were thought be crude and to upset civilized sensibilities.[7] Some of the revised translations also sought to censor overly nationalist messages, which were toned down to avoid offending foreign visitors. While this formalization of the city's linguistic landscape appeared as a necessary step in the city's access to global city status, making the city more accessible to foreign visitors, this sanitization and uniformization process eliminated colourful insights into China's culture, thereafter 'lost in translation'.[8]

These private initiatives have been very proficient enforcers of the Olympic civilization ideal, going as far as using public shaming as a deterrent, by posting pictures of offenders on their websites, and using the power of face and personal image as levers for social change. Active participation in the Olympic civilizing process also involved mutual policing and surveillance, as local governments offered rewards for denouncing uncivilized behaviour and reporting violations. It helped turn a simple state-led campaign into a panoptic surveillance apparatus, and maximize its impact with minimal state involvement.

Embodied means of reform

If much of the civilizing process rested upon propaganda and discourse, the main mode of operation of this reform movement was performative, and rested upon a vast public education programme predicated upon embodied practices. It was through active participation in state-sponsored activities that both the body and mind of the population would be transformed, and state-sanctioned norms of behaviour internalized.

The way people spoke, dressed, stood in line, cheered at events and carried themselves in public were all considered important measures of Olympic success. Anti-social behaviours such as spitting, queue jumping, jaywalking, hawking, swearing and smoking were specifically targeted. Focusing on manners, personal hygiene, civility, morality, sports ethics and respect for law and order, public education programmes instructed citizens on the correct use of public toilets, urged them to speak proper Mandarin, to learn English and to smile more.

In Beijing, embracing the Games therefore became an embodied engagement to transform oneself and embrace Olympic civilization ideals on a daily basis. People were encouraged to actively participate in the continuous flow of Olympic education activities that relentlessly punctuated the years leading to the Olympics. The acquisition of civilized behaviour was most often carried out through a mimetic learning process. It was through participation in activities involving bodily re-enactment, repetition and mimicry that the desired behaviour was most effectively learned.

To increase participation, Olympic education programmes and social reform activities were often of a festive or ludic nature: role play, simulation games and repetitive rehearsals were ubiquitous. Participation was also motivated by drawing upon people's competitive spirit, in televised quiz shows on Olympic knowledge, popular etiquette contests, in English language competitions and other forms of challenges held in schools, university campuses and neighbourhoods around the city. The pervasiveness of events that facilitated acculturation into civilized norms of behaviour allowed for this active participation to permeate everyday life.

Many civilizing activities were held in the classroom, where Beijing school children were regularly lectured on the need to improve their city's civic virtue and learned proper behaviour through repetitive enactment. The Beijing Education Bureau enjoined every elementary school in the city to teach 'A volunteer's smile is Beijing's best name card', an Olympic chant that would regularly be sung by pupils to remind them to smile. Children also learned and practised a series of key English phrases to be used to greet foreign visitors.

In order to sustain public interest, Olympic education programmes and social reform activities were ceaselessly renewed and often used humour rather than admonition to convey their message. Throughout the year, there were recurring thematic days, such as the 11th of each month, known as 'queuing day', when volunteers in satin sashes helped people into lines at busy subway stations, hospitals, banks, post offices and various other public places. Others, dressed as penguins, walked in line on city streets to promote lining up and orderly conduct. Some regular holidays were also reformatted to fit the Olympic civilization ideals: for example, Children's Day (1 June) was celebrated in 2006 under the slogan 'Befriend civilization, grow up with the Olympics' and included activities relating to the theme 'Civilized Olympics, Harmonious Families'.

A specific area of concern that was addressed through embodied practices and repetition was sport etiquette. Since 2005, etiquette and sportsmanship campaigns were conducted to educate the Beijing public. In the interest of promoting harmonious relations between China and the world, efforts were focused on keeping overly nationalist partisanship in check. These efforts came as a result of several incidents when local fans displayed anti-social behaviour against foreign teams in international competitions, which also led to the adoption of a law banning hooliganism in 2006.

Aware of the tenuous line between promoting national pride and fostering xenophobic behaviour, the party sought to control and channel nationalist impulses at Olympic events by civilizing popular expressions of support. Strict instructions about proper cheering behaviour (including cheering for non-Chinese teams) and appropriate hand gestures were thus issued. In the months leading to the Games, Olympic cheering practice sessions were held for workers around Beijing to learn the state-sanctioned '*Zhongguo Jiayou*' (Add Fuel China!) and a cartoon illustrating the official civilized hand gesture (clap twice, thumbs-up, clap twice again, punch the air with both fists) was widely distributed and published in local newspapers.

If mimesis and repetition were important means of facilitating the acculturation process and the assimilation of new social norms, the most effective way to encourage proper behaviour was to preach by example. Much of the Olympic civilization programme thus rested upon the production of idealized citizens, presented as social models to be emulated. Olympic propaganda made profuse use of movie stars, such as Jackie Chan and Andy Lau, and sports personalities, such as hurdler Liu Xiang and basketball player Yao Ming, to personify the ideals of Olympic civilization. These celebrities were ubiquitous in Olympic propaganda, both as proud bearers of the Olympic brand and as social models of virtue. Erected as national heroes and icons of civility, these popular public figures embodied the best of what China had to give.

Other, more ordinary heroes were also promoted to the status of iconic social models. Olympic volunteers held a central position in Olympic propaganda. Images of the selfless Olympic volunteer, idealized for his ethics, hard work and desire to make the nation proud, dominated Beijing's pre-Olympic landscape and sought to stir the national psyche and inspire citizens through their loyalty and dedication to the motherland.

In recent years, volunteerism has been on the rise in China, fuelled in part by the new humanist vision promoted by the harmonious society doctrine.[9] As the Olympics grew near, volunteering, an important Olympic tradition, became a trend, especially among university students, with record-breaking numbers of applicants.[10] Overall, 470,000 volunteers – selected from more than 2 million candidates – would serve the Olympic spectacle in Beijing, not only with their devotion, but with their unpaid labour as well, as Olympic village drivers, stadium ushers, field 'gofers', media runners, ceremonies performers, VIP escorts, tour guides, traffic supervisors and interpreters (Bonnot 2008). Required to be fluent in foreign languages, they were subjected to series of rigorous theoretical and physical examinations and received extensive training in proper etiquette, protocol, first aid and basic security. Among them, members of the elite 'Olympic brigade' were placed in the direct service of athletes, journalists, officials and other VIPs, while on city streets and outside official Olympic sites, members of the more mundane 'urban brigade' were charged with providing assistance to millions of visitors.

Members of the city's numerous local residents' associations (*juweihui*), which represent the lowest level of the Chinese party-state, constituted another important portion of urban volunteers. As a central feature in the Socialist tradition of voluntarism, these neighbourhood committees have long played a role in the maintenance of social and political peace inside residential neighbourhoods, while giving a social responsibility to the retired and the elderly (Bobin and Wang 2005). During the Olympic Games, older men and women, wearing Olympic shirts and hats with the red armbands indicating their volunteer status, acted as surveillance agents, patrolling the neighbourhoods, watching for trouble from protesters or dissidents, and reporting suspicious behaviour to the authorities, thereby contributing in their own way to the state's elaborate surveillance system.

The civilizing process helped constitute the ideal Socialist citizen as an iconic sign, contributing to both national and urban image construction. As the incarnation of the multitude, and representatives of the nation, Olympic volunteers played a central role in the creation of a particular vision of what a citizen of modern Socialist China should be. As a corollary, those who failed to comply with this ideal brought into question their very belonging to Chinese society. The Olympic civilization project thus allowed for the discursive construction of certain members of society as uncivilized, and therefore unworthy of citizenship. By introducing social norms that reinforced distinction, it justified the emergence of new hierarchies and power disparities at the local level.

In Olympic Beijing, if some model citizens embodied the key qualities of civilization and came to stand for the nation as a whole, others who failed to embody this ideal were denied representation as valued members of Chinese society. Absent from Olympic propaganda and marketing brochures were Beijing's mass of migrant workers, who were the main target of this civilization campaign, and were discursively constructed as a major threat to the image of civilized modernity conceived for the Games. Their crude manners, coarse language and unhygienic habits – often the result of their own destitution, lack of education and exploitation – were taken as proof of their need for reform.

Beijing is estimated to have a migrant population of about four million, mainly employed in the construction, manufacturing and service industries, within a total municipal population of around 17.5 million. In recent years, rural migrants (generally identified by the derogative *waidi* or 'outsider') have routinely been accused of tarnishing the city's image. They were blamed for the capital city's deteriorating civilization level and were portrayed in the media as the perpetrators of most violent crimes. The effects of this exclusion from collective representation are tangible. The ideological construction of the migrant as uncivilized, dangerous and pathological has helped naturalize his subordination and exploitation and devaluate his labour. Marking the migrant as structurally irrelevant also helped justify his further abuse and legitimate his exclusion from full citizenship rights. As the constitutive other of civilized urbanites, against whom the latter can construct their own cosmopolitan identity, the migrant was made unworthy of equal rights and opportunities in the city, even, as we will see below, of the right to be seen.

Discipline and order: social reform through control, surveillance and coercion

The two civilizing strategies previously discussed resorted to discursive construction, and on the acquisition of new behaviour through embodied practice involving direct participation, repetition and emulation. A third approach relied upon the imposition of disciplinary measures and regulatory controls. In this case, state notions of discipline and order were enforced through juridico-political means, by arresting or regulating movement, through the implementation of an elaborate surveillance system, and with the imposition of limits on personal liberties (freedom of movement and of expression). State attempts to achieve full population control also relied upon more coercive methods, including repression, internment and expulsion.

With harmony as the main code word for the Olympic Games, public order and security became a paramount focus of the Olympic civilization programme. Although Beijing has long been among the safest cities in the world, the idea that the media attention generated by the Olympics could make Beijing an ideal target for international terrorists seeking maximum visibility required outstanding security measures. The IOC-sanctioned emphasis on security as a paramount Olympic host responsibility combined with exaggerated reports about internal security threats, especially from insurgent groups such as the Falun Gong, Tibetan activists and Uyghur separatists, also validated diverse control and discipline mechanisms in the public eye and gave licence to the imposition of limits on personal freedoms.

China's desire to eliminate potential threats to the largest gathering of world leaders the mainland had ever hosted justified massive security operations. With a security budget of over USD 2 billion, double that of Athens 2004, Beijing mobilized a 150,000-strong anti-terrorism force that included commandos and other military units equipped with surface-to-air missiles and military aircrafts. Over 80,000 policemen, security agents and other peacekeepers also provided added security and increased surveillance (Bonnot 2008).

Weary of the international press's scrutinizing presence and of its eagerness to expose any cracks in Beijing's tightly controlled image, local authorities were determined to soften China's reputation as an authoritarian state. Police patrols were thus ordered to project a sympathetic image and to be on their best behaviour with foreign tourists. They were instructed to assist those in need, to avoid using their sirens to bypass traffic and to refrain from physically or verbally abusing anyone (Bonnot 2008).

Citizens were subjected to heightened scrutiny, with the use of over 300,000 surveillance cameras (Zhou et al. 2008). Sophisticated and innovative surveillance technology enabled the government to keep tabs on private electronic devices, allowing the surveillance apparatus to reach into the privacy of people's homes. Mobile phone and the Internet, which are often portrayed as instruments of liberation in post-Mao China, were thus turned into instruments of control. They also became instruments of civilization, as mass SMS messages were routinely sent to Beijing residents to remind them to adopt a civilized demeanour.

Security became a code word for social beautification initiatives and was used to justify the imposition of strict public order. Months prior to the Games, Beijing authorities launched a preventive anti-crime campaign to ensure that crime rates stayed as low as possible before and during the Olympics, targeting vagrancy, begging, prostitution and other illicit activities in the city. These actions clearly stemmed from an image construction imperative rather than from a real security threat, and were meant to ensure that undesirable social elements would not ruin costly image construction efforts.

Just as physical signs of poverty and backwardness, which could dispel China's newly proclaimed prosperity and modernity, were carefully camouflaged behind newly erected walls, shrubbery and fresh coats of paint, many local residents who did not fit the image of the civilized Beijing citizen were also carefully hidden from view. For the duration of the Olympic events, Beijing became a sort of closed city, protected by a series of filtering systems that controlled access to its urban fortress and determined who would be granted the 'right to be seen' (Mitchell 1995).

In order to create a 'safe and harmonious security environment' and to preserve the image of a city unburdened by poverty, Chinese authorities began enforcing residence permit (*hukou*) regulations, after having turned a blind eye to illegal residents for many years. Migrants were thus subjected to diverse forms of harassment, including mass identification checks, the confiscation of the tools of their trade, and the destruction of illegal schools and homes. Those found without Beijing residence permits were fined, forcibly expelled from the city, or sent to detention centres and re-education work camps. Police sweeps were also conducted on the streets of the capital. Beggars, street children, the homeless and other conspicuous indigents were picked up at train stations, pedestrian underpasses, railway bridges and other hideouts, to be placed in relief centres on the city's outskirts, or in custody and repatriation camps, before being exiled back to their hometown.

The criminalization of informal activities, with the banning of unlicensed taxis, sidewalk vending, peddling and hawking, furthered the victimization of migrant citizens – who were rendered 'illegal' in the name of 'order and security'. In spite of their great contribution to Olympic image construction, through their labour as construction workers, street sweepers or garbage collectors, most migrants were thus barred from active participation in the Olympic celebrations, even as simple bystanders. Innumerable informal businesses were closed, and for the duration of the event, the contribution of these workers to the city's economy was made blatantly visible by their absence.

Foreign nationals living in or entering the country were also the object of increased scrutiny and tighter state control. Eager to monitor those who lived in or visited the country during August 2008, state authorities tightened foreign visa regulations, imposing new restrictions on visa applications and renewals in the month before the Olympics. As a result, thousands of foreign residents whose visas were not renewed were forced to leave China, while others with valid visas were barred from re-entering the country.

This policy would prove to have negative consequences on the success of the Games, by triggering a sharp decline in the number of visitors to Beijing. Only 389,000 foreign tourists visited Beijing during the Olympic month of August 2008, far lower than the 500,000 guests expected (Anon. 2009). Overall, in what was meant to be a standout year, Chinese tourism experienced a 2.3 per cent decline in 2008, equivalent to a 2 million drop in the number of travellers visiting China, marking the first decline in visitors since the 2003 SARS epidemic. Beijing hotels reported a 32 per cent drop in occupancy rate between November 2007 and November 2008, recording the largest average hotel occupancy decrease for any major global city in 2008 (Bowerman 2009).

Crowd control was another major area of concern, for the IOC, local Olympic organizers and Chinese authorities. The Chinese Communist Party has long been suspicious of all forms of public gathering, especially in the national capital, where all forms of popular congregation remain tightly regulated and closely monitored. For the duration of the Olympics, Beijing's public spaces were kept under constant and vigilant surveillance. Access to several key sites, including Tiananmen Square, was tightly controlled by police check-points, equipped with X-ray machines.

The IOC's fear that the host city be overcrowded and overburdened during the event also justified the imposition of strict limits on Olympic site accessibility. Efforts to preserve status hierarchies and regard for prestige accumulation, rather than a simple security concern, often appeared to rationalize such restrictions, which gave the event the feel of an exclusive elite festival. The Olympic park and the area surrounding major Olympic venues were protected by a 'cordon sanitaire'. Entry was restricted according to a finely calibrated social ranking, which included people with tickets for specific daily events, athletes and their entourage, and holders of special privilege (accredited members of the press and VIPs, among others). As a result, the expansive park remained eerily empty during most of the Olympic weeks. Along the perimeter fence that protected the Olympic park, crowds of curious local residents and foreign tourists who failed to get their hands on elusive Olympic tickets could be seen stretching their necks to catch a glimpse of the world famous Olympic stadium and of the festivities they had come to celebrate.

Also empty during the Olympic Games were the three official Olympic protest zones that had been set aside by the Chinese government for legal protests. Chinese citizens had been told they would be free to protest during the Games, as long as they applied and obtained a permit from the Public Security Bureau. However, the Bureau claimed the right to reject all protests that could potentially 'harm national, social and collective interests or public order'. As a result, all 77 individuals who had applied for permits were either persuaded to withdraw their proposals 'after amicable settlement between the parties and authorities was reached', or had them rejected (Callick 2008, O'Brien 2008). Some of those who submitted applications were subsequently arrested and detained. International human rights NGOs such as Human Rights Watch denounced protest zones as political traps, set to facilitate the arrest of dissidents. They qualified the application process as a tactic to facilitate the suppression of protests rather than giving people greater freedom of expression (O'Brien 2008). In the end, the three official protest zones remained devoid of activity for the duration of the Games, except for the occasional foreign journalist who came looking for some form of political activity.

Reception and criticism

Two years after the 2008 Olympic Games, it remains difficult to assess the success of the civilization programme. If participation in civilizing activities was overwhelming, a more in-depth analysis would be required to measure their effectiveness in altering social behaviour. From both the state and Olympic authorities' point of view, security measures were a success. Olympic festivities went relatively smoothly and were disturbed by few incidents. Small protests were held by foreign human rights activists both at Tiananmen Square and near the Olympic park, but they were rapidly contained and their participants promptly arrested. A dozen local citizens were also detained after staging illegal protests or talking to the foreign media to alert international public opinion to abuses they suffered with regards to Olympic redevelopment.

Resistance to the civilizing process was also limited. Sixty years of socialism and repeated exposure to similar state propaganda programmes and social engineering campaigns may have made people impervious to such state intrusion into their personal lives. If the overwhelming majority approved of the civilizing campaign as a necessary step to facilitate social coexistence in the city and to raise the international status of both the nation and its capital, several members of the general public reacted to the campaign with a certain amount of cynicism and criticism. Many Beijing residents were sceptical about the effectiveness of yet another attempt to civilize society. Others admitted suffering from slogan fatigue, numbed by the omnipresence of messages in an urban landscape already saturated by competing propaganda. The impact of the programme was effectively weakened by the recuperation of Olympic symbols, icons and slogans by diverse commercial interests, which both confused and diluted its message. For example, one billboard advertising the Silk Alley shopping emporium proclaimed: 'One World, One Shopping Paradise', in a brazen reworking of the 'One World, One Dream' Olympic slogan.

Some harsher critiques privately condemned the Olympic civilization programme as a thinly veiled tool of repression and control. They denounced what they saw as the state's attempt at cosmetic social beautification and the leadership's excessive concern for face and appearances, when more pressing social issues are left unresolved. People denounced the criminalization of behaviours that did not fit globalized images of modernity through a process that turned social norms into legislation and punishable offence. Many also resented being used as props in the staging of this Olympic spectacle. They saw the programme as counter-productive as a means of promoting social unity and harmony, because of its tendency to produce social identities that reinforce social stratification and fragmentation.[11]

The heavy-handedness with which the Olympic civilization programme was carried out was similarly criticized. Foreign visitors deplored how state and Olympic authorities' concern for security had turned Beijing into an over-regulated and antiseptic fortress, which spoiled the festive spirit of the Olympics. Others feared that Beijing's sanitized landscape was self-defeating as an image construction initiative, as it confirmed representations of China as a police state, which the foreign media was eager to portray (York 2008).

Frustrations with excessive restrictions and security measures imposed on city residents and criticisms about Olympic civilization programmes were most often expressed in the form of jokes and satire, found on irreverent websites, and widely circulated via email and text messaging. One text message reveals the rising exasperation and widespread Olympic fatigue experienced by the general public. Styled as a newsflash, the message claimed that millions of Chinese had fallen into a faint after hearing that the IOC had decided to hold the 2012 Olympics in Beijing once again, because of its great hosting performance. Another message, widely circulated before the Games, advertised an 'avoid the Olympics' holiday package (*bi yun tao* – an expression that is homonymous with 'condom') to those wishing to escape the capital during the event (Blanchard 2008).

Conclusion

This chapter underlined the important role of mega-events in the planned management and control of human activity, especially in the implementation of social reform and disciplining programmes framed by global norms of order and civility. It demonstrated how the hosting of mega-events such as the Olympic Games not only acted as a catalyst for the physical transformation and beautification of the urban landscape but also helped intensify the processes of civilization process that have long accompanied modernization and globalization. This analysis of the elaborate and multi-faceted strategies developed in pre-Olympic Beijing to reform both the mind and the body of the city's inhabitants, sought to provide further insights into some of the social costs associated with mega-events.

In Beijing, the Olympic Games not only gave local authorities the licence to undertake social reform and control programmes, implemented through a mix of cultural ideology, embodied practices and coercive power, but it helped create specific conditions that facilitated the implementation of the civilizing programme and maximized its success with minimal state intervention. The campaign effectively managed to devolve the part of the enforcement of the civilizing norm to individuals, thereby alleviating state costs and efforts while promoting mutual surveillance and self-policing.

The chapter demonstrated how the Games were instrumentalized as a political tool of legitimation. The Beijing Olympics played a key role in the official orchestration of the civilizing discourse and represented both a means and an end in carrying out the ideological doctrine of the harmonious society. The Olympic civilization campaign not only served as a tool of persuasion to rally popular support for the Games and compliance with the associated programmes of social reforms and control, but it was also designed as an opportunity for a major propaganda effort to build national pride and gain popular consent for the continuance of CCP rule. The Olympics were thus instrumental in giving weight to a vast mass persuasion campaign that sought to maintain the core political status quo, and to persuade the masses that the party in power and the political system it represents are legitimate and should be sustained.

The patriotic framing of the Olympic civilizing process and its focus on the promotion of social harmony also served as an instrument of pacification, allowing diverging interests to converge and mobilize around a grand, collective and unquestionable goal. The Olympics thus acted as a great unifying moment at a time of growing social fragmentation.

The chapter also emphasized the role of mega-events in the organization of social stratification and domination, exacerbating pre-existing problems of social justice and inequality. It demonstrated how basic rights and freedoms can be compromised as result of the Olympics, including the right to the city and the right to be seen. By promoting a tightening of the social control apparatus, the imposition of limits on civil liberties and social exclusion, mega-events help redefine the rights to participate in society, and affect the constitution of citizenship.

Notes

1. One can think, for example, of the numerous 'hygiene campaigns' (*weisheng yundong*) initiated in the Mao years to educate the Chinese population about various healthcare issues.
2. See for example, Guo Shixing's 1999 play about disrespect, *Bad Words Street*.
3. See Simon Leys's (1997) introduction to his translation of the *Analects of Confucius*.
4. Beijing's Capital Ethics Development Office also published a series of handbooks and pamphlets on etiquette, civility and sports ethics, which they freely distributed throughout the city.
5. As translated in Meyer (2008).
6. The Green Woodpecker Project was started by Wang Tao, a worker at the Xicheng District Sanitation Bureau, who manages the Forbidden Spit website at Jintan. org.
7. One of the most often cited examples in the Western media is that of the 'Dong Da Hospital for anus and intestine disease', whose name was changed to the 'Dong Da proctology hospital'.
8. One can think of evocative dishes such as 'ants climbing up a tree' now translated into the more mundane 'sichuan pork with noodles'.
9. In the Mao years, volunteering formed an essential role in the social and physical construction of Socialist China. It fell out of favour in the 1980s, in reaction to the excesses of the Mao years. Since 2005, the CCP is again advocating volunteering, but in a state-managed form rather than through independent civil society movements. The outpouring of volunteer help in response to the May 2008 Sichuan earthquake, when people rushed to lend a hand rather than to wait for the party and its army to take command, testifies to this new trend (Bonneau 2008).
10. The previous record for the number of volunteers serving the Olympic Games was held by Athens (2004), when 45,000 volunteers were chosen from among 160,000 candidates.
11. These comments are based on a series of interviews conducted in Beijing between 2006 and 2009.

4
Istanbul's Olympic Challenge: A Passport for Europe?

Jean-François Polo

Over the past 15 years, Turkey has been increasingly present in the international sporting arena, either through the results of its teams and its athletes or through its numerous bids to host international sports competitions (see Appendix 4.1 for a comprehensive list). Prior to the 1980s, indeed, Turkey had only hosted wrestling competitions (in 1974 and 1977); perhaps unsurprisingly, the first successful bids were in sporting disciplines that are part of traditional Turkish sport practices and in which Turkish athletes have always been successful (wrestling, weightlifting). But in recent years, and as bids have become increasingly frequent, we have been able to observe an increasing diversity in the sports for which bids are made: not only sports which have a strong international and media impact, and in which Turkey has begun to have some success (football, basketball) but also in other disciplines (swimming, motor racing, archery, etc.). Of course, the most prestigious applications have been its bids to host the Summer Olympic Games.

Indeed, Istanbul has repeatedly bid to organize the Olympic Games since 1992, though without success; Turkey was also perhaps unlucky not to be awarded the co-organization with Greece for the UEFA Euro 2008 and 2016 Football Championships. Not all bids have been unsuccessful; in 2000, Istanbul hosted the 35th EKF European Karate Championships; in 2001 Turkey hosted the FIBA European Basketball Championship; in 2005, Istanbul hosted the final of the UEFA Champions League and the first FIA Turkish Formula One Grand Prix. And in the coming years Turkey will welcome, amongst others, the FIBA World Basketball Championships in 2010, the FISU Winter Universiade (World Student Games) in Erzurum in 2011 and the IAAF World Indoor Championships in Athletics in Istanbul in 2012.

Focusing on the numerous Olympic bids, and on the Galatasaray sports clubs, the central argument of this chapter is that the increasing number of these international sporting events is not accidental, but must be linked with the political stakes of sport in Turkey, and its specific relationship with Western countries and, more precisely, with Europe.

The question of the links between sport and politics is as recurrent in common reasoning as it is in the academic field (see special issues of *Géopolitique*
1999; *Politix* 2000; *Relations Internationales* 2002a, 2002b; *Bulletin d'Histoire
Politique* 2003, as well as Villepreux 2008). Despite occasional naive statements such as 'sports and politics are different' or 'sport is not war', and
despite the sporting world's affirmation of the neutrality of sport (Defrance
2000), sport, being a 'total social fact', has numerous effective and potential implications in political issues (Polo 2005). Sports mega-events raise the
question of the political dimension of sport which is obvious in international relations. According to Pierre Milza (1984), the international stakes
of sport cover three areas: sport as part of and reflecting the international
stage; sport as a means of foreign policy; sport as signifier of public feeling.
Thus, the organization of sports competitions, or merely the act of bidding
to organize such events, goes far beyond the economic benefits hoped to be
derived from hosting. International competitions remain a quest for recognition of national power (Elias & Dunning 1987: 307). There are numerous
examples in history where states have striven to demonstrate their power
through the success of their teams and athletes at international sports
events (*Géopolitique, Revue de l'Institut international de géopolitique* 1999;
Boniface 2002; Houlihan 1994). Even the reintroduction of the Olympic
Games by Baron Pierre de Coubertin can be partially explained by a drive
to enhance the national status of a France that had been humiliated in the
1870 Franco-Prussian war; as Milza (2002: 300) points out, the reinvention
of the Olympic Games at the end of the nineteenth century thus stems
from an explicitly political, national programme. Thus, international sport
is deeply impregnated with nationalism, which has often reached spectacular degrees. From the Nazi propaganda in 1936 (Brohm 1983) to the Beijing
Games in 2008 (Collectif anti-jeux olympiques 2008), there are numerous
examples of the political uses of sports results and events.

But international sport is not only a means used by states to show off
their power; it can also simply be a way of achieving international recognition of existence and legitimacy. International sports authorities, such as
the IOC, FIFA and UEFA, can act as an efficient parallel channel for international diplomacy. Obtaining the organization of an international sports
event such as World Cup Finals or the Olympic Games means international
recognition, amply demonstrated by the fierce competition between states
to host this kind of event. In authoritarian states, hosting an international
sports event has been used to give legitimacy to their political system, or
at least to make them appear less unattractive in spite of it (e.g. Argentina's
hosting of the FIFA World Cup in 1978, Moscow's of the Olympic Games in
1980, etc.). In other contexts, such sports events help to demonstrate that a
country has reached a high standard of development (the Olympic Games in
Seoul in 1988, Barcelona in 1992, Athens in 2004 or Rio in 2016). Similarly,
FIFA wanted to reward South Africa for its peaceful political transition from

apartheid by attributing the organization of the 2010 World Cup to this country.

Finally, the issue of sport as a sign revealing public feeling is linked more broadly with the meaning of these events for individuals or groups. Sport often helps express the feeling of belonging to a community, to a nation, whether by nationalistic means or by the proclamation of plural identities (Arnaud 1998; Dauncey & Hare 1999).

Sports mega-events may also give rise to social and political mobilizations, with two basic dimensions. On the one hand, they can simply exploit the media coverage of the event to publicize social and political concerns which have no causal links with the sports events themselves (Polo 2003). On the other hand, they can directly contest the mega-event per se, for political, social or cultural reasons. The passionate opposition to the Beijing Olympic Games is a recent and dramatic example, with calls to boycott the Games widespread. The Olympic flame's global tour through the five continents was the scene of protests against the Chinese political system, its violation of human rights, its authoritarian regime, the occupation of Tibet and the repression of minorities. In this volume, Xavier Renou describes in detail the organization of protest against the passage of the torch through Paris; in Istanbul, Uighur Muslims demonstrated and also tried to interrupt the Olympic torch ceremony (*International Herald Tribune* 3 March 2008). Protests can also address the direct effects of mega-events for different reasons such as their typically huge public cost, or their environmental impacts (such as in Albertville for the 1992 Winter Games; see also the chapter by David Whitson in this volume, on Vancouver).

Yet there has, so far at least been little evidence of this type of mobilization in Turkey. Indeed, Turkey's multiple applications to hold mega and other prestige sports events have not as yet given rise to any protest. Here, I argue that this lack of public opposition must be placed in the specific context of the country's relationship not only with the international stage but also with itself, and more broadly speaking, the complex relationships of Turkey with the West.

A loyal ally of the USA (at least until the war in Iraq), a member of NATO since 1952 and a candidate for EU membership since 1987, Turkey continues to claim its anchorage to the West, thereby pursuing the work of its founder, Mustafa Kemal Atatürk. However, this Western polarization is questioned both inside and outside its borders. Although Turkey applied to the EU before the new Eastern European members did, negotiations for its membership were only formally opened in October 2005, and are set to continue for least ten years, given the strong reluctance of some member states (including France and Germany) to admit Turkey to the EU. Hosting international sporting events is a means for the Turkish state, whose key founding ideas are 'modernity, secularism, nationalism', to demonstrate its modernity, its organizational abilities and thus the legitimacy of its application for membership of the European Union. In this perspective, hosting mega-events is a national cause enjoying comprehensive popular support.

The question of the consensus about both European stakes and sports mega-events raises another relevant issue for political scientists working on Turkey: is there a civil society able to be mobilized and to promote alternative points of view, challenging the decisions of political authorities? This is a very crucial and delicate question in a country marked by a strong state capacity, by which we understand the state's ability to 'penetrate society, regulate social relations, extract resources and use them effectively' (Migdal 1988: 4). Metin Heper (1991) uses the concept of 'Stateness' to describe the Turkish state's capacity to resist the institutionalization of political models which include the participation of interest groups within decision-making processes. The question of civil society and its capacity to be mobilized is very recurrent among both academic research on Turkey (Yerasimos et al., Dorronsoro 2005, Polo & Visier 2006), and in political debate: civil society strength is used as an indicator of the level of democratization by a number of international organizations, including the European Union.

This chapter aims to show therefore that sport in Turkey has specific political meaning based on Turkey's relationship with Europe. Sport helps us to understand how, in applying for EU membership, Turkey is striving to overcome European procrastination (often perceived in Turkey as disrespect) while simultaneously declaring its attraction to Europe. The analysis of the meaning of sport in Turkey will help us to improve our understanding of both the role of sport in the modernization of Turkish society, and of the strategies of elites, which blur the borderlines between sport and politics, civil society and political authorities.

In the first section, we show that Turkey's bids to host sports events are aimed at promoting Turkey in Europe, but that they stem from the same fundamental political goals pursued by Turkish elites (second section). Finally we will see that sport reveals, beyond the international stakes, the tensions and ambiguities of Turks towards their own identity and history, and thus the complex relationship of Turkey with Europe (third section).

Sport as a means of supporting EU membership

The question of the link between sport and politics in Turkey is a classic problematic of international relations. The multiplication of Turkey's bids to organize international sporting events is based on its determination to strengthen its international position. However, its ambition goes beyond this. Sport is perceived as a means of demonstrating that Turkey belongs to the West, even if the West is reluctant to admit it. In Turkey, sport is perceived as part of the 'modernity' of the country, which is one of the founding principles of Turkey's political project. For Turkey, EU membership will mark the ultimate recognition of its European identity, the materialization of its age-old attraction to the West, proving that Turkey is European. In this context, sporting mega-events can be perceived as one of several means (such as the victory of a Turkish singer in the 2003 Eurovision Song Contest, the successful organization of the United Nations International Conference Habitat 2000, or the designation of Istanbul as European Capital of Culture in 2010) of promoting its European ambitions, despite Turkey's poor image in Europe. This poor image is linked with the effects of the war against the Kurdish guerrillas, the still considerable role of the army in public affairs, and to a certain extent with the large Turkish migrant population in Europe (notably in Germany), reinforcing the image of the country as poor, rural and conservative. Finally, it is also inextricably linked to the fact that 98 per cent of Turks are Muslims, even if this link is generally denied by those opposing Turkey's EU membership application (see Visier 2009).

Sport as a symbol of political modernity

In the mid-nineteenth century, during the reformation period of the *Tanzimat* (1839–76), the Ottoman Sultans tried to modernize the Ottoman state, administration, army and education system as a means of resisting the decline of the Ottoman Empire. The political modernization process throughout the nineteenth century drew examples from European models, culminating in the promulgation of the first parliamentary Ottoman constitution in 1876. At the beginning of the twentieth century, as the Ottoman Empire became increasingly decadent, the 'Young Turks' and Mustafa Kemal resolved to make a clean break with the Empire and create a new state, the 'modern and secular Republic of Turkey'. Once again, the West was the model. For Atatürk, quoted by a former Turkish president (Özal 1988: 208),

> peoples who are not civilized are condemned to remain dependent on those who are. And the West is civilization. If Turkey wants to survive it must be part of the modern world. The nation has decided to adopt exactly and completely in both shape and content the way of life and the means that contemporary civilization offers all nations.

Though the European winners of the First World War tried to erase Turkish power through the Treaty of Sèvres in 1920, Turkey remained attracted to the European political model. The subsequent reforms undertaken by Atatürk aimed to transform and adapt Turkish society to Western values; the recurrent affirmation that *the Turks have always been walking to the West* remains crucial today, with the process of applying for EU membership. Like the principle of national independence, the European dimension is at the heart of the Turkish identity as defined by Atatürk and has never been questioned (at least theoretically).

The obsession with modernity has guided Turkey's various political reforms from the *Tanzimat* period to the birth of the Republic in 1923,[1] especially in educational reforms (as in the UK, where sport became a fundamental discipline in British elite education; see Holt 1989, Mangan 1988). The history of Galatasaray Lisesi (from *lycée*, high school), an institution in Turkey, shows us how much the country is attached to European modernity in all its dimensions, including the role of sport in education. Created in 1868 during the *Tanzimat* by Sultan Abdulaziz with the collaboration of Napoleon III, this French-speaking high school aimed to educate the new elite of a declining Empire. The new Turkish Republic continued to respect it, and its special aura remains even today. In the early days of the Galatasaray Lisesi, sport occupied a singular place in the school's educational and pedagogical project. It became the first high school to have modern sports equipment (1869), and many of Turkey's first sports clubs and teams were created within it, including the Galatasaray Football Club. For its founder, Ali Sami Yen, who became the first president of the Turkish Football Federation in 1914 and president of the Turkish National Olympic Committee (NOCT), the goal was to 'have our own uniforms, our own colors, to play together like Englishmen, and to beat the non-Turkish Teams'.[2] It also became the first Turkish multi-sports club: tennis (first court in 1911), sailing, rowing and so on. The club also won the first national championships in various sports, such as water-polo (in 1931) and handball (in 1956).

Although Galatasaray Lisesi is now only symbolically linked with the Galatasaray Sports Club, it houses the Galatasaray sports museum. The historiography of the high school stresses the importance of sport in student education. Turkey's sporting history is inextricably linked with Galatasaray Lisesi, and sport and education have long been perceived as a means of promoting the political modernization of the country. The development of sport and its organization are thus fully integrated into Turkey's political project, which is aimed at achieving a high level of political development, and the Western modernity that the EU symbolizes today.

Europe, an unachieved quest

The multiplication of sports event candidacies in the 1980s and 1990s was cognizant with a striking period of economic and political development marked by the will to draw closer to Europe. In 1980, in order to break a situation of growing political violence leading to civil war, a military coup overthrew the government, froze democratic institutions and organized severe political repression, especially against the left. The army then prepared a new democratic constitution, adopted in 1982, which strengthened military control. However, in 1983, it returned power to civil control after the elections were won by Turgut Özal, leader of the centre right (Motherland) party. Özal's victory inaugurated a remarkable period of increasing economic liberalization (see Akagül 1995). Turkey experienced important social changes, including rural depopulation, accelerated urbanization and growing popular demand for a more Western style of consumption. New entrepreneurs emerged, some of whom tried to organize or support sports events or teams, while the numbers of sports fans and spectators increased. This process encouraged an increase in the number of sports event bids and helped the country learn how to manage applications and events.

At the same time, the Turkish authorities tried to strengthen relations with Europe. After the signature of the association agreement (in 1963), Turkey officially applied for EU membership in 1987. Without actually saying no, the European members states slowed down the process. Since then, while the issue of European membership has become the leitmotif of Turkish national political debates, the relationship between Turkey and the EU has been marked by considerable ambiguity.

Indeed, from the refusal to make concessions to the acknowledgement of Turkey's progress in respecting EU requirements to new European demands, Turkey's membership application has been a long and tortuous process revealing misunderstandings and ambiguities on both sides. Although in 1995, the EU signed a Customs Union Agreement with Turkey, it then rejected the country's candidacy at the 1997 Copenhagen summit, provoking strong feelings of bitterness in Turkey, especially in the context of the agreement to open membership to ten new member states, eight of which were nominally 'Eastern' countries. The principle of Turkey's candidacy was finally accepted in 1999 at the Helsinki summit, but the date for the start of negotiations was postponed until December 2004. Finally, in October 2005, the EU opened accession negotiations. In the history of European construction, 'there is no country which has been so long pushed aside from the European project while it desires so deeply to be part of it' (Marcou 2002: 9).

Nowadays, EU membership remains the main goal for all Turkish political leaders (except perhaps the far right, and nationalist movements and parties). To be recognized as Europeans and be accepted by and in the EU is the major political ambition for a country still striving to negotiate its relationship with modernity. European membership, it is widely held, would sanction this deep-seated aspiration.

Sport as a parallel diplomatic channel

Turkey's numerous bids to organize international sports events are thus a means to demonstrate the nation's modernity and its organizational capacity, and thus the legitimacy of its application for membership of the European Union. The image that organizers want to portray of their country can be seen through the application files.

International sports authorities (FIFA, UEFA, IOC) define the requirements for all applications; these include such things as the quality of sports infrastructures, security, communications and media, financial guarantees, transport systems, environmental quality, accommodation and so on. Candidates typically highlight their assets in their applications. In addition to these more technical aspects, which do have some symbolic effect insofar as they are an indication of a country's level of development, it is important for candidates to underline the specificity and particular assets of their own candidacy. For bidding cities, this is an opportunity to build an image with a strong identity relating to Olympic ideals.

For example, the IOC strives to promote 'a policy that seeks to provide greater resources to sustainable development in and through sport at national, regional and international level, and particularly at the Olympic Games', also symbolized by the introduction of a clause in the Olympic Charter. Candidate cities must set out their environmental protection and sustainable development programmes. Therefore, one should be not surprised that the bid books of each Istanbul Olympic candidacy (1996, 2000, 2004, 2008) included a chapter on this topic.

More remarkable is the focus on the multi-cultural and culturally diverse aspects of the city, while Turkey's EU membership is still debated on the grounds of human and minority rights issues. Istanbul's Olympic Bidding Committee (IOBC), which is responsible for the city's candidacy, tries to portray an image of a tolerant and generous country. All Istanbul application files point out the city's unique geographical location. Since its first Olympic bid to host the 2000 Olympic Games (in 1994), Istanbul's logo has depicted two interlinked rings with the slogan 'The Meeting of the Continents'.

These two rings represent the continents of Europe and Asia linked by the two bridges over the Bosphorus, and invoke the stylized Olympic rings. Istanbul is presented as a 'bridge between the cultures and civilizations of Asia and Europe. [...] It is a city of culture where religions and languages have merged over thousands of years of co-existence.' It is claimed in the brochures that Istanbul 'is home to 26 ethnic groups and its people speak 10 different languages. The very existence of Olympist Istanbul is a challenge to prejudice and sectarian divide'.[3]

It is interesting to note that these statements are quite different from the official state position towards ethnic groups and multiculturalism in Turkey. The Republic of Turkey remains firmly based on a unitary national conception which doesn't recognize ethnic minorities; only religious groups are recognized as minorities. This means that Alevis and Kurds are not recognized as minorities because they are Muslims. Because of the strong fear of Turkey breaking up, the Turkish state opposes not only all separatism, but also any claims to differential ethnic identity. The war in the southeast against the PKK (Worker's Party of Kurdistan) between 1983 and 1997 led to thousands of people being killed (combatants and civilians), and the deportation of a million people from the combat zone. In spite of the fact that Turkey has had to make major concessions on this issue in order to fulfil European requirements (in March 2006, for example, the first private Kurdish television channel was allowed to broadcast four hours per week of Kurdish-language programmes), minority rights remain at the core of the dispute between the EU and Turkey. By insisting on diversity and pluralism, the authors of the Istanbul Olympic bid are striving to portray an image of a peaceful and tolerant country. But to do this they nonetheless draw on the image of the Ottoman Empire, which is represented as a multi-cultural state.

Bids to host sports events are also an opportunity to portray the image of a peaceful country not only with its own citizens but also with its neighbours. This was the goal of Turkey and Greece's unsuccessful joint bid to host the Euro 2008 UEFA football championships. In the context of the complex and conflictual (if improving) relationship between Turkey and Greece, this joint bid, initiated by the Turks, was presented as the sign of a new era for peaceful cooperation and 'greater mutual understanding'[4] between the two countries. Following the example of the 2002 FIFA World Cup Finals in Japan and Korea, Şenes Erzik (a member of FIFA's executive committee, of the Korea-Japan World Cup Finals Organization Committee, and vice-president of UEFA) considered that this event 'would help to build a better relationship between Turks and Greeks' (Interview, Istanbul, January 2003). The Turkish Football Federation's proposal, made to Greece in February 2000, was enthusiastically accepted by the Greek Football Federation and was supported by the foreign ministers of each state. In less than 18 months, a joint bid committee was created and prepared the official bid for submission to UEFA in November 2001. The change of majority in the Turkish Parliament following the general election victory of the AKP (the moderate Islamic *Justice and Development Party*) in November 2002 did not change the strong political support for this project. On the contrary, R. T. Erdoğan, AKP leader, future prime minister (and a former professional football player) recorded a speech in favour of the joint bid which was delivered in front of UEFA on the day the host country for Euro 2008 was chosen.

From a more technical perspective, even if bids are a means of promoting a country or city's assets, the various bidding committees have not tried to avoid the difficulties that Istanbul may face in hosting events. In Olympic Games bids, transport, security, environmental issues and so on are not denied but rather have been consciously analysed to demonstrate problem-solving capacity. Indeed, bid committees have taken into account the criticisms made by the International Olympic Evaluation Committee in relation to previous bids. The experience of the members of the NOCT and especially of its former president, a member of the IOC, is integral to the evolution of each new Olympic bid. The library at Olympic House in Ataköy, headquarters of the NOCT, contains every bid document filed by rival cities over the last 40 years as well as the IOC's reports and evaluations of those bids.

The IOBC is aware that the information they give in the bidding document has to be credible and reliable if it is to be taken seriously by the IOC. Documents for the 2000, 2004, 2008 and 2012 bids were produced by sociologists and economists at the Social Research Centre (Interview with Cenap Nurat, director of the Social Research Centre and Ferhat Kentel, sociologist, Istanbul, January 2003). It is fair to see in this approach a will for transparency and realism, which parallels with the approach adopted by the Turkish governmental authorities towards EU entry requirements: according to the head of the 'Turkey' desk at the European Commission, the most notable change in relations with new Turkish government following the November 2002 elections was a more shared perception of issues for the Turkey membership (Interview, Brussels, June 2003). Turkish sports bids promote the image of a European country on the verge of becoming an EU member: 'At present, Istanbul is the largest city in a country that is preparing for membership of the European Union' (bid document for 2008 Olympic Games).

'Now, we are Europeans!' claimed Turkey's former president, Suleyman Demirel when Galatasaray won the UEFA Cup in 2000 (*Le Monde* 9 June 2009). Hosting an international sporting event or the victories of Turkish teams in international events are presented as true national successes and a further step forward to Europe. Indeed, these victories are all the more celebrated in that they are achieved on the sports field. According to Ehrenberg (1991), the sports field is a condensation of the democratic society ideal in which competition is equal and the winner the best. In other words, sport establishes a hierarchy based on merit rather than on human ranking or categorization. The focus of Turkey's policy is that, through sports victories, the nation can achieve the European recognition that European politics refuses to grant it.

Hosting sporting mega-events: a consensus project promoted by elites

If we consider sport only as a means of diplomatic action, or as an instrument serving political strategies or ideologies, then we are forgetting the complex relationship that exists between sport and politics in Turkey. Of course, political actors have to refer to sport: sport is a valuable resource. Sport can glorify youth (which is one of the most popular themes in speeches on the Republic of Turkey) and patriotism in a country where the 19 May is a national day for 'sport and youth', with ceremonies organized in stadiums full of children, featuring poems and speeches made to the glory of young people, the Republic and Atatürk (see Yurdsever 2003). The political authorities in Turkey have always supported bids and tried to draw benefits from victories (whether sporting victories or victories in hosting sports events). However, sport is not used in this way in Turkey to the same extent as it was in the former USSR or in China. Even if the relationship between sport and politics is particularly close in Turkey, we cannot reduce it to mere political exploitation. Sporting elites are close to the political elite, share the same republican values, and know that they can draw support from them. In this section therefore, we underline the process by which the sporting sphere exploits the politicization of sport for its own benefit. This phenomenon blurs the borderline between sport and politics in Turkey.

Sporting directors: a specialized elite expecting political support

Although it is obvious that politicians support sports bids, sport organizations are managed by specialized elites and sports federations. These directors acknowledge that they want to see their teams winning competitions, but also that they want Turkey to become a member of the EU. Demirel's exclamation following Galatasaray's 2000 UEFA Cup Final victory is not an isolated example, but an assumption shared by sporting and political elites: 'now, I hope you understand that we are Europeans!' claimed the president of NOCT following a traditional Turkish dinner (of fish and *rakı*) to IOC members during an official visit to Istanbul in 2000.

However, these elites also aim to support other specific goals, such as the development of sport in Turkey. Istanbul's bidding efforts are also a means of building national sporting infrastructure, as defined by the IOBC, the General Directorate of Youth and Sports, and the Greater Istanbul Municipality, as well as private bodies. Through urban policy, Turkey is also able to fulfil European standards in infrastructure terms and to justify expenditure in this area. Sporting elites also want to promote sport among young people, not only from a health perspective but also with the aim of producing future champions. However, even if sporting elites think that their sport project may help Turkey's application for EU membership, they are convinced that the opposite is more likely: Barcelona or Athens hosted the Olympic Games only *after* their countries had been accepted into the EU (six years later for Spain, 22 years later for Greece).

Sporting elites consider themselves to be in the vanguard, using sport to build the image of a modern Turkey in order to receive vital support from the political elites in preparing serious bids to host international sport competitions. Furthermore, the sports sector draws important benefits from the support of political elites. An astonishing example of this goodwill is the Turkish Olympic Law. Istanbul's bid is endorsed by a special law, passed near-unanimously by the national parliament in April 1992, making Turkey the first (and so far the only) country to have enacted such a legal instrument. The Turkish Olympic Law (No. 3796) established the Istanbul Olympic Games Preparation and Organization Council, or IOBC, and authorized it to 'take all necessary action in the pursuit and organisation of the Games'; it further 'recognises and respects the supremacy of the IOC in all Olympic matters'. The law requires that all public institutions and agencies, as well as all local government bodies, give 'priority to the requests of the IOBC in relation to the pursuit and organisation of the Games'. The Olympic Law guarantees a continuous flow of funds, both for the pursuit and the organization of the Games. These include: 1 per cent of football betting revenue; 5 per cent of the preceding year's net income of the National Lottery; 1 per cent of the Housing Fund receipts; an annual appropriation from the Consolidated Budget, the amount left to the discretion of the legislature; 1 per cent of the budget of the Greater Istanbul Municipality; and 1 per cent of the Horse Racing Joint Wagers ticket sales.

Thus, thanks to the political stakes of sport, sporting elites have received continuing political support for their project, which has given rise to a greater accumulation of experience and further developed the country's ability to prepare for hosting international sporting events.

Political and sporting elites: a close family

Whilst sporting elites develop their own independent projects, they also enjoy very close ties to political elites. They were educated in the same high schools or universities, and share values and social experiences. This closeness between elites reinforces the confusion between sport and politics. Almost all NOCT members graduated from high schools such as the Galatasaray Lisesi.

The biography of Sinan Erdem, the former president of NOCT and former member of the IOC, is a significant example of this kind of sport elite career. Erdem was educated at Galatasaray Lisesi where he played volleyball, becoming a member of the national volleyball team and taking part in the Olympic Games. Afterwards, he chaired the Turkish Volleyball federation and worked at NOCT, of which he became President in 1987 until his death in 2005. Hosting the Olympic Games in Istanbul was his cherished dream; bid by bid, he worked to make it increasingly plausible. Invaluable help came from Galatasaray's alumni club: an important social network of Galatasaray's former students is dispersed throughout the world and of course throughout Turkey. The network has a common view of the Republic of Turkey, its modern project, its European aspirations, and of the role of sport in improving Turkey's image in the world. But Erdem's ambition also needed strong and continuing state support. Thanks to his contacts within the political sphere, he was able to suggest the adoption of the Olympic Law, itself prepared by Prof. Erdogan Teziç, a former student of Galatasaray Lisesi, who played volleyball in the national team once coached by Erdem. Afterwards, Teziç became director of the Galatasaray Lisesi, and then rector of Galatasaray University (established as a secular, Francophone institution by agreement between France and Turkey in December 1991). He is also a member of the NOCT and in 2004, was appointed as president of the YÖK (Council of Higher Education).

The confusion between the interests of these elites is patently obvious. Sport may serve to promote national and international political strategies, but sporting elites also exploit this political use of sport in return to gain more support. Furthermore, the shared conception of a common political project to modernize the country gives both elites the conviction that they are working towards the same aim, only by different means. In other words, the borderline between sport and politics is thin, even vague. Sport lies within the political modernization process that the elites have been trying to achieve for decades.

Sport as revealing a complex relationship with Europe

Sport as a social practice also bears a symbolic dimension and political meaning which reveals the confusion of interests that exists between sport and politics. In Denis-Constant Martin's terms, we can perhaps see sport in this sense as a 'non-identified political object' revealing the ambiguities and multi-dimensional nature of politics, where

> Ambiguity is a fundamental characteristic of power. [...] Power is always the instrument of a domination whose forms are more or less blunt. [...] In turn, it creates ambivalent attitudes, which are not only varied and ambiguous, but shared and contradictory; attitudes which combine a fascination with, fear of, challenge to and rejection of power; attitudes which can lead to behaviours which are apparently contradictory and illogical, with unforeseen recalibrations and changes in direction. (2002: 22)

Indeed, sport in Turkey appears a reflection of the country's passionate relationship with the EU and European identity, constructing the stage to play out the ambiguities of this relationship. The hazardous route taken by the negotiation process and the still unfavourable prejudices of Europeans towards Turkey reinforce Turkey's ambiguous feelings towards Europe; but becoming an EU member is all the more fundamental for Turkey in that it is based on the complex issue of Turkish identity. Sport appears as a domain where success and failure affect the image that the Turkish state wants to project to the world, but also the one Turks have of themselves. The resentment varies in intensity and formalization as a function of the social origin of the groups.

Bitter setbacks for sporting elites

We have seen above how Turkish victories in European competitions have been perceived by the political and sporting authorities as giving greater weight to Turkey's EU candidacy. They grant material and symbolic resources. But defeats or failures are felt as a rebuff which echoes European reluctance to admit Turkey to the EU. After the defeat of the Turkish football team by Latvia in a Euro 2004 qualifying match, the president of the Turkish Football Federation, criticizing a biased referee claimed 'the Europeans do not want us'. Of course this kind of statement is rare, because sporting authorities generally control their tongues. Yet failures are perceived with strong bitterness.

In December 2002, UEFA's decision to grant the organization of Euro 2008 to Austria and Switzerland, in preference to the Turkish-Greek bid, fostered deep disappointment and frustration. This feeling was made all the more acute by the fact that during the same period, the 15 European member states taking part in the Copenhagen summit had postponed Turkey's request for a date to be set for the start of admission negotiations. The Turkish coordinator of the bid committee, who was very disappointed by the final vote (the Turkish-Greek bid lost in the last ballot), underlined the extraordinary political dimension of this candidacy. According to him, the success of their rivals could be explained by the fact that both countries were better integrated into Europe, either from a sporting perspective (the president of FIFA, Sepp Blatter, is Swiss), or from a political one (Austria is already an EU member, but Switzerland does not wish to be). In addition, he suggested that while these countries' teams were neither competitive nor strongly supported by fans (in contrast to Turkey and Greece), they offered the reassuring financial guarantees of rich countries. In other words, although from a 'sport logic' Turkey and Greece had deserved to organize the Championship, because of the political and financial resources of their opponents, they failed at the final hurdle (Interview, Istanbul, December 2002). Sport is thus constructed in this argument in terms of its neutrality (as a space of a fair play competition between individuals or states), in contrast to a political space of dubious legitimacy. Thus, Turkey and Greece, in spite of the strong symbolism of this joint bid and in spite of the sporting and technical qualities of their bid, were relegated by the UEFA decision to the ranks of secondary zone countries, to the periphery of Europe. Finally the Turkish coordinator claimed that if Turkey had been an EU member the decision might have been different.

The same level of disappointment was perceptible when in May 2004, the IOC announced the shortlist of the five cities bidding for the 2012 Olympics, without selecting Istanbul. For the director of the bid committee, 'Istanbul was accepted as a candidate city for the 2008 Games – and not for the 2012. Were there any negative developments? No. On the contrary. What has changed?' (Interview, Istanbul, May 2008). For the secretary general of NOCT, this vote was highly competitive, with the intrinsic qualities of the candidacy insufficient to obtain the Games (Interview, Istanbul, May 2008). In September 2007, NOCT decided for the first time since 1993 to refrain from bidding for the 2016 Olympic Games – but will nonetheless apply to host the 2020 Games.

The failures of Turkey on the sports field echo its failure in the political arena. Both reinforce the feeling of exclusion of a country which is already on the margins of Europe.

Sport as a sign revealing political frustration:
the nationalism of Turkish fans

Certain Turkish football fans and articles printed in sports papers express the extent of the ambiguity of Turks towards Europe, through the glorification of nationalism.

While 75 per cent of Turks were claiming that they were in favour of Turkey's membership of the EU (*Le Monde* 27 March 2002), football matches between Turkish teams and teams from other countries are frequently the scene of tough demonstrations of nationalism. The stadiums often offer a large array of slogans and nationalist symbols, sung or written on banners such as: 'Avrupa Fatihi' (the Conquerors of Europe); 'Tremble In Fear Europe, We Have Arisen!'; 'Hail Our Country, The Turk Is Coming' and so on. This attitude is particularly widespread when Turkish teams play against foreign teams from the former Ottoman Empire. Very often, matches against Greek teams are an opportunity for the most radical fans to remember the domination of the Ottoman Empire over Greece. In 2002, the Turkish football club Fenerbahçe played against the Greek club Panathinaikos. Turkish fans raised the provocative banner, 'Greek, Can You Hear The Steps Of The Soldiers Of The Conqueror Sultan Mehmet?'. Such aggressive slogans and the violent behaviour of fans reinforce the already negative opinion Europeans have of the Turks, who are perceived as violent and extremist.[5]

This ambiguous relationship with Europe is also perceived in the media and especially in two sports dailies, *Fanatik* and *Fotomaç*, which altogether have about half a million readers every day. The logo of *Fanatik* (the meaning of the word is more to do with passion and exaltation than religious fanaticism) represents a Turkish flag with the sentence: 'This Land Is Ours'. According to the writer and journalist Tanil Bora, nationalism on the Turkish football scene increased in the 1980s and 1990s, when nationalism was thrust onto the political stage:

> Following the military coup of 12 September [1980] attention paid to national and international football games focused and grew around the goals of national unity. During the transition from the 1980s into the 1990s, the purportedly reflexive growth of Turkish nationalism in responding to the Kurdish national movement, brought to the stadiums a nationalist agitation of until then unparalleled proportions. (2000: 378)

Nowadays, sport nationalism expresses frustration and humiliation at being repelled by an arrogant Europe, to which Turkey would still like to belong. For the leader of the Ultra Aslan (a Galatasaray fan club), the organization of an international event or Turkish sporting victories would not really do anything to help Turkey's EU candidacy because 'they don't want us anyway, because we are Muslims' (Interview, Istanbul, July 2002).

Unlike sports executives, who have to control what they say in public, sports fans express their social and political frustration. For Christian

Bromberger (1995: 9), the stadium remains 'one of those rare areas where the display of collective emotion is tolerated, within certain limits'. But sporting and political authorities are aware that the images of violence in and around the stadiums harm Turkey's image in Europe. In December 2002, the University of Galatasaray and NOCT organized a symposium on 'Violence, sport, and fandom', bringing academics together with police officers and club executives (Yarsüvat & Bolle 2004). However, this kind of attempt to address the problem is far from sufficient: in 2002, the Turkish and Greek foreign ministers sought to turn a match between Turkish and Greek football clubs into a symbol of reconciliation and friendship between the two nations in order to support the joint bid to host Euro 2008. But football fans threw objects and plastic bottles at them while they were entering the stadium. This kind of event shows us the limits of the potential power of sport as an instrument serving governmental strategies. However, it demonstrates that sport is highly revealing of social and political frustrations. It allows symbolic revenge on the real world of social relations.

For Kozanoğlu (1999), the ambiguity of Turkey's relationship with Europe reveals a kind of social schizophrenia in Turkey, between surges of nationalism and Islamism on the one hand, and the desire for Europe on the other. Indeed, despite the denunciations and warlike declarations, Europe remains the supreme reference point. During a Turkish Championship match, Galatasaray fans raised banners against the rival team reminding them that the Galatasaray Football Club was the only one to be playing in the UEFA Champions League: 'We Are Europeans, We Are Playing In The Champion's League! And You, Where Do You Play?'. This ironic provocation testifies once again to the fact that Europe remains synonymous with superiority in Turkey and that it is important to be recognized as Europeans. It also shows the uneasy concern with 'not existing beyond Edirne'.[6]

Thus, Europe appears as a remote horizon which can be criticized and defied, but that is still the common aspirational standard. In reflecting the ambiguous relationships that exist between Turkey and Europe, the sporting arena confuses even further the line between sport and politics.

The multiple overlaps between sport and politics in Turkey make it difficult to draw the borderline between them. Sports mega-events are a political message sent to Europe; a message that can produce high hopes or bitter disillusions, depending on the subsequent response. Turkish bids to host events have to cope with a variety of political stakes that we can certainly analyse, but whose understanding has to be sought in the dynamic interpenetration of this reasoning and the historical relationship between Turkey and Europe. In some contrast to the dynamic of 'deterritorialization' identified by Sandvoss to lie at the heart of the cultural globalization of football (2009: 89), the European stakes of Turkish sport and mega-event bids perform a 'reterritorialization' of a country set within narrow borders, its own traumatic history, and its – as yet – unachieved ambition of European acceptance.

Appendix 4.1   A comprehensive list of sports event bids and event hosting

In the 1980s, Turkey multiplied its bids to host international sports competitions. I have recorded more than 20 major official competitions i.e. European or World Championships (regional or youth competitions are not included).

	World Championship	European Championship	Bids
1971			– Mediterranean Games in Izmir
1974	– Wrestling		
1977		– Wrestling	
1985		– Taekwondo	
1989		– Wrestling	
1990		– Judo (teams)	
1991	– Skiing (on grass)	– Wrestling (juniors)	
1992	– Freestyle and Greco-Roman wrestling (young men) – Wrestling (handisport)	– Basketball (club)	
1993	– Archery – Volleyball (young men)	– Freestyle and Greco-Roman wrestling – Swimming (juniors) – Basketball (juniors)	– OG 2000 bid
1994	– Weightlifting – Boxing (juniors)	– Sailing (optimist) – Athletics Nations Cup – Greco-Roman wrestling (juniors) – Tennis in teams (women) – Volleyball (juniors)	
1995		– Sailing (Laser and Finn) – Basketball (junior, women) – Tennis (Ch. Des Clubs Champions) – Tennis (teams, men)	
1996	– Equestrian World Cup		
1997	– Equestrian World Cup	– Wrestling (junior)	– OG 2004 bid
1998	– Offshore 4th Stage – Taekwondo (juniors)	– Sailing (470) – Basketball (junior, women)	
1999	– Freestyle wrestling	– Swimming – Tennis (teams, men and women)	
2000		– Karate – Archery	
2001	– Weightlifting	– Basketball – Greco-Roman wrestling – Triathlon European Cup	– OG 2008 bid
2002		– Taekwondo – Volleyball (women's champions league)	– Football Euro 2008 bid (with Greece)
2003	– WRC Rally of Turkey – Wrestling (junior)	– Basketball (young women) – Volleyball (women) – Athletics Nations Cup (2nd div.)	– OG 2012 bid

Continued

Appendix 4.1   Continued

	World Championship	European Championship	Bids
2004	– Windsurfing – WRC Rally of Turkey	– Freestyle wrestling Seniors	
2005	– Istanbul Formula 1 Grand Prix – WRC Rally of Turkey – University world championships (Izmir) – Volleyball: Under 20 World Championship (women)	– Football: Final Champions' League – EuroBasket Women 2005	
2006	– Istanbul Formula 1 Grand Prix – WRC Rally of Turkey – Archery: World Cup Circuit (Antalya)	– Youth European Championship Freestyle, Greco-Roman and Women's Wrestling	
2007	– Senior World Cup Greco Roman Wrestling – Fencing: Junior and Cadets World Championships – Karate: 5th World Junior & Cadet Championships – Formula 1 Grand Prix (Istanbul)	– 2007 Kuşadasi ITU Triathlon European Cup	
2008	– Fencing: Male Senior Sabre World Cup – UCI Mountain bike World Cup – Marathon – Sailing: IODA World Sailing Championship – WRC Rally of Turkey – Formula 1 Grand Prix (Istanbul)	– Taekwondo: European Qualification Tournament for Beijing Olympic Games – Taekwondo: 7th WTF World Junior Championships	
2009	– Fencing: Senior World Championships		
2010	– Judo: World Team Championships – Basketball: World Championships		Football Euro 2016 bid
2011	– Winter Universiade (Erzurum) – Archery World Cup Final (Istanbul)	– European Youth Olympic Festival (EYOF) in Trabzon	
2012	– Athletics: IAAF World Indoor Championships – Swimming: 11th FINA World Championships (25m)		
2013	– Archery (Antalya) – Mediterranean Games (Mersin)		

Note: The major senior competitions are in bold; events and categories are in parentheses

Notes

1. Among the main reforms of the first ten years of the Republic were the abolition of the Califat (1924), the deletion from the Constitution of the reference 'the state religion is Islam', the proclamation of secularism, the abolition of (Muslim) religious schools, courts, and uniforms, the adoption of the Christian calendar, language reform, the adoption of the Latin alphabet and new civil codes inspired by European codes, the introduction of women's suffrage and so on.
2. 'Bizim Hedefimiz bir forma ve renklere sahip olup, Ingilizler gibi top oynamak, ve Türk olmayan takımları yenmektir'. See, amongst other sources, *The Offside* Galatasaray blog, http://galatasaray.theoffside.com/team-news/4.html, accessed 12 June 2008.
3. *Istanbul 2008*, Application file, p. 2.
4. *Bid Document Greece-Turkey Euro 2008*, p. 24.
5. Several similar events have occurred in the past ten years or so. In 1999, two English fans were fatally stabbed in Istanbul before a match between Turkish and English teams. In November 2005, violent incidents took place after a football match between the Turkish and Swiss national football teams. In February 2006, FIFA's Disciplinary Committee imposed sanctions on the two teams, players and executives. One of the players involved complained that, in Europe and the world at large, Turks were perceived with a different eye: 'They see Turks as barbarians. If I had been Totti or Zidane, I wouldn't keep on being called here to give a testimony'. Quoted in *Turkish Daily News*, 8 February 2006.
6. Edirne, former capital of the Ottoman Empire (c.XIV) is the westernmost city of Turkey, just 10 km from the Bulgarian and Greek borders.

5

The Failed Bid for Lyon '68, and France's Winter Olympics from Grenoble '68 to Annecy 2018: French Politics, Civil Society and Olympic Mega-Events

Hugh Dauncey

The failure of Paris to host the Summer Olympic Games of 2012 focused attention on France's relationship with the IOC and on the internal functionings of her Olympic Movement. Paris also lost out to Barcelona for the 1992 Games, but France did stage the 1992 Winter Games in Albertville, echoing the success of the 1968 Winter Olympics, held in Grenoble during the heydays of Gaullism. France's winning bid for the 1968 Winter Games and the general national popularity of the event obscured the fact that Lyon had been a rejected candidate for the 1968 Summer Games. The sporting, political, diplomatic and economic background to what became a forgotten element of France's Olympic history sheds interesting light on how Gaullist governments of the 1960s aimed to harness the developmental impetus of Olympic mega-events to their project for France, and how French cities and regions hoped to instrumentalize the Olympics in their own 'regional' agendas. Both the failure of Lyon's bid and the success of the Grenoble Games affected France's relationship with the IOC and set patterns of French thinking about sports mega-events at municipal, regional and national levels. This chapter analyses Lyon '68 as a failed bid for the Summer Games eventually awarded to Mexico City, Grenoble '68 and Albertville '92 as successful bids, and the prospective candidacy of Annecy for the Winter Games of 2018, examining bidding processes and municipal and national politics and the hosting and event staging of France's two Winter Olympics so far.

The diachronic perspective offered by an analysis of how France has related to the function and concept of mega-events over the five decades which separate the failure of Lyon '68 and the failure of Annecy '18 allows us to suggest that the case study of France demonstrates a changing frame of reference for mega-events within national and regional political contexts, itself reflecting changes in the relationships and discourses which have underpinned the interactions between local conditions of organization of the Olympics and the global contexts of the IOC. France's interactions with the IOC around the Olympics over the past 50 years or so seem to reveal a complex hybridity of the Games in the French context at least, where they entertain an at times confusing dual identity as simultaneously national and city-specific events. This shift from an understanding of mega-events as primarily national to more city-specific has been discussed by numerous analysts in recent years (e.g. Searle & Bounds 1999, Shoval 2002, Hall 2006), with emphasis being placed on how the function and concept of the Olympics has changed the frame of reference from symbolic displays of state power to vehicles of entrepreneurial rivalry between cities of aspiring or established 'world-class' status within a globalized economic framework of neo-liberal competition. France's experience of the Olympics since the 1960s seems to demonstrate how this trend coexists and interacts with enduring concerns about national prestige.

Lyon '68: what does a failed bid show?

Lyon's failed bid for the 1968 Summer Games is linked to a key juncture in France's post-war development: in 1958, the ailing Fourth Republic (founded in 1946) gave place to the Fifth Republic, under the presidency of Charles de Gaulle. The new regime (whose constitution gave more effective government, and ultimately the presidentialism characteristic of contemporary French politics) transformed the French state and French approaches to France's 'place in the world'. Gaullism heavily emphasized French 'grandeur' and independence in all domains, and elite sport was identified by the Gaullist state as an instrument of French national prestige. Lyon's unsuccessful rivalry with Mexico City for the 1968 Games – less as a 'world-class city' than as a representative of a nation with an agenda of 'prestige' and of a continent – was played out within the context of a new importance for sport amongst France's governing elites. Lyon also was undergoing political change during the late 1950s and early 1960s, stimulated by a 'generational shift' in pivotal agents such as the city mayor, regional prefect and president of the regional council, engendering new thinking and fostering ambitions for redevelopment of the city and region. New manifestos at state level to instrumentalize elite competitive sport in support of national prestige thus intersected with new networks of governance in Lyon, which saw opportunities to build bridges between central government 'sports policy' and municipal/regional development.

Institutional histories of French national sports policy are complex, but 1958 always figures in analysis of state interest in sport and associated portfolios. The Gaullist inflection to 'sports policy' focused on sport *and national grandeur* through Olympic success, but interest in medals also came with realization of the significance of sports mega-events as showcases not only of national athletic prowess but also of the organizational genius of the state and the growing modernity of France. In September 1958, the new Gaullist government created a 'High Commission' of *Jeunesse et Sports* (youth and sports, directed by Maurice Herzog) which progressively instrumentalized elite sport and mega-events. Throughout the Fourth Republic, state interest in sport had been rather dormant, as governments rebuilt the economic system, decolonized the Empire and managed the difficult socio-cultural modernization of the beginnings of France's '30 glorious years' of prosperity, 1945–75. Any will to employ mega-events as vehicles of national prestige was slight in comparison, say, with the UK's celebrations of post-war internationalism through sport in London '48 (Dauncey 2004) or of British know-how in the 1951 Festival of Britain, both of which were much about (re)affirming Britain's 'place' in the post-1945 world. The preoccupation with sport of early governments of the Fifth Republic did however continue institutional initiatives of the Popular Front (1936–8), when left-wing governments promoted sport for all as healthy leisure for the working classes. It also echoed more problematic attitudes and structures of the collaborationist Vichy Regime (1940–4), when sport and physical exercise had served the more reactionary sociopolitical goals – 'Travail, Famille, Patrie' – of the *Etat Français*. From the Liberation in 1944–5 until 1958, French sports policy remained in some limbo, both because interventionism in sport was tainted by Vichy, and because dire public finances precluded expenditure: any plans for improving sporting infrastructures within the five-year national planning system came to naught, as education took priority over sport/leisure. Any will of some governments to promote sport was vitiated by these financial problems, by antagonistic understandings of 'sport' as elite or mass, professional or amateur, by a reluctance of the refounded 'Republican' state to 'direct' civil society, and perhaps, ultimately, by a lack of national self-confidence which undermined the state's ability to conceive organizing a mega-event for purposes of French self-assertion. In a period when the Olympics were about symbolic displays of state power, France's polity lacked the self-confidence to host them.

The 1958 regime-change meant that 'sport' would be taken seriously. Gaullist emphases on sport for 'le prestige de la Nation' *and* for 'la santé de tous' ('the health of all') potentially resolved the elite/mass dichotomy in 'practice' by positing that success in international competition would foster grassroots uptake, and by hoping, conversely, that mass sporting activity would create a 'pyramidal' upwards provision of elite athletes to represent France. Behind sport as instrument of national prestige lay both an implicit conceptualization of sports mega-events as potential demonstrations of national strength, and the beginnings of practical preparation of athletes and of infrastructures that could help France in future Olympic bids. The IVth Plan (1962–5) – whose preparation was ongoing as Lyon's bid was drawn up, and which was to have laid some of the bases for Lyon '68 – started to address the needs of sport and physical education with ambitious infrastructure developments intended to encourage mass participation and identification of sporting champions.

Emphasis on elite sport was arguably the defining feature of 'sports policy' in the late 1950s and early 1960s, and essentially it has been interpretations of sport and national prestige through success in international competitions – the Olympics especially – which have dominated the French state's attitudes towards sport in the Fifth Republic. This has created resentment amongst some politicians of the left and left-wing analysts of sport who see success in medals tables as a distraction from investment in 'sport for all' in the tradition of earlier French thought on 'éducation physique' – creating a fit and healthy population (e.g. Brohm 1997). France's bid – through Lyon – for the '68 Summer Games was intended as an instrument of national prestige through the staging of a sports mega-event, glory through hoped-for sporting victory, and (for Lyon) the – accessory – creation of sporting infrastructures for the general public. The Lyon bid was thus an early product of the Fifth Republic's new-found interest in mega-events (in terms of medals, and in staging them), combined with the particular conditions of the city's municipal politics, infrastructure development and relations with Paris.

The 'Olympic turn' in French sports policy appeared most strikingly in the aftermath of Rome '60, when the French team returned with a humiliatingly meagre medal haul. In the climate of national pride and confidence newly engendered by the Fifth Republic, and against the ideological backdrop of 'grandeur' developed by de Gaulle, France's tally of five medals and not a single Gold, prompted a celebrated caricature of de Gaulle (by Faizant for *Le Figaro*) depicting the septuagenarian head of state – in tracksuit and plimsolls – setting off to compete, with the grouching caption 'Do I really have to do everything *myself* in this country?'. Such satire emphasized how the team and the administrative structures to identify and support athletes had failed the Nation, or at least Gaullist understandings of how sporting *performance* should serve French 'prestige'. For a state as 'volontariste' as that of Gaullism, it was a small step to realizing that France might need to organize the Olympics herself, as the Third Republic had staged the 1924 Games (Summer and 'Winter') and the 1938 FIFA World Cup in a period when France had been a 'destination of choice' for international sports spectacles (Tumblety 2008: 83).

The Rome debacle reinforced the Gaullist state's opinion that the government needed to be as *dirigiste* in elite sport as it was in other fields. In 1961 a special General Delegation responsible for preparing France's participation in Olympics was set up, and the sports 'High Commission' was transformed into a minor ministry in 1963, administering relations between the state and national sporting organizations (the 'sports federations'). The first Programme-law developing sporting and educational infrastructures invested 600 million francs in public and private sports facilities during 1961–5. A thousand swimming baths and 1500 stadiums and playing fields were created, and this continued during 1966–70 with the construction of 2000 sports grounds and 1500 gymnasiums. This national effort by the state to enhance facilities (for all, at local level, to feed athletes to higher levels in the sporting pyramid, as well as elite training centres) coalesced with local municipal/regional ambitions to develop sports infrastructures, and, maybe even, to harness a mega-event to their own urban/regional development needs.

Also during the early 1960s – as Lyon worked on her bid for '68 – discussion arose about the need, naturally in Paris, for a 'national' sports stadium seating 100,000 spectators able to host the largest national and – crucially – international competitions. The non-existent national stadium became a standing joke and an appropriate stage for mega-events was only finally provided in preparation for the 1998 World Cup, in the form of the Stade de France in the northern suburbs of Paris as the central, capital element in nation-wide stadium renovation (Dauncey 1997, 1998). But the timidity with which Paris and successive governments dealt with the vexed dossier of the national stadium contrasted with successful infrastructural improvements made in the 1960s by the state in partnership with munici-palities and private sports clubs (30–40 per cent of costs were subsidized by central government). France in the 1960s was in full 'modernization' as volontarist, *dirigiste* policies addressed perceived failures of the Fourth Republic in adapting French socio-economic structures to modern reali-ties, and sport, variously, but most noticeably in terms of 'equipment' was at the forefront of the process. This, in brief, was the 'politics of sport' which contextualized Lyon's ambitions to host the '68 Games.

Local governance and global ambition

A further contextual element of the pre-event processes of Lyon '68 is the relationship between Lyon and Paris. Lyon is France's second city, vying with Marseille for second rank in the urban hierarchy. Provincial cities in France have traditionally suffered strained relationships with the capital, in the highly centralized 'Jacobin' administrative system characteristic of France since revolutionary times. Lyon has struggled to preserve its iden-tity and 'independence' against Parisian dominance, but until 1977 at least, when President Giscard d'Estaing finally authorized Paris to regain a full municipal dimension with an elected mayor, Lyon benefited from a strong tradition of municipal/departmental/regional politics, and exploited advan-tages brought by strong mayors, often important national politicians, who protected Lyon's interests in, and against, Paris. While France's candidate for '68 was being selected, Parisian 'identity' was 'diffuse' because of the absence of an elected mayor; the capital's resistance to Lyon's candidacy, although exploiting Parisian links with central government, lacked clear municipal/regional definition and backing, a precise geographical identity, and strong 'champions' to lobby for it.

Contrastingly, Lyon's champion of the Olympic bid was Radical mayor Louis Pradel, an unexpected victor in the 1957 and 1959 municipal elections, and successor to the long-serving nationally famous Edouard Herriot (a Radical, three-times prime minister in the interwar period, and Fourth Republic minister). Herriot was mayor from 1905 to 1940, and from 1945 until his death in 1957, and did much to develop sporting infrastructures and activities. As early as World War I, Herriot had used his national visibility and influence to lobby for Lyon's sporting interests: Pierre de Coubertin visited Lyon during the Great War, inspecting sports facilities and, in particular, the omni-sports stadium being constructed by Herriot in the Gerland quarter (now Stade Gerland, home to football club Olympique Lyonnais). In private correspondence with Coubertin, Herriot negotiated that Lyon be considered for the 1920 Games (if Antwerp should decline) or the 1924 Games (eventually awarded to Paris, after Lyon withdrew). Thanks to Herriot, and his ability to exert influence at state level to increase the 'visibility' of Lyon, the city's bid for the 1968 Games was thus its third attempt.

New-found state support for sport arising during the early Fifth Republic catalysed initiatives already implemented in Lyon during the late 1950s under the guidance of Mayor Pradel. Pradel also had strong interest in sport, and under Herriot in the late 1940s and 1950s had led a city council committee focusing on sporting and leisure activities. Under Pradel, as well as general encouragement to sporting activities in the city and region, the idea of Lyon staging the Olympics centred on building administrative teams and networks to produce the bid, and building of actual facilities. The process centred on Pradel and his aide Tony Bertrand, whose expertise in organizing large-scale sporting events led to his involvement in a number of major Francophone-world competitions during the 1960s. In the context of presidential wishes for 'une France qui gagne' in international sport, Pradel was increasing the urban governance capacity of Lyon through the motivational project of bidding for the Games, and simultaneously renovating urban infrastructures in general and sports facilities in particular.

Bidding for the Olympics as rejuvenation of city governance and infrastructures can be observed in an early Pradel initiative: an ambitious 'festival' marking Lyon's 2000th anniversary. Held in 1958, these bi-millennial celebrations included cultural and political activities and a programme of sports events, based on the work of the *Office municipal de la jeunesse et des sports* (initially created in 1946, building on structures left by Vichy). The event brought many visitors to Lyon, almost as a dress-rehearsal for future enterprises, and planning the 'festival' shook up the procedures of city administrators, which had drifted during the later years of Herriot's mayorality. Pradel was responsible for municipal sports policy under Herriot, and as mayor he appointed Bertrand to the responsibility: it was this partnership which from 1958 concentrated resources on sports infrastructures in the city and region, and from about 1960, piloted Lyon's Olympic bid. Supported by the regional prefect of Rhône-Alpes, Alain Ricard (credited with the idea for the Games and an example of the 'generational shift' in decision-makers, 1957–60), and with more implicit backing (from Paris) of the national High Commissioner for Youth and Sports, Herzog (Lyon-born and hoping to obtain a seat as a Gaullist deputy in Lyon in the 1962 legislative elections), Pradel and Bertrand constructed a support-coalition for the Olympic bid.

After Herzog's visit to Lyon in January 1961, discussion in the council meeting of 20 February about the preparation of a bid allowed Pradel to answer minor concerns, and Lyon's intention to be a candidate was declared to the French Olympic Committee and passed to the IOC. But nationally, the city's ambition caused jealousy and gave rise to some scepticism: in reflection of France's steep urban and political hierarchy, Lyon's bid was judged by some in the capital as 'pretentious', despite little ambition in Paris – at least initially – to consider organizing the Games. Notwithstanding rumbling Parisian criticisms, in April 1962, a visit to Lyon by Marceau Crespin, the national official managing France's 'Olympic preparation' resulted in his support, and in July, Armand Massard, IOC vice-president, municipal councillor in Paris and president of the French Olympic Committee also gave his backing. Massard's endorsement was perhaps surprising: although he had a *national* responsibility for Olympic matters and might thus be expected to be impartial, he also held elected office in Paris, and in August 1962, the head of the Paris Municipal Council Pierre-Christian Taittinger announced that Paris would also bid to host the Games, giving the French Olympic Committee a difficult choice. In late November 1962, however – after the legislative elections involving Taittinger in Paris and Herzog in Lyon – the Parisian candidacy disappeared as the French Olympic Committee decided to back Lyon, by 19 votes to 5, at Baden-Baden in October 1963. The short-lived Parisian bid reflected longstanding centre-periphery rivalry and short-term contingencies of party politics and electioneering. One of the rare studies of Lyon's failed bid for the 1968 Games (Terret 2004) emphasizes the difficult relationship between Lyon and the capital, and other more general studies of Pradel stress how he struggled to balance his coalition support and was sometimes obliged to make concessions to Gaullists at national and local level (Sauzay 1998). But study of the preparation of the Lyon bid suggests that it was the quality of Pradel's local political leadership which carried the process. One of Pradel's weaknesses was his lack – contrasted with Herriot – of experience and influence in national politics, but in compensation he benefited from a network of contacts and 'fixers' facilitating access to ministries and Parisian officials, and despite occasional compromises to avoid antagonizing Gaullists at all levels, the bid itself was significantly the product of local politics and municipal vision, exploiting a general context of state support for mega-events and an uncharacteristically disorganized central/national rival in Paris.

What opposition there was in Lyon centred mainly on the concern that the costs of staging the Olympics would outweigh income and the government subsidies promised by what was criticized as a recently created and doubtless impermanent Ministry of Youth and Sport. Worry was expressed at the deficit of the 1960 Rome Games, and criticism of feared costs tended to neglect completely, or at best discount as intangible and unmeasurable, any secondary benefits of hosting the Olympics. Local opposition was focused on the impact of the Games on Lyon itself, and this is a clear reflection of the significant dimensions of planning which were specific to the city. As has recently been suggested (Liao & Pitts 2008), 'Olympic urbanization' can be periodized into four eras: founding origins stressing a unique urban context as a 'Modern Olympia' (1896–1904); an era where the Olympic stadium dominated planning (1908–28); where the Olympic 'quarter' was the principal focus (1932–56); contemporary interpretations (1960–2012) of the developmental impact of the mega-event which stress thoroughgoing urban transformation. The 'non mega-event' of Lyon '68 confirms this periodization. Planning for Lyon '68 occurred essentially during the period when mega-events were 'about' Olympic quarters, but was also on the cusp of a more ambitious interpretation, as the authors suggest Rome 1960 as the 'first paradigm' of an Olympics catalysing a wide urbanistic programme (Liao & Pitts 2008: 152). Lyon's long tradition of developing sporting infrastructures at city scale (although somewhat fallow post-1945) was a factor which perhaps contributed to tying planning during the late 1950s to relatively restricted municipal/quarter levels, but Rome 1960 and drives for regional/national development from DATAR (the national regional development planning delegation, founded in 1963) pulled discourses and conceptualization towards more intensive and extensive urban transformations characteristic of 1960–2012.

Allowing the development of purely local facilities in an 'Olympic quarter' *as well as* within the large-scale Lyon conurbation, governance networks in Lyon saw the bid as a bold initiative to enhance Lyon's status nationally as France's second city, and also *internationally*, within the context of the new Common Market, and even globally: the global scale of the Olympics could refocus attention on Lyon not just as also-ran to Paris, but as a successful contemporary centre of industry and business located at the heart of European transport corridors. The issue was described in terms of Lyon remaining either just 'une petite ville de Province' or aspiring to the role of a major European metropole.

Thus already one of the key discursive and economic frameworks for the Games in Lyon seems to be a global one, also linked to the shorter-term, smaller-scale catalysing effects of the competition itself or the intermediate-term legacy of infrastructural or entrepreneurial development. The official bid document somewhat naively stressed Lyon's geographical position at what was claimed as the 'crossroads' of Europe, Africa, East and West, and the promotional film produced for the campaign was proudly but over-ambitiously entitled 'Lyon, grande ville européenne'. In retrospect, Lyon's bid-planners should have anticipated the attractiveness of the 'risky' choice for the IOC of awarding the Games to developing Mexico, and elaborated a narrative of Lyon's place in the world which was stronger and more persuasive than what amounted essentially to a vague restatement of traditional Eurocentrism, as yet unsupported by a developing centrality of the city to European flows and markets in the EU, or even at a more global level.

Since the late 1940s, the French state had been concerned at over-concentration of economic activity, wealth and power in Paris and the surrounding region. Thinking on regional development during the late 1950s and early 1960s aimed to re-balance regional disparities, but Lyon's political and business elites saw that beyond this national planning perspective, their city needed to be more than merely the regional capital of Rhône-Alpes. Lyon's past prosperity had been based on traditional industries, but as France's economic modernization proceeded apace, transferring the working population from agriculture and industry towards tertiary activities, Lyon's planners saw a need to encourage industry, but also to focus on infrastructures favouring the city in the new tertiary economy: hosting the Olympics was seen as a catalyst for urban development in transport, hotel accommodation and other facilities. Concretely (Pradel was nicknamed 'Mr Concrete' for his enthusiastic redeveloping), Lyon looked to the Games for upgrades to road, rail and airport facilities, planned during the mid-1960s to open the city to the rest of France, Europe and world. Within the city, the Gerland Stadium (1919) needed renovation, and improvement of other sports facilities was required. The renovation of entire 'quartiers' such as Gerland was a contribution to new urban planning, and new road systems and parking were important for business and industry. In particular, new hotel accommodation, and future use of the Olympic village were welcomed to boost Lyon's tertiary activities in tourism, conference hosting and other non-traditional activities. The mega-event of Lyon '68 would have merged national, state priorities of DATAR-led regional development (rebalancing the relationship between Lyon and Paris within an increasingly 'European' and even global context) with longstanding municipal objectives for improving infrastructures, at the same time as demonstrating French organizational genius and sporting prowess to the world.

Above and beyond Lyon's relationship to Paris and the city's strategic centrality (or otherwise) to developing European spatial dynamics, inflected by the process and after-effects of bidding for the Olympics (Benneworth & Dauncey 2011), it was Lyon's 'place in the world' (in a broader sense) which seemed very relevant to Pradel's team in Baden-Baden as they digested the IOC's decision: 30 votes for Mexico City; 14 votes for Detroit; 12 votes for Lyon. The 'Old World' in the shape of France's ancient Roman 'capital of Gaul', had lost out to the New World in a contest which seemed to have stressed the geopolitical significance of a nation's hosting of a mega-event rather than more technical aspects of the bid dossier itself.

Grenoble '68 and Albertville '92: national grandeur and regional development

The Winter Games of Grenoble '68, although naturally tied to the specific context and conditions of the Olympic Movement and French politics in the 1960s, also dialogue with the 'proto' mega-event organized in Chamonix as part of the 1924 Olympic Games. Arnaud and Terret (1993) suggest that Grenoble '68 can be understood, in some senses, as a transition between Chamonix '24 and Albertville '92. Their study of Chamonix and Grenoble (undertaken as research to inform the planning for and expectations of Albertville '92) locates these Games of France essentially within a context of utilitarist exploitation of the events for national, regional and local agendas. The sporting events organized in Chamonix in 1924 (the official title of 'Winter Games' was not authorized by the IOC until 1925) were arranged something at the last minute as the contract between Chamonix and the Comité olympique français (COF) was signed only in February 1923, 12 months before the events. What was described as a 'winter prelude' to the 'Summer Games' of Paris '24 was rather an adjunct to the main point made by the IOC in awarding the Games to France, namely that France, six years after the end of the Great War, was capable of staging a sports mega-event on a world scale and was fully recovered in her political and economic life. Mega-events were thus understood in discourse and concept as imprimaturs of development and of global integration.

Chosen precisely because existing infrastructures were well-enough established to enable the staging of the activities of the 'grande semaine internationale des sports d'hiver' without need for too much development, the Chamonix Games were nevertheless a financial catastrophe, costs being shared by the town and by COF. Arnaud and Terret (1993: 111) suggest that the impact of Chamonix '24 was overall rather difficult to evaluate: as a 'side-show' to the larger-scale activities later in the year in Paris, Chamonix nevertheless contributed towards France's renewed image as a nation capable of organizing international events, and Chamonix itself derived positive benefits in terms of publicity and facilities. But more importantly, it would seem that in this period, it was more for effects on mentalities and practices, rather than on infrastructures, that the 'Winter Games' of 1924 were significant: first, because Chamonix fostered development of skiing in France as a recognized autonomous discipline (the sport was 'institutionalized' through the creation of governing bodies); second, because building on dynamics already operating, Chamonix helped usher in the boom years of French/alpine winter sports in the interwar period; third, because along-side Paris '24, Chamonix publicized both the Olympic Movement itself and the concept of linkage between (international, Olympic) sport and national life. To borrow the analysis of a recent summary of Olympic urban lega-cies, Chamonix was thus a special (Winter Games) case where the Olympic stadium dominated planning in practical terms, as well as a (proto) mega-event with geopolitical significance for France and (already) entrepreneurial spin-offs for regional economics.

In contrast to the almost 'accidental' Games of Chamonix '24, those of Grenoble 1968 were prepared, planned, organized and evaluated in more thorough manner. As we have seen in discussion of the Lyon '68 candidacy, France in the late 1950s and 1960s was a context propitious for ambitious plans of municipal/regional development, and increasingly in the 1960s especially, for projects instrumentalizing sport and sporting achievements to the greater prestige of France. Francillon has suggested (1991: 14) that the Games of 1924 were 'improvised', those of Grenoble 'programmed' and those of Albertville 'calculated'. We will return later to the notion of Albertville '92 as a (mis?) 'calculation', but here we need to consider how easily and completely the ambitions of Grenoble, of the Isère *département* and of the Rhône-Alpes administrative region meshed with IOC requirements for the Games of 1968 and how Grenoble '68 represents a further 'node' in France's developing interpretations of the concept and function of the mega-event.

Grenoble '68 was indeed 'programmed' in the dual sense that, geopoliti-cally, France was due the favour of approval of Grenoble by the IOC after the disappointment to the Gaullist administration of Lyon's failure, and in terms of event staging, the full force of French technocratic management – at all levels of government – was gradually deployed to organize the event's success. To take up the distinctions made by Hoberman concerning the dif-ferent stages of Olympic sport defined by evolving balances between civic, public and commercial logics (Hoberman 1995) and discussed as part of a more recent collection of reflections on 'post-olympism' (Eichberg 2004), Grenoble '68 was a 'public' event (a taxonomy represented at the extreme by Berlin '36). Eichberg suggests that as 'public' mega-events, the Olympics were prone to state rationality turning them into a 'half-fascist festival', and if one substitutes the word 'Gaullist' in this striking term, then the Games of '68 were indeed representative of Olympism before it entered into its (essen-tially) commercial logic in the 1980s.

As with Lyon '68, the preparation of Grenoble's candidacy involved lit-tle questioning of the seemingly intrinsic advantages in terms of 'develop-ment' of hosting the Winter Olympics. In the late 1950s and early 1960s, Grenoble's governing elite – Gaullist Mayor Albert Michallon, prefect of the Isère Francis Raoul, and sports administrator Raoul Arduin (president of the national Fédération française de ski) – were conscious of the developmental deficit of Grenoble compared to the rest of France. Distant from Paris, marginalized both in the 'spatial imaginary' of politicians and officials in Paris and in terms of increasingly inadequate transport links, Grenoble wanted to alert the central government to its needs, and to the problems of its *département* and region, whose more 'alpine' areas were suffering economic and social problems. Elected mayor in 1959 (until 1965), Michallon was disappointed that the first Gaullist government (1958–62) of the Fifth Republic did little to support Grenoble and realized that both the failing economies of the alpine areas and rapid industrialization, congestion and transport problems in Grenoble required decisive action, preferably with state aid. Made public in late 1960, Michallon's plan was to bid for the Winter Games by federating as wide a coalition as possible behind the project, including the state, which would see the advantages of hosting the Games in terms of national prestige.

Grenoble was selected to stage the Games at the IOC's meeting of January 1964, and the hitherto relatively passive support of the COF, of Herzog and the Gaullist state, became more active, eventually leading to the famous description by de Gaulle of the Games being 'les Jeux de la France'. More so, perhaps, than the 'Games of France', Grenoble '68 were the Games of the French state, whose subsidies covered some 75 per cent of funding requirements, and whose oversight of event preparation extended to a inter-departmental working party chaired by Prime Minister Pompidou. As planning proceeded in 1964–8, the process of enhanced 'governance capacity' that was partially created in Lyon by its *failed* bid brought together a range of stakeholders in Grenoble whose uncomplicated ambition was to further their own concerns. Arnaud and Terret (1993: 150–8) suggest a number of often overlapping 'discourses' which informed and overlaid Grenoble's planning, and whose status at the time as unchallenged tenets of what a (proto?) mega-event should contribute to its host city and region show how little reflection existed about 'resistance' of any kind to IOC requirements or questioning of the intrinsic, teleological value of 'development'. These 'discourses' concerned political values of Gaullism and of national unity around France's aspiration towards 'grandeur', stimulation of winter tourism in the Alps, accelerated development of Grenoble and its economic base through infrastructure projects and enticement of foreign companies, boosting of French winter sports equipment industries and finally the fostering of achievement in French sport at all levels. As Arnaud and Terret conclude on Grenoble '68, some commentators wondered whether this total 'instrumentalization' of the Games in favour of municipal and regional development and of national Gaullist prestige was not in some sense a betrayal of the 'essence' of the Olympics: sport as a pretext for improving transport links to the Isère seemed to sell out on the Olympic ideal (ibid. 1993: 159). For all that they were in many senses geopolitically 'the

Games of France' showcasing France's modernity and genius for organization, Pompidou could also be prosaically down-to-earth in his view of how
the state was using the Olympics; speaking in February 1968, he explained
that the funding for Grenoble '68 was as much about economic development and infrastructures as about sport (Pompidou quoted in *La Gazette
Olympique* 11 February 1968).

The 40th anniversary of the Grenoble Games, coinciding with the city's
campaign to represent France in the final selection of a host-town/region for
the 2018 Winter Games elicited some popular revisionism about the 'success'
of Grenoble '68. This popular feeling against the legacy of the efforts made
in the 1960s to develop the city and region and of the event itself echoed
more academic reconsiderations of the overall balance sheet of Grenoble
'68: as Pierre Kukawka et al. suggest, the Games of '68 were paid for dearly
(Kukawka et al. 1991: 62) and over a long period (the debt was being paid
for by Grenoble inhabitants until 1995!). A representative example of the
retroactive dissatisfaction over the '68 Games (which also underlined many
citizens' actual indifference in 1968) is given by the Comité anti-olympique
de Grenoble, who under the counter-Olympic banner of 'Moins vite, moins
haut et moins fort' ('slower, lower, weaker') gave prominence on its website
to Pierre Frappat, a Socialist Grenoble city councillor in 1968, whose demys-
tification of self-congratulatory Gaullist and municipal discourses on the
'Jeux de la France' and of successful infrastructure development dates back
to the late 1970s (Comité anti-olympique de Grenoble 2008, Frappat 1979).

Announced initially in January 1983, France's bid for the Albertville Games
was led by a hero from Grenoble '68, the triple Gold-medallist Jean-Claude
Killy, and culminated in final victory in October 1986, when Albertville,
the Savoie *département* and the Rhône-Alpes region beat final competition
from Sofia (Bulgaria) and Falun (Sweden). Perhaps significantly fronted by
the highly successful businessman Killy – in illustration of a more 'commercial' turn to priorities – France's third Winter Games were generally perceived to be a success in sporting and political terms, although, as usual,
estimates of the overall finances were negative, and the most recent studies
conclude that the Games did not provide the durable hoped-for regional
economic development benefits (Terret 2010). Preparations for the Games
involved improvements to transport networks, sporting infrastructures and
telecommunications links. Francillon's image of Games which were 'calculated' seems apt, in contrast to the almost casual organization of Chamonix
'24, and the rigidly planned and programmed Grenoble '68, typical of contemporary technocratic procedures. A study undertaken by local researchers
drew up a detailed summary of what had been done to prepare the region
for the Games and what their 'imprint' had been (Dailly et al. 1992), com-
pleting a previous partner study that had assessed the mechanisms devised
to deliver the event (Kukawka et al. 1991); both analyses emphasize the
extreme complexity of designing and organizing a sports mega-event of the

nature of the Winter Games over the multi-site area defined by the Comité d'organisation des Jeux Olympiques (COJO).

The difficulty of conciliating the often conflicting local interests of the ten towns and villages hosting different elements of the Games – all keen to benefit from development and business – proved a conceptual and 'political' stumbling block in preparations for the 'Albertville' Olympics: the scale of the event and the number of disciplines involved required the facilities of a whole region – 'La Tarentaise' – to be *selectively* mobilized, with Albertville as central 'hub' of the network. Although the individual *communes* (or local authorities) shared a desire to see the overall success of the Games, local rivalries and selective lobbying complicated COJO's work, even leading to the resignation of Killy from his role as chairman of the organizing committee, in January 1987. In what became known as the Saint-Martin-de-Belleville 'incident', Killy concluded – too hastily – that conciliating the interests of all the localities hoping to benefit from the Games would prove beyond his powers and contradict his own understanding of how the event should be structured.

Informed by his past as a winter sports athlete, Killy favoured the siting of competition as much 'in the mountains' as possible and at the best possible facilities, and also within a manageable overall area, thus making travel for competitors less inconvenient. Saint-Martin-de-Belleville had hoped that it would figure as one of the sites, but an early decision to redefine the central area of competition ruled the locality out, precipitating bitter protest that the Games should benefit the whole of Savoie, and not just selected communes. In March 1988, Killy reassumed his post at COJO, now in partnership with the more 'political' Michel Barnier who, as president of the *Conseil général* of Savoie, was better able to manage conflict between competing towns, but the municipal elections of 1989 – when many communes changed political allegiance – further complicated compromise. Predictably, little mention is made of these difficulties in the official report on the Albertville Games (Blanc & Eysseric 1992), but it would seem that in the overall context of the conceptualization and functioning of the Winter Games as a 'mega-event', they demonstrate how conflicting 'entrepreneurial' expectations – especially when they emanate from para-public entities – can be hard to manage.

Annecy '18: re-engaging with the IOC?

On 18 March 2009, the French National Olympic Committee (now the Comité national olympique et sportif français, CNOSF) decided that Annecy (Haute-Savoie) – whose slogan had been 'The purest lake in Europe, and Mont-Blanc' – would represent France in the IOC's final deliberations for the 2018 Winter Games. It had only been in September 2008 that CNOSF had confirmed France's commitment to apply for 2018, in an international context where various cities had already expressed their desire to host the Games. Grenoble, Nice and surprisingly Pelvoux-les-Ecrins had presented their candidacies to CNOSF in January 2009, after a hurried drafting process. CNOSF's vote was eventually unambiguous: 23 votes for Annecy, 10 for Nice, 9 for Grenoble and none for Pelvoux, in only a single round of voting. Although CNOSF president Henri Sérandour congratulated all the competing bids on their quality, the CNOSF report evaluating the proposals – upon which final voting was substantially based – was severe, considering that only Grenoble and Annecy had devised plausible dossiers (CNOSF 2009), designating Nice's project as 'risky' and that of Pelvoux as unlikely to reach preliminary IOC standards.

In November 2008, CNOSF had briefed the candidate cities/regions to help optimize the 'fit' between their dossiers and IOC criteria for subsequent rounds of bidding. This briefing stressed IOC requirements in terms of environmental concerns, sustainable development and regional development in the medium- and long-terms and the necessity of avoiding 'white elephant' projects. CNOSF also underlined that any successful French bid should ensure that the Games would leave a legacy not only in material infrastructures, but also in attitudes and cultures towards sport, Olympic values, social cohesion and education, in contradistinction to the mixed benefits of Grenoble '68 and Albertville '92. CNOSF stressed the positive effects that bidding – even unsuccessful – can generate for cities aiming to host the Games, describing the possible 'virtuous circle' of governance capacity to be generated by collaborations, reflections on regional planning and joint infrastructure projects that result from bidding. Partly inspired by the disappointing failure of Paris 2012 (see Hayes, chapter 9, this volume), this innovative briefing session and the increased involvement of CNOSF in the early stages of France's candidacy, and by implication in subsequent later elaboration of the bid before presentation to the IOC in 2011, reflected a desire to align form and content of any application as closely as possible to IOC wishes. Some – on the left, particularly – viewed this 'accommodation' to the IOC unfavourably, as both tacit acceptance of an often criticized international Olympic Movement, and as implicit approval for a stronger role to be played by a national Olympic committee whose internal politics were deemed less than transparent (*L'Humanité* 18 March 2009).

Comparison of Annecy with rival bids is instructive. The Pelvoux bid was curious, being a collaborative venture led by the ski-resort village of Pelvoux (400 inhabitants) and a grouping of winter sports destinations in the surrounding area of Les Ecrins, including Italian resorts and facilities used in the Winter Olympics of Turin 2006. Relying on Marseille and the Provence-Alpes-Côte-d'Azur (PACA) region for logistical issues of transport, communications and infrastructural support, Pelvoux's candidacy stressed encouragement of the Alpine economy as a 'projet montagnard' for the whole 'massif' (*Libération* 22 October 2008). It was distinctive in its stress on returning the Games to a 'human scale', linking candidacy to a village rather than a city, and emphasizing that collaboration by small 'communes' in Les Ecrins could provide a sporting and spectating experience of modest infrastructures but high quality. Nice, in contrast, stressed that it alone was an urban centre of sufficient significance to rival the major cities of other countries – Munich was often cited – likely to be considered by the IOC in 2011. The city's loyalist Gaullist Mayor Christian Estrosi even described Nice's bid as 'geopolitical' (*L'Humanité* 21 January 2009).

The Grenoble candidacy was based partly on (fabricated?) nostalgia for 1968. But just as in the late 1950s and 1960s, the ambition to host the Games was tied to perceived developmental needs of accommodation (both tourist/business and social housing) and transport (improved train links and road access to Grenoble airport and to ski stations): the Socialist mayor of Grenoble, Michel Destot, argued that the 2018 Games would accelerate Grenoble's urban development by 20 years (*L'Humanité* 25 September 2008). Perhaps the defining feature of the Grenoble bid, as 'spun' by managers and publicists, was that it was supported by city and region sporting communities. Rather than stressing its 'environmentalism', Grenoble linked its candidacy to the CNOSF requirement for a bid which would federate the 'sporting movement' (CNOSF 2008). Grenoble also exemplified a sceptical response to the Rogge doctrine of environmental Games, as local Greens contested the idea that the Games – or any mega-event? – could not have an unfavourable impact on local ecology. In partnership with a local anti-Olympics association (the Comité anti-olympique de Grenoble), the Grenoble *Verts* were solidly against their city's accommodation with what they saw as token IOC environmentalism (Kempf 2009) and produced a comprehensive rejection of official histories of the impact of Grenoble '68 and of justifications for bringing the Games to Grenoble again (Groupe Ecologie et Solidarité 2008).

Contrastingly, Annecy's campaign stressed the desire to organize Games centred on and conceived by athletes within a context of financial and environmental 'excellence'. Mayor Jean-Luc Rigaut was supported by the president of the National Assembly, Bernard Accoyer (also a *député* for Haute-Savoie and leading member of President Sarkozy's UMP party), and the success of the Annecy dossier was interpreted by some as political favouritism in the best traditions of centralized Parisian opacity and establishment protection of colleagues' interests. However, Annecy's success against competition from Nice, Pelvoux and Grenoble is also explained by the bid's 'compromise' nature. Whereas Grenoble had already benefited from hosting the Games in the very different political, economic and environmental context of the late 1960s, Annecy and Haute-Savoie could argue their case for the developmental boost that 2018 would bring to an Alpine city of medium size and a region whose economy needed constant support. Nice, in contrast, was a significant city in France's urban hierarchy and with domestic politics similar to those of Grenoble, located in the more prosperous southern PACA region. The Pelvoux candidacy was that which most explicitly tailored itself to IOC requirements for sustainability, but failed to gain support of CNOSF delegates, perhaps precisely because it seemed too close to nature and 'back-to-basics'. In June 2009, the Annecy organizing committee was instituted, charged with developing the bid, and initial thinking seemed to increase the importance attached to environmentalism. Indeed, alongside commitment to close partnership with CNOSF, and pious intentions to make the Games those of exchange (culture, youth, accessibility, hospitality) and to put athletes at the heart of events and their organization, the organizing committee also promised to make optimal use of existing infrastructures and to guarantee carbon neutrality (Ville d'Annecy 2009).

Whether France was seriously interested in hosting the 2018 Games was frequently questioned by the media and occasionally undermined by CNOSF: Sérandour seemed to play down the likelihood that the IOC could not choose Pyongyang or Munich, or, indeed, give the Games to Europe following Sochi 2014 (*Libération* 19 March 2009). The suggestion seemed to be that bidding for 2018 was a means of re-engaging French politics and the national Olympic Movement with the IOC, after the partial breakdown in relations following the traumatic defeat of Paris in 2005 and attacks on the Olympic flame in France in 2008 (see Renou, chapter 12, this volume). Thus CNOSF may essentially only have been going through the motions with Annecy's bid; more important was re-establishing a working relationship which could facilitate a French application for Paris to host the Summer Games in 2024, a century after Paris '24.

France's probable playing of a 'long game' potentially reflects interesting considerations about how the French state perceives the Olympics. Most basically, France is more interested in the Summer Games than in the Winter Games, despite the arguably greater impact of Winter Olympiads on regional and entrepreneurial development in rural areas. Second, the Games are still perceived strongly as 'national/geopolitical', as well as in terms of their facilitation of world-city competition. Third, as a founding force in the Olympic Movement, France's identity and prestige in Olympism is most effectively served by a re-inscription of that identity within the long-term of Olympic history, in other words by celebrating the centenary of Paris '24: the Games have long been Eurocentric (Guttmann 1994) and Paris 2024 would be the strongest marker of that in the very different conditions of the twenty-first century.

In May 2009, elections designated Sérandour's successor as head (since 1993) of CNOSF, making his last substantive act the selection of Annecy. For some, Sérandour was a man of the past, too comfortable with the modus vivendi between the state and sport which had developed since the 1960s and the Gaullist emphasis on elite achievement and national prestige which had failed spectacularly in 2005 when the Paris bid (arguably the final 'evolution' of this model of sport-state relations) lost out to what the French saw as a more free market ('liberal') and politically independent organizing framework. With the election of President Sarkozy in 2007 and new inflections given to public policies, CNOSF was also keen to redefine its relationship to politics through a new president who would lead the French Olympic Movement and French sport into the future, towards Annecy 2018 or Paris 2024. The choice of Denis Masseglia, the only candidate to explicitly suggest including the world of business in CNOSF decision-making (*Le Figaro* 19 May 2009), but still a figure steeped in the internal politics of French sports associations, suggested that change would occur slowly, albeit with increasing opening of state and political mechanisms to business and civil society.

Conclusions

It is clear that France's bid through Lyon to host the 1968 Olympics lost out to an Olympics which were 'political' on a number of levels, from the IOC choice of the Third World rather than France, through the Black Power protests of US sprinters Smith and Carlos, to the massacre of 300 Mexicans protesting against their regime by the Mexican army. Lyon '68 would have been a mega-event organized by a Radical mayor of Lyon for Lyon and its region, but also with and for a Gaullist state wedded to the concept of national grandeur. It would have been a 'political' phenomenon which would have rivalled the events of the student and worker revolts of May–June '68 expressing the discontent of many in France at the results of ten years of Gaullism. Although the examples of Grenoble '68, Albertville '92 and most recently Annecy '18 seem indeed to show a growing degree in which the Olympics can be seen as vehicles of entrepreneurial rivalry between cities with aspirations to compete in a neo-liberal globalized economy, looking at the Olympics from a French perspective it is nevertheless still hard to disagree with the suggestion that the Olympics are enduringly 'political' (Hoberman 2008), most particularly in terms of national rivalries. The bitter disappointment felt in France when the 2012 Games were awarded not to Paris but to the capital of her traditional rival was representative of more than just neo-liberal commercial competition.

The suggestion that the bid for the Annecy Games in 2018 is to be understood as a means for France of 're-engaging' with the Olympic Movement illustrates another dimension to how states use mega-events, namely their long-term identification with the values and traditions of international organizations which can confer credit and validity to national representations of domestic politics and society. In this sense, France's interest is both to host the 2024 Games in Paris for the twenty-first-century reasons of entrepreneurial rivalry with competing world-class cities, as well as to demonstrate French national organizational genius, and crucially, to celebrate in a commemoration of the Summer Games of 1924, France's national identification with the Olympics and the positive elements of the Olympic 'brand'. If (as has been recently discussed) Los Angeles has in many ways a privileged rapport with the Games (Dyreson & Llewellyn 2010), Paris and the French state are likely to be keen to enjoy the same kind of special relationship. Since 2007 the French state has been negotiating the changes brought by President Sarkozy's new and more neo-liberal approaches to government and society which will inform future bidding for and hosting of mega-events, and as the replacement of Sérandour by Masseglia as head of the CNOSF demonstrated in 2009, there is also change afoot in the French Olympic Movement and in French sport in general which will further modify traditional French approaches to the Olympics.

Part II

Mega-Events, Environmental Impacts and Sustainable Development

6

The Role of Environmental Issues in Mega-Events Planning and Management Processes: Which Factors Count?

Pietro Caratti and Ludovico Ferraguto

Mega-events produce a number of long-term consequences on the host territory, both positive and negative (Roche 1994). An event has the capacity to radically change the surrounding territory, in a manner unparalleled by other factors in modern society (e.g. Roche 2003). The opportunities for development provided by mega-events, however, are contrasted with the fact that the concentration of a high number of people and activities within very limited areas (often a single site) constitutes a source of profound stress for the surrounding environment, as demonstrated, for instance, by the 1992 Albertville Winter Olympics (Cantelon & Letters 2000).

The environmental appraisal of events can be traced back to the mid-1990s (see Chalkley & Essex 1999a, Katzel 2007), when the 1994 Lillehammer Winter Olympics formally adopted the first environmental planning processes. In combination, the awareness of the environmental damage which could result from mega-events (May 1995) and new approaches to mega-event planning stressing environmental considerations (Chernushenko 1994) have however created the basis for a greater concern for the environment. Experience so far suggests that this has been expressed with various degrees of commitment at different mega-events, but it can nonetheless be seen as a continuous, cross-sectoral process.

Our chapter provides a review of the most significant environmental initiatives undertaken for events on a world-wide scale, including: the Sydney 2000 Olympic and Paralympic Games; the 2006 Turin Winter Olympic Games; the Kananaskis G8 Summit (held in Alberta, Canada, on 26–27 June 2002); the 2000 World Youth Day (WYD), held in Rome, 15–20 August (including the 2000 Jubilee); and the 2005 World Expo (held in Aichi, Japan, from March to September).

Starting from the definition set out by Spilling, for whom 'a mega-event is an event that generally attracts a large number of people, for instance more than 100,000, involves significant investments and creates a large demand for a range of associated services' (1996: 323), our objective here is to investigate how environmental concerns have been identified and dealt with through mega-event planning and management cycles. Discussion of the cases has been extended to include different categories of events with a global relevance (see Guala 2002), as focus on event iterations which have different dimensions, durations and activities enables us to better understand both the extent of stakeholder involvement in the event cycle, and the relevance of the adoption of environmental planning and management tools for addressing the impacts related to the concentration of people and infrastructures, often in sites previously unsuited for withstanding such concentrations. For this reason, one of the five events we have chosen to study is not strictly speaking a 'mega' – alongside the Olympics, World Expo and World Youth Day, the Kananaskis summit is perhaps conspicuous for its short timeframe, restricted elite participation and deliberately isolated location. We contend however that given the potential for mass protest (Kananaskis was the first G8 summit following Genoa in 2001), its comprehensive security operation and its global mediatization, it does share some key features of megas – as well as demonstrating the wider high-profile event context in which the environmental management of mega-events now takes place.

Our analysis develops as follows. In the first section, we explain the methodology, before focusing in the second section on a comparative analysis of the five events which make up our corpus, with reference to the environmental measures adopted. In the final section, we assess the experience of each event, highlighting aspects such as stakeholder involvement in the process, the inclusion of environmental considerations within general event planning, the consideration given to post-event management, and what factors are most influential in the adoption of environmental practices.

Methodology

The environmental considerations of mega-event hosting will be analysed through two different lenses. First, we examine the adoption of environmental management and planning tools over the course of the whole event cycle. The same processes will be then scrutinized to determine the extent of the involvement of relevant stakeholders in the decision-making process regarding the environment. These aspects are particularly important since disregard for the physical environment and lack of relevant stakeholder involvement have in recent years undermined the outcome of the events as key drivers for creating public acceptance and success (Bowdin et al. 2006).

Mega-events usually represent an exception to the bureaucratic routine of planning (Hiller 1998). Event planning processes are characterized by an initial lack of knowledge over which activities will be performed, which usually become clear only as the project progresses further (O'Toole 2002, Getz 2009). The principal planning efforts are therefore limited to devising strategies, evaluating potential risks and establishing a programme for controlling them (Kurscheidt 2000). In this context, the definition of a systematic planning and management approach becomes particularly relevant to ensure a successful event outcome (O'Toole & Mikolaitis 2002): the presence of a strategic approach to environmental issues, indeed, may constitute an important indication of how reasonably the process is structured.

Alongside normative compliance, the degree to which environmental concerns are taken into account through the adoption of environmental tools may be attributed to the influence of three main factors: institutional commitment, infrastructural impact and location. We discuss each of these in turn below.

The first driving factor is *institutional commitment*, which relates to the willingness of event organizers to address environmental issues. Events are usually developed under a special 'regime', conferred by the power coalition drawn from the private and public sectors, and which is capable of overcoming fragmentations in local policy (Burbank et al. 2002). Though these coalitions are usually established to alleviate the burden of event organization on the public, it is important to verify how the coexistence of different interests among the organizers influences the adoption of environmental practices. The motives underpinning environmental considerations may stem from various reasons, such as the close involvement of public institutions and other environmentally aware stakeholders, or the importance of environmental/sustainability issues to event promotion.

A second factor concerns the *infrastructural impact* of the event. The setting up of infrastructures, on a temporary or permanent basis, is directly linked to the influx of people to the event (see, for example, Essex & Chalkley 2002); nonetheless, infrastructures constitute an indirect but relevant consequence of the event (e.g. the construction of secondary infrastructural projects such as tourism facilities) within the planned activities of the event as a whole. Our aim is to understand how infrastructural development may lead to the inclusion of environmental measurement and management tools in the event planning cycle.

The third factor is the *location* of the event. Events can be located in very different sites, including multiple locations and environmentally sensitive ones. It is reasonable to think that the specific territorial dimension will influence the degree to which environmental concerns are taken into account. This happens through the combination of two factors, one being the risk of causing significant changes to the environmental consistency of a sensitive spot (see, for example, the case of the G8 at Kananaskis, below);

the other is the opportunity to reclaim previously derelict or contaminated areas, as was famously the case with Homebush Bay for the Sydney Olympics, thus making a 'proactive' use of the environment (Yeoman et al. 2003).

Given that these three factors are likely to influence the degree of environmental concern, it is important to assess the extent to which concerns are taken into account by observing which environmental tools have been adopted for the event and by evaluating their effectiveness in the general planning and decision-making processes. In particular, we focus on the adoption of two main groups of environmental tools. The first includes planning and decision-making support methodologies, such as Strategic Environmental Assessments (SEAs) and Environmental Impact Assessments (EIAs), which are commonly adopted for the evaluation of strategic acts or projects with an impact on the environment; the second includes management support tools such as Environmental Management Systems (EMS), and environmental reporting. First, though, we provide methodological clarification to enable better understanding of the specific contextual functions of each tool.

Though traditionally seen as separate activities (the first related to strategic level decision-making, the second to the project approval phase), SEAs and EIAs are considered here under the same category of planning instruments. SEAs are in fact a natural development of EIAs, since they can be deployed on the basis of the procedural stages of an EIA (Fischer & Seaton 2002, Pope et al. 2004). More importantly, the commonly adopted deployment into Policy, Plans and Programmes followed by Projects can be seen as a simplification; a planning system can be neatly subdivided into separated and consequential phases only arbitrarily (Fischer 2002). It ensues that the 'strategic' meaning of an SEA should refer to its ability to influence policy and to orientate decision-making, rather than simply being applied to a number of specific actions, a task which is also performed by an EIA (Bina 2007). SEAs and EIAs, more importantly, are able to provide preliminary assessments, thus enabling the identification of possible alternatives whilst the decision-making process is still ongoing.

The second category of tools concerns systematic approaches and methodologies which can be grouped under the heading of 'management support tools'. These approaches enable the combination of environmental protection with typical business and company demands, providing a schematic set of general requirements for management (MacDonald 2005). Though Environmental Management Systems and reporting are usually enacted on a voluntary basis, once they are enacted and their principles and targets are fixed, they provide a defined structure of activities that must be performed to comply with system requirements, by including them into an iterative cycle (Robèrt et al. 2002). At the end of these activities, reporting of the outcomes of the cycle and of their responsiveness towards previously fixed targets provides the management with a detailed monitoring of the system, enabling putative changes to the system (Eccleston & Smythe 2002).

Each of the tools therefore serves different purposes during the process, but can be regarded as necessary for achieving environmental objectives within the organization of an event. Their respective contribution is to be seen as complementary and not exclusive, since they intervene for different purposes and at different stages of the process. In particular, SEA/EIAs and Environmental Management Systems/monitoring tools are often combined, where the first apply at a planning level, and the second at a more operational level (Keysar 2005).

None of these instruments can however be considered as discrete from the more general planning and management context of the event. For this reason, following a commonly accepted subdivision of SEA phases (Gambino et al. 2005), we have identified three specific moments of an event, where environmental considerations should be properly integrated into decision-making:

- from event conception to the planning phase: in this phase, which includes the formalization of the event idea, the proposal and feasibility studies, it is important to see how strategically the environment is considered in order to define the design of the event, including opting for environmentally sensitive alternatives, through the correct use of strategic tools;
- from planning approval to the event conclusion, which includes fundamental steps such as application of the approved plans, monitoring and reporting (Bowdin et al. 2006). This is where the contribution of management tools should be highlighted, and the correspondence between them and the decision-making process as set out in the previous phase. Planning tools, indeed, can provide a basis for monitoring conclusions through dedicated plans;
- post-event management and follow-up: this third phase ensures the ability of both planning and management tools to act beyond the confines of the event, to inspire and guide further policy processes (Partidário & Arts 2005), related to legacy management and the continuous monitoring of environmental indicators.

Our approach for examining an event from an environmental perspective is therefore a procedural approach, aimed at identifying 'windows of opportunity' in decision-making for introducing environmental concerns into the whole process (Caratti et al. 2004). The focus of the work shifts from the simple evaluation of the environmental impacts linked to a plan, to the procedural integration between decisional processes and environmental evaluation as a means of producing a Policy or a Programme (Caratti 2004). We now give a brief summary of the specific contours of the application of these tools for each of the five case studies.

The Sydney 2000 'Green Games'

The Sydney 2000 Summer Olympic Games may be considered – along with the 1994 Lillehammer Winter Olympics – as the first attempt to encompass environmental considerations in the planning of a hallmark event. The commitment of the Olympic Coordination Authority (OCA, responsible for site management) and the Sydney Organizing Committee for the Olympic Games (SOCOG, responsible for the management of the event as a whole) was evident from the bid launch in 1993; the bid included detailed Environmental Guidelines delineating goals and areas of intervention.

The guidelines regulated numerous sectors of interest (e.g. energy, transport, waste, water), listing requirements to be fulfilled and expressing an explicit commitment to the integration of environmental concerns into all phases of planning. This document constituted the main reference framework for the adoption of environmental practices, strengthened by its adoption by the New South Wales Government as an official act for compliance by the two organizing entities. However, the subsequent legislative acts and policies that were issued by the NSW Government ended up creating loopholes and conflicts with the requirements of the guidelines, producing what Green Games Watch 2000 has called 'selective compliance' to environmental requirements (Green Games Watch 2000, 1999). As a consequence, the organizing committees had the opportunity to 'choose' which requirements were followed, producing less ambitious goals than had been set out in the bid (McGeoch 1999).

Consistently with the strong commitment to environmental impact mitigation set out in the Olympic bid, comprehensive plans were developed for the Olympic locations: remarkably, for the first time in Olympic Games planning, management and assessment tools for the infrastructural programmes were used. For each of its operations, the OCA adopted an Environmental Management System based on ISO 14001. Comprehensive environmental requirements for tendering (including the adherence to EMS principles by contractors) were also instituted. As part of this 'green' tendering process, an LCA (life-cycle assessment) was introduced for the environmental proposals of all contractors, which were then integrated into a general model detailing all the impacts of the Olympic Stadium.

Among the most interesting features was the decision to locate a significant number of Olympic activities at the Homebush Bay site, partly a former landfill for industrial and toxic waste, a move which called for remediation work before facilities could be built. Unlike the sectoral plans, which were limited to adding environmental specifications to general projects, this was the closest the Sydney Games came to integrating environmental concerns into general planning, since the successful bid gave major impetus to the development of a master plan for the entire area in 1995.

The process of environmental reclamation and the construction of the Olympic facilities in Homebush Bay were not exempt from criticism. The strongest opposition focused on the statement that Games infrastructures were not subject to the obligatory Environmental Impact Statement (EIS) procedure. This was probably due to time and budgetary constraints, which led to the abandonment of the lengthy requirements of the official procedure in favour of a more flexible one (McGeoch 1999). Less stringent legislation was consequently adopted, the main requirement being that any new infrastructure must receive prior authorization from the Ministry for Urban Affairs and Planning (New South Wales Department of Planning 1993).

Criticism of this decision focused on the exclusion of the consultation phase which is part of the EIS process. More critically, the exclusion did not affect NGOs (such as Greenpeace or Green Games Watch 2000) as much as the host communities, which had a potential interest in the operations in Homebush Bay. The local authority, in fact, was only involved in a very superficial way by the OCA, which led to very little information being disseminated to citizens (Owen 2001). This exclusion was reinforced by the State Environmental Planning Policies (SEPP) exemption clause no. 38, which prohibited the bringing of legal proceedings for any failure in compliance with the requirements of the guidelines, for any activity carried out by, for or on behalf of the OCA.

According to most organizations, it was thus clear that the commitment of the NSW Government and the organizing committees was driven by time and budgetary constraints, which took precedence over 'green' issues (Luscombe 1998). At a more general level, further interests and considerations by NGOs were not included in the original planning process instances (Waitt 2001). The overall involvement of environmental groups in the whole process was consequently less significant than they had hoped (Kearins & Pavlovich 2002).

A key element leading to the decision to identify Homebush Bay as the Olympic site was the opportunity of providing a reclaimed and perfectly operating facility as a legacy (Waitt 2001). The OCA reported on a regular basis on construction progress in the area, ensuring that environmental performance monitoring be independently verified by bodies such as the Australian Environmental Protection Agency, Earth Council and Green Games Watch 2000, who provided detailed evaluations of the projects. Furthermore, in some cases it was possible to see how the integration of environmental concerns led to an adjustment of the original projects: this happened, for example, with the discovery in the Olympic zone of the presence of a rare species of frog, which led to a modification in the Masterplan.

However, the management of the post-event phase was far from complete. Following the Games, a new Olympic Park Authority was created in order to manage the Homebush Bay area in accordance with the Environmental

Guidelines of the event. Among the various achievements, the Authority has developed a Masterplan for the area which is valid until 2015, including transport and precinct plans. But this initiative is still insufficient to address all the environmental implications arising from – above all – the construction of venues for the Games, for which a plan of sustainable post-event use remains – more than ten years later – missing (Mackenzie 2006, Sadd & Jones 2009).

The Turin 2006 Winter Olympics

The second Olympics event that we discuss here is the 2006 Winter Games, held in Turin. The environmental outcomes are the product of interaction between organizational and political actors such as the Regional Government, the Organizing Committee (TOROC) and the Italian Ministry of the Environment, joined in an ad hoc coordinating institution. This high-level institutional involvement enabled a potentially higher degree of integration of environmental aspects into the general planning process.

As lead event organizing body, TOROC was able to present from the beginning a detailed set of commitments which comprehensively matched the requirements set by the IOC for candidate cities. It immediately decided to carry out a preliminary environmental assessment for the whole of the Olympic plan, and develop an Environmental Management System to cover Olympic operations. The SEA was then approved together with the Olympic plan by the Regional government, which also assisted in the implementation of the planning process and supervised the legacy management of the event. At the top of this organization, the Italian Ministry of the Environment took part in the final approval process for the SEA.

From the outset, this strong commitment undoubtedly helped bring together Turin's environmental strategy within the Olympic Plan: the bid document contained a 'Green Card' which listed the environmentally sensitive areas which were to be given special consideration during the planning of the event's activities. The same is to be observed for the approval of the preliminary Environmental Compatibility Assessment, and for the definition of the monitoring plan. However, it must be noted that the SEA could not cover all the activities which had to be implemented in order to address all the environmental aspects considered in the preliminary assessment, which became a primarily descriptive instrument (Segre 2002). Most of the initiatives implemented after the SEA, in fact, had not been considered within the original Olympic Programme, and thus could not be assessed prior to the bid. The water use plan, which was adopted by the competent authority only in 2002, is a pertinent example of this.

As a potential remedy to these weaknesses, a comprehensive monitoring phase was developed by the organizers. In fact, one of the most interesting procedural features of the environmental plan for the Winter Olympics was the drafting of a 'balanced' framework to evaluate sustainability perform-ance, which included every potential impact and the related mitigation/compensation strategy. Furthermore, the main areas for assessment were identified from the conclusions of the compatibility assessment phase of the SEA, thus providing a linkage between the preliminary and monitoring phases. The monitoring plan was developed through a system of indicators with the help and support of the Regional Environmental Agency, and also covered territorial and community development.

One of the main achievements of this phase, indeed, was the adoption of the EIA procedure to assess the environmental compatibility of the locations chosen for at least two especially sensitive facilities. This process eventually led to the relocation of one of the facilities – the ski-jump in Pragelato – to a different site, while at another facility (the bobsleigh track in Cesana Torinese) asbestos was detected, necessitating remediation action.

TOROC also decided to provide 'green' management for all its operations, through the adoption in 2004 of an Environmental Management System, and the approval of three annual reports documenting the evidence of the interventions implemented. The adoption of the reporting measure can be seen as a major achievement of the Games environmental plan, since it provided a synthesis and combined assessment of all the potential issues to be tackled by the organization. Furthermore, this process seems to have strongly benefited from the application of an EMS, which created a well-defined framework for the operations related to the event's management (Frey et al. 2007).

Yet this deployment of environmental tools has not proven to be com-pletely effective when it comes to post-event management. While the origi-nal intention was to disperse the Olympic facilities throughout Turin and its surrounding territory in order to facilitate the integration of the Games into a single territorial area, what has emerged in the post-event phase is a lack of decisiveness about the final use of many of the facilities – some of them with a high environmental impact, such as the ski-jump (Guala 2009). Not even the establishment of a foundation specifically dedicated to dealing with Olympic heritage has contributed to improving overall man-agement in the post-event phase, since it has focused mainly on urban facilities.

As for participation in environmental decision-making, the SEA proce-dure ensured openness, transparency, and most of all various opportuni-ties for interaction with both stakeholders and the general public. However, more in-depth analysis has shown how, since the bidding phase, partici-pation has not been integrated into a strategic framework but limited to consultation phases where the discussion was either limited to generic terri-torial impact or focused on more technical issues, but without enabling the parties involved to see their issues addressed by the final decision-makers (Caratti & Lanzetta 2006). In this respect, the dearth of communication on environmental and territorial issues is a symptom of the lack of a deeper involvement of the general body of citizens.

The 2002 Kananaskis G8

The 2002 G8 summit in Kananaskis (Alberta, Canada) represents the first time an event of this type has included a detailed environmental plan for the site hosting the event. Two factors seem to have paved the way for this commitment: first, the application of exceptional security measures in order to forestall potential violent disturbance during the meeting; second, the nature of the site hosting the event (G8 Strategic Environmental Assessment 2002). The Kananaskis region is in fact a highly sensitive, protected area hosting various endangered wildlife and plant species.

Environmental management was conferred to an ad hoc Environmental Affairs Directorate at the G8 Summit Management Office. The directorate was charged with overseeing the environmental protection of the Kananaskis area, coordinating the work of other interested partners and providing essential tools for managing the event in an environmentally sound way.

From a planning point of view, the G8 Summit Management Office had agreed to a strong linkage between the general planning process and the definition of environmental measures. In fact, the preliminary environmental assessment was conducted at the very beginning of the planning process, in an attempt to assess whether the site and the plans presented specific issues which would require attention. The preliminary evaluation recommended that the event be subject to the legal framework of an SEA. Furthermore, it enabled an understanding to emerge that the effects of the event had impacts beyond its strict sectoral and temporal limits, creating potential social and economic impacts, and affecting the physical environment after the event itself had finished (G8 Strategic Environmental Assessment 2002).

The subsequent evaluation phase, conducted with the support of the Government of Alberta and the City of Calgary within an Environmental Advisory Committee, provided a site characterization enabling impacts to be determined and subsequent mitigation measures to be designed. These measures addressed a wide array of dimensions, including infrastructures, wildlife and terrain disturbance among others. Perhaps more importantly, the SEA process attempted to bring together all the relevant factors of the timing, scale, reversibility and cumulative effect of the impacts. The predicted impacts were then categorized according to their capacity to produce effects within the timeframe of the event itself, by the location of their effects, and their predictability and manageability within the planning process (G8 Strategic Environmental Assessment 2002).

The SEA for the Summit, finally, prescribed some initiatives for the monitoring and follow-up of the measures it set out. For each site, an inventory of the relevant environmental conditions was carried out, and their assessment referred to the Director of the Environmental Affairs Directorate. Further evaluations in the follow-up phase included continuous inspections over the six months after the summit and the following year, in order to evaluate the presence of any residual effect, and to decide whether further interventions were needed.

Alongside the production of a SEA process, two main sectoral plans were successfully adopted for the event's management. The first of these was the 'Envirosafe' Program, designed to train the large number of security staff employed during the event to minimize environmental damage while at the same time contribute to the prevention of any hazardous situation for humans, animals and habitats. The second was the 'Environmental Stewardship Program', designed for all partners and staff of the Summit Management Office, consisting of a series of meetings with the other Directorates of the Summit Management Office and the other relevant partners to help them identify the impacts related to their activity and the associated mitigation measures.

The initiatives for communication and stakeholder involvement, finally, were structured mainly through the creation of an EnviroNetwork including most relevant local and international NGOs to discuss the steps taken by the Summit Management Office to protect the environment during the event. Although scheduled to meet on a monthly basis, the network actually held only one meeting, subsequent contacts being maintained exclusively by email. Further communication initiatives were then developed for the media and the general public, including a website, newsletters and interviews. The particular feature of these initiatives was the coordination between the Environmental Directorate and the other parties, such as the Communication Directorate activated within the Summit Management Office. This ensured that the information provided would be effectively detailed and precise.

However, it must be noted that these initiatives lacked two essential features: integration into the decision planning process, and the presence of two-way communication between stakeholders and decision-makers. In fact, while the EnviroNetwork undoubtedly represented an opportunity to form partnerships with relevant stakeholders, it was developed as a one-way-only communication tool to inform the organizations about what had been developed, without involving them in the process of establishing their needs and defining strategic responses to environmental problems.

World Youth Day 2000

The first religious event to be informed by environmental considerations was the World Youth Day held in Rome in August 2000. The World Youth Day is a youth-oriented Catholic Church international event of culture and spirituality, which is held every two or three years. The event we discuss here – which culminated in a Holy Mass led by Pope John Paul II on 20 August – involved approximately 2.5 million participants. While at that time there was no mandatory normative requirement in Italy for applying a strategic environmental assessment of plans or programmes, Rome City Council nonetheless required an Environmental Impact Assessment of the full series of activities included within the planning of the event, given the transformation of the surrounding environment that some of them would cause. The SEA procedure was thus activated in order to provide decision-makers and the public with this kind of information, especially considering that some of the actions carried out (such as the construction of access routes to the site) entailed a permanent physical transformation of the area (Comune di Roma 1999).

A complication in the organization of the event was the simultaneous presence of different institutional authorities (Rome City Council, Regional Government, the Province of Rome), none of which was formally in exclusive charge of the organization and management of the event. Creating a strategic evaluation document also therefore meant establishing a unitary 'technical' reference point for the coordination of the different projects included in the planning for the event (Karrer 2002). In this sense, the procedure implemented for World Youth Day enabled the creation of a strong link between environmental analysis and general planning.

The procedure followed can be divided in two main phases: an initial evaluation phase, identifying the normative prescriptions that could create an obligation to adopt an EIA procedure for the planned works, along with a biological and hydro-geological classification of the event site, and a second evaluation phase of individual interventions, their impacts and potential mitigation. It emerged from the initial evaluation phase that none of the planned activities would have to be subject to the mandatory environmental assessment required by Italian law, but rather only to discretionary assessment. However, given the 'systematic' nature of all the potential interventions, Rome City Council decided to proceed with the application of an environmental assessment framework.

Concerning the evaluation of the impacts of the event activities, as anticipated, the organization of the event consisted mainly of management activities, such as providing essential services, or managing the flow of participants and related risks and safety. These measures did not have a direct environmental relevance, or interfere with the ordinary use and normative regime of the concerned area. An accurate description was instead attached to the infrastructural programme, which consisted of both temporary and permanent systems and structures. Both direct and indirect impacts were taken into account, evaluated with respect to both the needs of the local population and to the overall value of the planned interventions. Impact evaluation demonstrated that many of the planned interventions could be matched with a more general restoration of the area, creating an improved, rehabilitated environment for subsequent use (Comune di Roma 1999).

However, the evaluation process also shows that few impact minimizing measures were considered, especially in the utilization phase of the permanent infrastructures. Furthermore, for most of the WYD events no alternative was actively considered, either in terms of their mode of operation or of their geographic localization, if one excepts the general evaluation of the benefits of locating the event in the Tor Vergata district of southeast Rome instead of another district in northern Rome. In this case, we can infer that SEA planning was a key factor in justifying the final decision to locate the event in Tor Vergata, with environmental considerations among the decisive factors (Karrer 2002). This conclusion is strengthened by the evidence that no sectoral plans were drawn up for dealing with waste, or with water and energy consumption, apart from short-term, specific interventions addressed by the SEA.

Participation in the SEA process, finally, was guaranteed by the meeting of two Services Conferences, that is round tables with local district councils which provided high-quality analysis of the needs of the local community, as well as providing a forum for bottom-up contributions to environmental evaluation. In particular, a hydrological site classification was requested by one of the parties involved (a local citizens association), pointing to a well-developed degree of citizen participation from the very first steps of the process.

The Aichi 2005 World Expo

Although no specific environmental planning or assessment measures were required by the Bureau International des Expositions (BIE), the body which governs and awards the hosting of World Expos, the Aichi Expo master planning document presented at the time of the bid included an environmental commitment for the site's development and management, in order to promote environmental protection beyond the context of the event (Expo Environmental Report 2006). Indeed, environmental concerns were included at an early stage of Expo preparation. The official application approved by the Japanese government was guided by an 'environmentally conscious' approach consistent with the chosen Expo theme of 'Nature's Wisdom'; this approach acted as a framework for the three phases of the event (site construction, Expo hosting and facility dismantling). The main organizational responsibility for the Expo was conferred on the Japan Association for the 2005 World Expo, created by an alliance of Japanese economic and political actors to promote national industrial development and culture.

Consistent with this approach, the original master plan was evaluated under a specific EIA procedure, which highlighted how the site initially identified as the main destination for the Expo – the Kaisho forest, a particularly rich area of biodiversity, home to a number of endangered animal species, including goshawks – was a highly environmentally sensitive location, leading the organizers to change the original plan, identifying in collaboration with environmental organizations a reduced occupation of the Kaisho forest. The EIA procedure was further integrated into the revised master plan and the follow-up biannual monitoring reports. This led to a substantially 'two-tiered' approach: an initial impact assessment was carried out on the general site plan, in order to assess the general compatibility of the event with natural sensitivity of the site; subsequent assessments were targeted at incorporating event development within the plan's environmental compatibility aims (Expo Environmental Report 2006).

The Expo 2005 strategic environmental vision was not, however, limited to this compatibility assessment, but also tried to promote specific plans at various stages of the event's development as well as an EMS, based on an environmental policy expressing the commitment of the organizers, and above all, a comprehensive set of environmental objectives for the event itself. Furthermore, specific environmental conservation guidelines were issued for each exhibit at the Expo site.

The EMS developed for the Aichi site was comprehensive, including aspects related to site development and management, modulating objectives as a function of the continuously changing operations and contents of the Expo. The EMS was designed both to act as a framework for activities, linking environmental policy and objectives to multiple sectoral plans, imposing a single plan and clear distribution of responsibilities among the institutional actors involved, and to measure the capacity of the implemented actions to respond to previously set objectives. Within this framework, the sectoral plans adopted for transport and mitigating global warming were developed according to a 'continuous efforts' model, which constituted one of the main achievements of the programme (Expo Environmental Report 2006).

As for the involvement of stakeholders, at each stage the EIA procedure included a very detailed plan for open participation, directing the process in close cooperation with central government and relevant local authorities, while at the same time organizing opinion exchange meetings with citizens and NGOs – in particular, national and local organizations concerned with environmental issues, such as Japan Environment Association, Wild Birds Society of Japan and WWF. These organizations were particularly active in pressurizing BIE for action; BIE, in turn, urged the organizing committee to make changes to the original construction plan. Most interestingly the outcome of this process was also the establishment of an environment forum, the 'Aichi Expo Review Committee', with decision-making power equally divided between NGOs and governmental authorities (Gu 2007). NGOs were thus able to effectively push for moving the site to a less sensitive area which only partially coincided with the previously chosen one.

Conclusions

The experiences of adopting environmental tools in the events we have examined give rise to a number of conclusions, concerning their role in the event, and how their development can be explained through the observation of the driving factors considered. Table 6.1 below illustrates how environmental issues were dealt with during each event through the planning processes.

As can be noted, it is in the event management phase that environmental concerns were most effectively integrated. Event organizing committees typically concentrated their efforts on the definition of management and reporting tools, and these efforts proved quite effective. EMS were seen and adopted as primary tools for establishing an environmental commitment, even though they were in some cases restricted to facilities and not to the entire process. Finally, environmental reporting enabled dissemination of the SEA and EMS processes, providing public evidence of the measures implemented at specific moments.

Our study shows that the most critical phase is in fact post-event management. The case studies have in fact generally pointed to a lack of environmental monitoring or potential further measures in legacy management planning, with few exceptions, mainly related to the continuation of activities within the framework of Environmental Management Systems. This shortcoming confirms a trend in the analysis of mega-events, that is the prevalence of ex-ante assessments which are not confirmed by subsequent phases at the end of the event (Roche 1992). Further and lasting interactions, such as those with an environmental impact, are not commonly taken into account through planning and management instruments in the period following the event.

In terms of the effectiveness of the separate environmental tools, it is interesting to highlight that two of the events (Rome 2000 and Turin 2006) point to a striking feature of the SEA process: its capacity to bring the different institutional actors involved within a single coordinating framework,

Table 6.1 Dealing with environmental issues in the planning process

Integration of environmental concerns in the event management cycle			
Case/Phase	Event planning	Event management	Post-event
Sydney 2000	😐	☺	☹
Turin 2006	☹	☺	☹
Kananaskis 2002	☺	☺	☺
Rome 2000	☺	😐	☹
Aichi 2005	😐	☺	😐

enabling the full series of event interventions to be implemented in compliance with event planning, regardless of their environmental nature. We therefore conclude that SEAs and EIAs can enable procedural harmonization in the planning processes, and can create a defined context for planning where this is not available. This is particularly important where various actors are involved without a defined hierarchy and with potentially juxtaposing competencies. Furthermore, this helps in creating a proactive climate of cooperation with stakeholders, avoiding shortcomings where consultation phases are only nominal and where the contribution of stakeholders is not usually considered.

The adoption of management tools has proved very useful in the phase from planning approval to the event, enabling the proper consideration of environmentally sound practices, especially concerning the management of facilities. Management tools have also been useful in raising awareness among those actively involved in the organization of the event.

It is now possible to assess the influence of the factors which have been identified as driving forces behind the adoption of environmental tools. Concerning *institutional commitment*, it is first important to stress that is that the power of organizing committees is usually limited to the management of activities and sites directly related to the development of the event. Secondary infrastructural initiatives, for instance, are not under their jurisdiction, and are usually excluded from the event's environmental programme. The influence of institutional commitment, then, can mainly be related to the central management phase of the event cycle, without significantly affecting the two other phases (event planning and post-event). The same can be said of *infrastructural impact*. Our case studies suggest that those events with a higher number of infrastructures are characterized by the adoption of a wide array of environmental tools, especially those specifically related to the physical projects – such as EIAs, which proved to be quite effective in all the cases under consideration. However, the presence of infrastructural development did not produce particular focus on environmental impacts in the post-event phase, with the exception of the maintenance of EMS in some of the areas previously adopted as event sites. The final element under consideration is *location*. Our research suggests that the particular environmental sensitivity of a site is in fact an influential element in focusing attention on the design of environmental tools. Thus in the Kananaskis case, for instance, the specific character of the physical environment increased the importance of environmental tools in the first and final phases of the event cycle, with more comprehensive post-event monitoring.

Although local communities and NGOs have a common interest in environmental protection, the events analysed here show that their objectives and role can diverge in significant ways. In the course of mega-events, local communities are also concerned about issues of territorial development, which require a comprehensive planning process and a general commitment for the post-event phase. These issues have been neglected by organizers through the non-comprehensive application of planning instruments; host communities have generally been considered only as the target of one-way, top-down communication or have even been excluded from the whole process, with the remarkable exception of Rome 2000. NGOs have been more easily considered in the planning and management processes, though this involvement has often proved to be only superficial (as demonstrated by the cases of Sydney and Kananaskis). The case of Aichi however highlights that where NGOs manage to create political pressure and build an interest coalition, they can find significant space to see their interests discussed and evaluated.

As a final remark, our research suggests that though the events examined mostly demonstrate a potentially useful approach towards the application of environmental tools, they lack however an integrated environmental strategy covering all three phases of the event. It is evident that the events are mainly considered discretely, even where such an approach has an impact on the planning process, despite initial intentions (Owen 2001), as was the case with Sydney 2000. The risk is the introduction of shortened environmental approval procedures, or the exclusion/limitation of consultation and environmental analysis, with the creation of an 'extraordinary' regime. This leads us to conclude that the quality of environmental strategies will depend on the ability to integrate the event into more general local development strategies and measures. If planning processes are designed in a short-term, reactive fashion, their consequences can be both unpredictable and unstable (Searle 2002). The present analysis, therefore, clearly suggests that event organizers should avoid an episodic conception of the event, and should seek to integrate event planning within existing territorial needs and plans in a much more detailed and sustained way.

7
Sustainability as Global Norm: The Greening of Mega-Events in China

Arthur P. J. Mol and Lei Zhang

In our (post-)modern times of high fluidity, with rapid turnover of fashionable ideas, where 'everything that is solid melts into thin air', it is remarkable that such an old-fashioned concept as sustainability – launched over 20 years ago – is still so solidly present in debates and practices. This chapter investigates whether sustainability has become a global concept, idea or norm that increasingly redirects institutions, practices, structures, norms and ideologies globally. If that is the case, sustainability becomes a shared notion that can no longer be ignored, not even by the powers-that-be, which now have to pay more than lip-service to sustainability in order to legitimize their behaviour.

Global mega-events are interesting cases to study this 'global norm' claim. The argument would be that, as high-profile and very visible happenings that attract world-wide attention, mega-events can hardly ignore common global norms on – among others – environment, democracy, transparency and equality. Mega-events are points of convergence (Close et al. 2007: 2), or crystallization points, for a cluster of major developments at and between different levels of social life. Moreover, mega-events are almost by definition global: the site selection is done globally, the reporting on the event is global, the participants are global, and the politics, economics and culture around it are global. So: do we see sustainability emerging and taking central stage at such global mega-events in discursive *and* material ways? And how and to what extent does sustainability become crystallized, institutionalized and 'fixed' in material and social infrastructures, to have some permanency beyond the mega-event, when things seem to return back to business as usual? These questions are particularly important in mega-event locations where sustainability seems to be far from an accepted norm, such as in China.

China is at the moment seen as one of the leading polluters and widely perceived as an undemocratic state, with limitations in transparency. 'Global norm' advocates would claim that through global mega-events organized in China – such as the 2008 Olympics and the 2010 World Expo – the country would strongly improve its environmental and democratic profile. With the world watching such mega-events and thus China, China could not but modernize – ecologically and politically. But others have been less optimistic, using two arguments. First, while ecological modernization and political modernization might emerge discursively with the Olympics and the World Expo, this does not necessarily result in processes of institutionalization, crystallization and 'materialization', making improvements short-term public relations accomplishments at best. Second, the mega-event-related advances in environmental quality and democracy/transparency in China are claimed to be rather meagre in comparison to the sheer size of the problems China faces in these respects. Here we use the 2008 Beijing Olympics and the 2010 Shanghai World Expo as interesting cases to investigate (1) whether we indeed witness central staging of sustainability during (the run up to) such mega-events, and (2) to what extent sustainability has some permanency beyond the mega-event, both locally at the place of the mega-event and globally through the redirection of networks, flows and infrastructures that structure these events.

In this chapter we therefore use the 2008 China Olympics and the 2010 Shanghai World Expo as case studies for investigating the concept of sustainability as global norm (second and third sections). In the first section we develop a theoretical perspective of the centralization of sustainability in social processes and institutional developments, drawing on ecological modernization theory and the sociology of networks and flows. The fourth section will interpret these two case studies against more general developments of ecological modernization in China. In all this, the main focus is on the environmental dimensions of sustainability, and only transparency and democracy issues related to that will be addressed. That means that we will not touch upon issues of, among others, Tibet, Falun Gong, the Uighur minority, unequal development and general media openness (which are all key issues for democracy and transparency as such in China). While these can indeed be considered as closely related to sustainability in a wider definition, for reasons of space we concentrate primarily on a more restricted, environmental definition of sustainability here.

Attractors, sustainability and mega-events

The notion of attractor stems from system theory, as primarily used in the natural and complexity sciences but now also in social sciences. If over time a dynamic system does not occupy all possible parts of a space randomly, but instead occupies only a restricted specific part of that space, a pattern emerges. The emergence of such patterns within a system then stems from attractors. In complex systems numerous iterations through time are drawn in certain directions, also through various feedback loops. Social scientists and sciences have paralleled this language and conceptualization of natural and complexity sciences, using similar concepts, ideas, metaphors and 'laws' to frame and understand the dynamics of social systems and society-nature interaction patterns. Within sociology, John Urry (2000, 2003) is arguably one of the better known examples, but we see it also emerging in notions of resilience and adaptive management and governance (see Folke et al. 2005, Lebel et al. 2006), following the work of the systems ecologist C. S. Hollings. It goes beyond the scope of our argument here to analyse the advantages and weaknesses of such disciplinary border-crossing, especially between the social and natural sciences. Our use of the complexity science notion of attractors is rather instrumental. In using the attractor concept, we're not suggesting that a wider merging of natural and social sciences into one overarching complexity science is a (let alone, the only) road to be taken. Nor are we implying embracing systems theory as a privileged theoretical position in the longstanding social science debate on agency and structure.

The current emerging global order is no longer structured centrally, but through multiple interdependent organizations and institutions that together demonstrate the 'capability to "orientate" to macro-level properties' (Gilbert 1995: 151). In contrast to the view of anarchic global disorder, as well as to the idea of a single ordering structure of the global system (e.g. capitalism, the UN system, the clash of civilizations), Urry (2003) introduces the notion of pockets of ordering: processes involving a particular performance of the global and operating over multiple time-spaces and through various feedback mechanisms. These pockets of ordering involve various networks, flows and governance mechanisms and produce similar shapes and 'results' at very different places and scales across the world. Attractors can be seen as the 'governing' signifiers or properties that structure the pockets of ordering, demonstrating similar and repetitive shapes and regularities at distinct places, processes and practices. Attractors act as magnetic forces, drawing complex (social) systems into specific trajectories during iterative loops in the system.

Sustainability as attractor

If we interpret sustainability as a so-called strange attractor (see Gilstrap 2005) we mean that across time and place, and through multiple networks and flows, institutions and practices are structured, 'ordered' or patterned following notions and ideas of sustainability. Pockets of ordering in a sea of disorder can be identified, when practices and institutions are infiltrated with sustainability claims, norms and interests, strengthened through various feedback mechanisms of interest groups, politicians, media, businesses, citizens and competitors who react upon short-falling as well as successful attempts to restructure practices and institutions along ideas of sustainability. Information flows play a crucial role in restructuring conventional processes and practices into more sustainable ones. Information flows do not just communicate ideas and interest over large stretches of space and time; they have also turned into a constitutive force in the Information Age, also with respect to the environment (see Mol 2008). The current information scape enables these pockets of ordering to have global outlooks.

To some extent the idea of sustainability as attractor resembles ideas of ecological modernization (Spaargaren & Mol 1992, Mol 2006). Ecological modernization refers to the growing emergence of an ecological rationality in processes of production and consumption, next to and partly independent from economic, political and other rationalities. The articulation – and subsequent institutionalization – of an ecological rationality can be witnessed in politics (e.g. ministries of environment, environmental laws, Green Parties, multilateral environmental agreements), in the life world (e.g. environmental NGOs; environmental periodicals, magazine and television programmes; widely shared environmental norms and values), as well as in economics and markets (e.g. green accounting, company environmental reporting, green investment funds; environmental standards and labels; environmental industries). This emergent ecological rationality enables societies, social groups and organizations to not only analyse and judge, but also shape and design production and consumption processes by independent ecological criteria (alongside other criteria of efficiency, fairness, equity, etc.). This idea of the ecological modernization of production and consumption especially fitted the analysis and understanding of contemporary processes of environmental reform in a relatively ordered world, constituted of nation-states and societies as units of analysis. And numerous authors have used this concept/idea in analysing (the failures of) environmental reform in political, economic and life world institutions around the world (see Mol et al. 2009).

But with globalization and the increasing complexity of the global order, an increasing number of social theorists argue that the 'zombie concepts' of states and societies have to be replaced by networks and flows as the key analytical concepts. Networks and flows have become the true architects of global modernity. While, as argued elsewhere (Mol 2008, Mol et al. 2009), this does not render the ecological modernization paradigm out of order, it does call for a further 'update' of environmental reform analysis and theory to make it better fit current times. Until further notice, various attempts to rethink environmental reform under conditions of globalization can be brought together under the banner of an environmental sociology of networks and flows (Spaargaren et al. 2006). Introducing the notion of attractor into environmental reform analysis should be understood in this context.

How should we understand sustainability as an attractor? In multiple networks and flows of different kinds, which are – each in their own way – formative for global modernity, sustainability has become included and is a point of direction. This means that actors, rules and resources; the socio-material infrastructures/scapes; and the flows themselves articulate and include sustainability considerations and interests. With the inclusion of sustainability in multiple networks and flows, these architects of modernity have changed. For instance, increasingly commodity networks and flows of (green) products, financial flows (of ethical investments as well as more conventional ones), and tourism mobility include environmental conditionalities in some ways and to some degree. Consequently, sustainability inclusion in networks and flows pursue a different 'outcome' at the places, institutions and systems they connect to. Hence, sustainability can be identified as becoming a 'point of orientation', a commonality that reroutes flows and reconstitutes networks, be it to varying degrees for different networks and flows. And the processes and dynamics through which this happens are of course very diverse for the different networks and flows; they include a variety of resources, governing actors, mechanisms, dependencies and power relations. Global advocacy networks, with their information flows of blaming and shaming, and commodity or financial networks of business and investors articulate sustainability in different ways; both different from the political networks of states, cities, and international organizations that are involved in international standards setting and regime formation. Interpreting sustainability as a global attractor basically brings a common focus to the diverse ways in which sustainability norms, interests and considerations are articulated in these global networks and flows. Sustainability is incorporated in and gives direction to these very diverse networks and flows, be it in different ways, to different degrees, and through different mechanisms. No matter how different each of these networks and flows is, together they direct a pattern (or pocket of ordering) of sustainability in systems, places and institutions. As with the notion of ecological modernization, also here sustainability should be interpreted in (interdependent) material *and* discursive ways; as norms and ideas, but also as actual/materialized outcomes at places, systems and institutions.

Mega-events: the Olympics and World's Fairs

Mega-events originate from the nineteenth century, during times of nation-state and empire building. They are short-lived collective – usually cultural – actions, which have long-lived pre- and post-event social dimensions, impacts and effects (Roche 2003). Planned, but also unexpected, mega-events – World Expos/Fairs, Live Aid concerts, Earth Summits, Olympic Games, World Cups, Diana's death and funeral – bring together various signifiers of an emergent global order. Such signifiers reflect and perform a global imagined community, and provide a united frame for different people, generations, genders and classes. On occasions of mega-events, global images and signifiers are (re)produced, circulated, recognized and consumed, not only at the space of place where the mega-event physically takes place, but also world-wide through the media of screens and bits. As such mega-events are time-space condensed hubs, around which the economic, cultural and even political flows and networks interact and exchange, and thus structure social life. But they are a special, extraordinary kind of time-structuring institution in modernity, because of their large scale, the temporal cycles, and lack of a fixed, national location.

Mega-events have changed under a progressing globalization. Although since its origin the Olympic Games always presented itself as a global event, its expansion has gone through various stages, only to turn more recently into a truly global mega-event. In 2008, 205 NOCs were members of the IOC. With its close linkage to global advertising, global corporate sponsoring and selling broadcasting rights to the global media since the 1980s, the Olympics became more profitable, more closely linked to global capital and turned into a global media event. It also resulted in growing interest from cities – especially global cities (see Sassen 1994) – in the bidding process. A similar development can be seen in the organization of World Expos, where the number of participating countries and international organizations increased over the last decades. While global television broadcasting is less important for World Expos compared to Olympics, advertising, corporate sponsorship and global capital made these expos truly global events. Organizing such mega-events became increasingly 'a significant opportunity to forge new and improved links with the wider world that plug the city more effectively in the global flows of capital, people and ideas' (Short 2004: 108). Hence, city authorities and leaders develop what Andranovich and colleagues (2001: 113) label a 'mega-event strategy' to position their city in global economic competition. By themselves, mega-events constitute exchange hubs in the global economic, political and cultural networks and flows of social life (Roche 2003), not unlike how Saskia Sassen has analysed the position of global cities in world financial flows.

The facts that the Olympics and World Fairs (1) are truly global events and embody globalization, (2) have an overall positive connotation throughout the world and (3) are closely related to many vested interests mean that major corporations have a significant interest in getting Olympics and World Fairs in places with yet underdeveloped but fast growing markets. But it is much too simple to draw a straight line between the selection for China to host the Olympics and the World Expo, and the interest of large corporations. Besides economic mega-events, Olympics and World Fairs are also political and cultural mega-events, as Close and colleagues (2007: 24) rightly argue. These mega-events are also important for their political and cultural impacts, and these dimensions play a full role in – among others – decisions regarding proposing cities as candidates, the bidding process, the decision-making process to grant cities the organization and location of the event, and the final implementation of the event. This prevents us from being trapped in economic determinism in analysing such mega-events.

If indeed sustainability has turned into a global attractor that reroutes and restructures the flows and networks of global modernity, it should 'materialize' and become visible and identifiable at and around mega-events such as the Olympics and World Fairs. There we should be able to identify and point out the proliferation, impact and institutionalization of sustainability in networks, flows and institutions that structure and construct these global happenings. The 2008 Olympic Games at Beijing and the 2010 World Expo at Shanghai are a key proof of that, given the poor sustainability record of China at the start of the first decade of this millennium (World Bank 1997, Carter & Mol 2007).

The Beijing Olympics and the quest for sustainability

China's ambitions to stage the Olympics started in the late 1980s (Ong 2004: 35), in essence to increase international prestige, to build an image of national strength and unity, to strengthen domestically the position of the China Communist Party (CCP), and to use the Olympics as a development engine. Compared with other hosting cities, the objectives for Beijing to host the Olympics were rather on a national than on a city level, and more political than economic. The mega-projects of Beijing's Olympics reflect China's ambition to transform its image of backwardness, tradition, red tape and corruption and to reclaim its position as global leader with an image of progress, efficiency and economic success (Broudehoux 2007). And in current times, good environmental quality is definitely part of such a new, modern image.

It is only since the 1990s that the environmental dimensions of Olympic Games have started to become articulated, basically in two ways. First, especially in the 1990s, environmentalists – both global NGOs and ones based locally to the Olympic Games – became very active in discussing the environmental consequences of the Games, in all its dimensions. Second, an environmental 'pillar' was slowly but steadily built into Olympic institutions. This process followed a wider development, with the 1987 Brundtland report and the 1992 UNCED as the symbolic milestones of increasing sensitivity around the world of the environmental agenda. In a rather timely move, in 1986, IOC president Samaranch declared the environment to be the third pillar of Olympism, alongside sport and culture. But only in the 1990s did the IOC begin to operationalize this third pillar, resulting in increasing precision of environmental requirements in the manual for candidate cities (MCC) (which guides the bidding process) from 1996 onwards. The 1994 Winter Games at Lillehammer provided a first model for taking environmental conditionalities into consideration, followed by Nagano (Cantelon & Letters 2000). But it was especially the 2000 Sydney Games that moved 'green Olympics' from the IOC agenda into practice. Through Olympic Games Impact Studies (introduced in 2003), the impacts of Olympic Games are now assessed on over 100 economic, socio-cultural and environmental indicators.

Beijing first bid in 1993 for the 2000 Olympic Games, losing in a very close finish to Sydney. In that bidding process, the environment hardly played any role, either in (criteria for) the bidding document prepared for IOC decision-making, or in the discussions and decision-making process itself. This aspect was notably different in Beijing's bidding in 2001 for the 2008 Olympic Games. Beijing used Sydney as the example for budgeting their Olympics, including the latter's significant, path-breaking attention to the environment and a green Olympics. In the 2001 final promotion of Beijing during the IOC decision-making event, Wang Wei, the secretary general of Beijing's Bid Committee, indicated that with respect to the environment Beijing had 'come a long way since its last bid in 1993' (BOCOG 2008), emphasizing a major commitment to the environment in Beijing's bid for the 2008 Olympics. Beijing articulated three themes: a green Olympics, a High Tech Olympics and the People's Olympics. While there remains much to be said and analysed on the latter two themes (see Ong 2004, MacDonald 2003), here we concentrate on the first theme.

The green profile of the Beijing Olympics

Beijing started Olympic environmental preparations as early as 1999, integrating the environment into the design of the Games at an early stage. This was, of course, a necessity for a successful bid, as China's environmental problems were of concern to environmentalists, medical specialists and athletes in many countries. Although within China and Beijing the environment was gaining increasing attention from the public authorities, scientists and the public, it was far from being a priority area for the state and the CCP at that time. With double-digit economic growth figures, outdated technology and industrial systems, limited investment in environmental protection, a short-falling environmental state and the sheer absence of a countervailing environmental movement, China witnessed ongoing increases in environmental deterioration in all sectors. 'Clear water, blue skies' (World Bank 1997) was put on the agenda at the turn of the millennium, but was far from reach (Carter & Mol 2007). But with the example of the then forthcoming 'green' Sydney Olympics, BOCOG understood the growing importance of the environment in (the various global economic, political and cultural networks and flows that constitute) the Olympic Movement. Turning Beijing's/China's poor environmental profile and record into an advantage, BOCOG argued that staging the Olympics would inspire, facilitate and disseminate further environmental improvements in Beijing and across China: by introducing global standards and benchmarks for urban development; by bringing in foreign technology and expertise; by raising environmental awareness;

by changing political priorities and domestic and foreign investments; and by triggering institutional innovations. Beijing's Olympic Action Plan (OAP) for the 2008 Games clearly reflected this. It identified 20 key projects for improving Beijing's environment. The OAP covered how ecological infrastructure development, energy saving and the use of renewables, and environmental protection would be integrated into Olympic facilities, systems, processes and the organizational structure.

What has been the actual environmental impact of the Olympics on Beijing and wider China? Liao Xiudong, member of the Chinese Olympic Committee and consultant for BOCOG, stressed the opportunities the additional environmental budget provided (the planned USD 12.2 billion over 1999–2007 (see also UNEP 2007) increased to USD 17.5 billion; UNEP 2009a); the significant institutional strengthening of Beijing's environmental authorities; and the special opportunities this mega-event offered for environmental experimentation and setting new routines and rules (Interview, Beijing, 20 May 2008). While at other large events, such as the 50th anniversary of the People's Republic of China, temporary environmental improvements were achieved by shutting down factories, many of the Olympic efforts involved some sort of long-term, permanent and institutionalized clean-up. The environment was part of all Olympic processes: the design and construction of facilities, refurbishment, marketing, procurement, logistics, accommodation, transport, office work, publicity and operational affairs. Environmental Impact Assessments (EIAs) were organized for all projects, even though they were often not legally necessary. Fourteen new wastewater treatment facilities were put into operation. Five new public metro transport lines were built for the Olympics, doubling the network to over 200 km. Acceleration in leapfrogging vehicle emission standards was achieved, to Euro IV standards in 2007. Around 90 per cent of all buses and 70 per cent of all taxis were converted to natural gas by early 2008. A subsidy programme helped to accelerate the replacement of old taxis with new ones, and a city-wide vehicle inspection, testing and green labelling programme was carried out. The rapid transition of coal-fired power plants (all major ones (>20 tons), nearly all 16,000 plants under 20 tons, and 44,000 boilers under one ton) towards gas fired power plants was achieved by late 2007. Some 200 industries were relocated – sometimes combined with environmental upgrading – between 2000 and 2006, including two of the largest industrial polluters, the Beijing Capital Iron and Steel Group and the Beijing Huaer Company (chemicals). Rapid phase-out of ozone-depleting CFCs took place in the city of Beijing, six years ahead of the national deadline. Procurement guidelines for the Olympics reflected sustainability. And several other deadlines of the already existing and approved Beijing Environmental Master Plan (1997–2015) were moved forward, to have their effect before the Olympics. In addition, expansion of the city greenbelt in Beijing (from 36 per cent to 43 per cent urban surface coverage between 2000 and 2008; UNEP 2009a:

64) and reforestation of the mountains and fields surrounding Beijing were meant to reduce sand and dust storms (UNEP 2009a). Major national anti-erosion and reforestation programs – although not specifically designed for the Olympics – had similar land coverage effects in more Western provinces, which together affected the sand and dust storms that reach Beijing. Still, in spring 2006, 18 dust storms were reported in Beijing.

But did all these efforts result in improved environmental quality? The crucial and most debated issue in the run-up to the Olympics was air quality. Most monitoring data did indeed show improvements in air quality indicators in the years up to the Olympics, such as ozone, CO, NO_x and SO_x. Figure 7.1 provides annual average monitoring data for NO_2, SO_2, CO and PM_{10} for a number of years. Perhaps except for PM_{10} (on which most of the discussion focused), all indicators showed slow but steadily decreasing tendencies. And the number of days where air quality was better than the national standard II increased, from 100 in 1998 to 246 in 2007 (see Figure 7.2). Some critics, however, accused the Beijing municipality of providing only average data for its 23 monitoring stations, and not the site-specific ones (in the Olympic zone). Others accused the authorities of moving the monitoring stations out of the most polluted areas, thus causing environmental improvements in monitoring data (Elliott 2008, Gang 2009: 88–102); this was denied by Beijing EPB (Environmental Protection Bureau) authorities (*China Daily* 27 February 2008). NASA satellite data seem to confirm air quality improvements (UNEP 2009a: 32ff). The general assessment on air quality is that, although significant sustainability measures were taken to reduce air pollution, a number of reasons prevented more far-reaching air quality improvements: dependency on other (up-wind) provinces (most notably Hebei, Shanxi, Shandong and Inner Mongolia where fewer clean-up activities of coal mining, power plants, cement plants and steel and iron plants took place), the specific geographical location of Beijing and the fact that air quality improvements require more than six years (see Streets et al. 2007). As UNEP (2007; see also 2009a) concluded one year before the Olympics, 'despite the relatively positive trends of recent years, air quality remains a legitimate concern for Olympic organizers, competitors and observers, as well as for the citizens of Beijing' (UNEP, 2007: 18). In the end, a combination of emergency measures and a fortunate change of the weather caused relatively low air pollution levels during the Olympics and Paralympics. From 20 July 2008 onwards, a driving ban for half of the 3 million registered cars was implemented as one of the emergency measures. In addition, most construction in Beijing was halted. And over 150 companies in Beijing and in the neighbouring province of Hebei – and Shanxi, Inner Mongolia and Shandong provinces – were temporarily closed two weeks before the Olympics started, until the end of the Paralympics. A tropical storm that hit southeastern China on 28 July brought strong winds, some rain and lower temperatures, which cleared the skies just before the Olympics started, helped a little by the launch of over

1000 rockets with artificial weather modification and control chemicals. So, in August 2008 levels of SO_2, CO and NO_2 in Beijing were 45 per cent lower than in July, falling back to levels normal in developed countries. August 2008 also counted an extraordinary number of days with the highest air quality (14 days of level I), and only one day with the lowest air quality (level III) (BEPB 2009).

Besides air quality a number of other environmental improvements have been achieved. Through the installation of 3.0 million m^3/day wastewater treatment capacity by the time the Olympics started, surface water quality in Beijing improved considerably. In addition to water pollution prevention measures, a variety of water-saving schemes, rain water collection technologies and water re-use systems were implemented. Many of the improvements in waste re-use and recycling will have similar environmental legacies beyond the Olympics. The Beijing Olympic Village received a LEED (Leadership in Energy and Environmental Design) award from the U.S. Green Building Council for, among others, its use of environmentally friendly paints, used building materials, applied energy systems and waste recycling.

The general assessments made not only by the Beijing Environmental Protection Bureau (BEPB 2009) but also by independent organizations such as UNEP (2007 and 2009a) and Greenpeace (Zhang 2008) are overall positive on the actual environmental improvements following the Olympics.

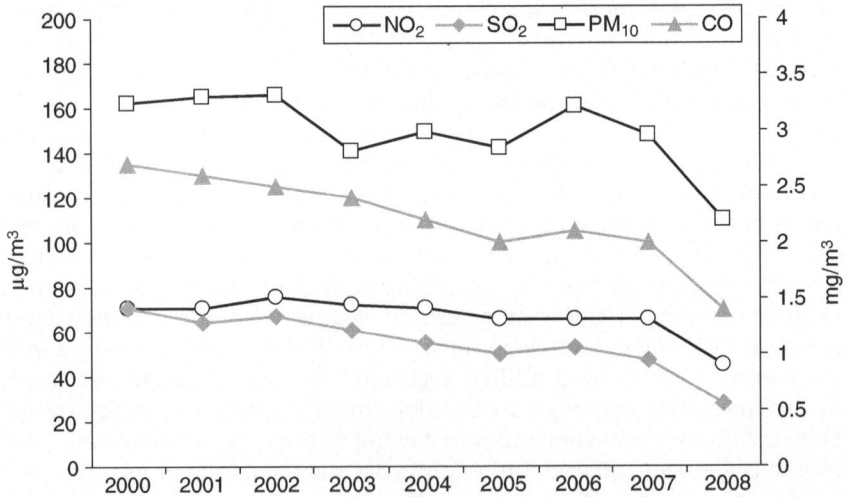

Figure 7.1   Annual mean NO_2, SO_2 and PM_{10} (in $\mu g/m^3$; left axis) and CO (in mg/m^3; right axis) concentrations in Beijing, 2000–8

Source: BEPB 2009, UNEP 2007, 2009a

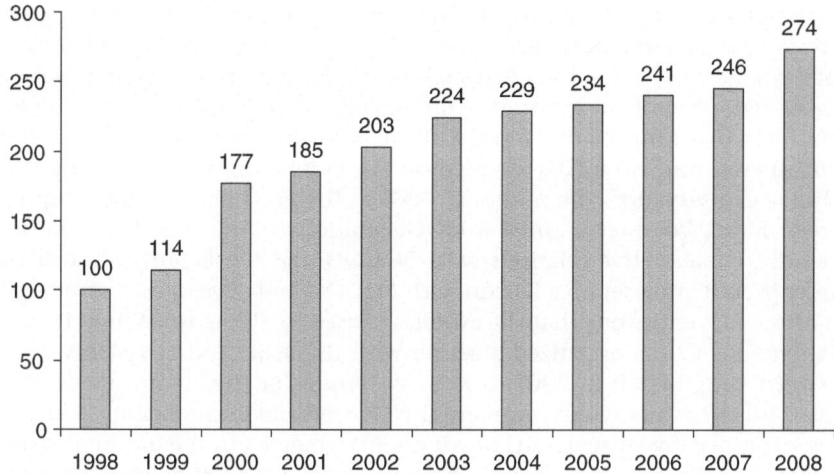

Figure 7.2 Number of days with air quality index equal or better than national standard II, 1998–2008

Source: BEPB 2009, UNEP 2007, 2009a

In that sense, the green Olympics was more than just a set of exercises in public relations and successful bid planning.

Environmental criticism

In the run-up to the Olympic Games, most host cities have experienced environmental concerns and criticism from local, national and even international NGOs over the potential environmental (and other) consequences, often leading to reduced support for or even opposition to the event (see Del Corpo & Dansero 2007). Polls in Beijing and China (by Chinese and independent organizations) witnessed continuously high levels of public support for the Olympics (all above 90 per cent). And very limited criticism was raised on the environmental dimensions of the Olympic facilities within China. This can be explained by (1) the significant environmental projects planned and implemented by Beijing, in a context of generally poor environmental quality; (2) the limited experiences with and opportunities for environmental activism and criticism in China (see Xie 2009, Carter & Mol 2007). The Beijing authorities did stimulate citizens to complain about environmental misbehaviour through the offer of financial incentives, resulting in some 2300 complaints being lodged in 2006 (with 59,000 RMB paid on incentives to complain; Beijing 2008 (2007a)).

But in the environmental progress assessment commissioned by BOCOG, UNEP (2007, 2009a) was least enthusiastic about efforts regarding collaboration with 30 domestic and international environmental NGOs. Only limited activities were undertaken in this area, and opportunities to further raise environmental awareness (during and after the Olympics) and mobilize civil society have not been seized. Especially considering the (expected) negative media reporting of the environmental profile of Beijing and China, UNEP considers this a major omission. Still, it is surprising that the (local and global) environmental NGOs active on the Olympics, such as Greenpeace China, Conservation International, WWF, IUCN, Global Village Beijing, Green Earth Volunteers and Future Generations/China, developed rather friendly, collaborative relations with BOCOG and the Beijing authorities. Greenpeace China set up a liaison with BOCOG and developed a campaign to offset CO_2 emissions that the athletes caused by flying to Beijing; Future Generations/China organized together with the Beijing Forestry University a Green Long March in 2007 to raise awareness for the environment. It is illustrative that many environmental NGOs defended the Beijing authorities when negative discussions on air quality emerged in the (international) media. And Greenpeace (Zhang 2008) produced a very friendly assessment of the Beijing Olympics, although it also stressed the lack of transparency, the difficulties of independent verification of environmental data and the poor use of international certification systems such as LEED and FSC (Forest Stewardship Council). While – certainly compared to issues such as Tibet, Taiwan, Falun Gong, the Uighur and the military – environmental reporting has become less restricted in the media over the past decade (see Gang 2009, Mol 2009), the high profile Olympics did limit the room for manoeuvre for environmental NGOs: 'At least two leading environmental organizers have been prosecuted in recent weeks, and several others have received sharp warnings to tone down their criticism of local officials. One reason the authorities have cited: the need for social stability before the 2008 Olympics, once viewed as an opportunity for China to improve the environment' (Kahn & Yardley 2007). And in the run up to the Olympics, environmental scientists were warned to be cautious in their media contacts on environmental quality in Beijing. Often official permission was required for environmental scientists to talk to the foreign media.

Shanghai World Expo and the quest for sustainability

With its 150 years of history, the World Expo is considered the Olympics of the economy, science and technology. Article 1 of the Convention relating to International Exhibitions (BIE 1928 (2009)) defines an exhibition as 'a display which, whatever its title, has as its principal purpose the education of the public'. The World Expos from 1851 to 1938 delivered the message of 'knowledge is power' and focused on trade and technology. From 1939 onwards the World Fairs had cultural themes and focused on cultural exchanges, although technology remained important. From 1988 onwards countries started to use World Expositions more widely and more strongly as a platform to improve their national images through their pavilions. Pavilions became advertising campaigns, and the Expo a vehicle for 'nation branding'. Apart from cultural and symbolic reasons, organizing countries (and the cities and regions hosting them) also use the World Exposition to brand themselves. Presently, there are two types of World Expositions: 'registered' and 'recognized' (sometimes unofficially known as 'major' and 'minor' fairs, respectively). Registered exhibitions are the largest events. At registered (or universal) exhibitions, participants (usually countries and major international organizations) generally build their own pavilions. They are therefore the most extravagant and most expensive expos. Their duration may be between six weeks and six months. Since 1995, the interval between registered expositions has been at least five years. Smaller recognized expos are organized in between (such as in 2008 in Saragossa, Spain, and in 2012 in Yeosu, South Korea) and have fewer participants.

The Bureau International des Expositions (BIE, established in 1928) regulates the frequency and quality of exhibitions. With 155 member countries, it organizes two meetings a year. Environmental concerns were first fostered as part of the World Expo by the United States in 1974 in New York. Since then, the idea of sustainable development has gained prominence, especially in the themes of some of the World Expos. But it has not become so fundamentally institutionalized in the BIE structure as with the IOC. There is no committee for environment, there are no environmental indicators that bidding documents need to pay attention to, and it is not clear whether and how environmental issues play a role in evaluating the bidding documents.

China appeared at the first London Expo in 1851 but lost interest in these exhibitions for quite some time. The return of the New China at the Knoxville, US, World Expo in 1982 (at the end of the Cultural Revolution and the start of the opening-up and reform era) marked a turning point for China. China has participated in 12 World Expos, culminating in the bidding for the 2010 Expo. On 3 December 2002, after four rounds of voting, Shanghai won the bid over the World Expo 2010 and became the first developing county to host a registered World Expo in history. Like the 2008 Beijing Olympic Games, this World Expo is not only a local event for Shanghai, but an event with national impact and outreach, lasting half a year. Together, these two mega-events serve as markers for the new global position of China at the start of the twenty-first century, a strategically critical period for China.

In May 1995, when Shanghai decided to apply to host the World Expo, a research team on Expo themes was established. Based on studies, surveys and expert consultations, over 30 themes were proposed. During the theme-selection process, Shanghai communicated intensively with BIE. In November 2000, 'Better City, Better Life' was chosen as the theme for Shanghai Expo 2010. In the application report (2001), Shanghai answered

58 questions posed by BIE, of which 10 questions were related to the theme (Bureau of Shanghai World Expo Coordination 2009a).

China started to promote the Shanghai Expo as a green expo in 2000, with the theme launch. Shanghai, a mega city facing all kinds of 'modern diseases', aimed to present during the Expo not only the challenges facing today's cities but also possible solutions. The idea of a green expo aims not only at 'educating' the 70 million expected visitors, but also at leaving a green legacy for the citizens of Shanghai and contributing to world-wide initiatives of making cities more sustainable. The Shanghai Expo has 242 official participants (191 countries, the others were international organizations); in comparison, the 2005 World Expo in Japan had presenters from 122 countries and territories, as well as many other organizations, under the theme 'Nature's Wisdom'.

The green profile of the Shanghai World Expo

Since 2000, when preparations for the Expo 2010 commenced, the municipal government of Shanghai began upgrading the city's infrastructure, strengthening its pollution control measures and introducing more renewable and energy-efficient technologies. Shanghai has taken the Expo as an opportunity and catalyst for the introduction of new, and the acceleration of already planned, measures for turning Shanghai into a more resource-efficient and environmentally friendly city, focusing on environmental targets set for 2010 (Bureau of Shanghai World Expo Coordination/SEPB 2009a, UNEP 2009b). A 'Three-Year Environmental Action Plan' approach was adopted to attain the set goals. Since 2000, three rounds of Three-Year Environmental Action Plans have been implemented, and the fourth round started in early 2009. Together, over 900 environmental projects have been implemented under these plans by 2011.

To coordinate the environmental governance of different departments at different levels, the cross-departmental Shanghai Environmental Protection Committee (SEPC) was established in 2003, headed by the mayor. The SEPC is responsible for coordinating, reviewing and realizing these three-year action plans. To implement the ambitious plans and its projects, the city invested heavily in environmental protection, especially after winning the Expo bid. Since 2000, investment in environmental protection has tripled, reaching 42 billion Yuan (4.4 billion Euros) in 2008 (UNEP 2009b). By then, environmental expenditures made up 3.1 per cent of the city's GDP, with over half of all the accumulated environmental investments since 2000 coming from the government.

With the Expo approaching, Shanghai introduced a series of new environmental measures, both for the city as a whole and for the exhibition site. Cutting city air emissions traffic measures included: imposing restrictions on polluting vehicles within the mid-ring-road area; the major upgrading of emissions norms for public buses, taxis and new cars, using similar standards (Euro class III and IV) to Beijing for the Olympics; encouraging the phasing out of old vehicles; and significantly extending the public transport system from three to eight lines. To reduce air pollution during oil/gas storage, transportation and refuelling, oil and gas recovery projects were to be completed in all (more than 800) gas stations in Shanghai. Major investments were made in extending wastewater treatment capacity from 55 per cent in 2000 to 75 per cent in 2008; in transforming almost 6000 coal-fired energy plants in the city to cleaner energy sources and adding desulphurization devices to the remaining coal-fired plants above 10 million KW; in increasing the green area in the city; in closing, relocating and/ or significantly upgrading over 2000 heavy polluting industries since 2006; and in modernizing the urban solid waste collection and re-use system (with collection rates at almost 80 per cent of municipal waste). Recently renewable energy has taken off (with 27.3 MW wind power capacity and 200 KW solar by 2008).

Did these intensifications of environmental measures and investments result in environmental improvements? The green coverage in the city increased (from 25 per cent in 2000 towards 38 per cent in 2008; UNEP 2009b), and moved from 4 to 13 m^2 of green land per capita between 2000 and 2008. Several air (see Figure 7.3) and water quality indicators showed some improvements between 2000 and 2008, but not as dramatically as in Beijing. National air quality standards for CO, SO_2, NO_2 and PM_{10} have been met during these years, except for PM_{10} in 2002. But when compared to WHO standards and with health effects (Cao et al. 2009), further improvement of

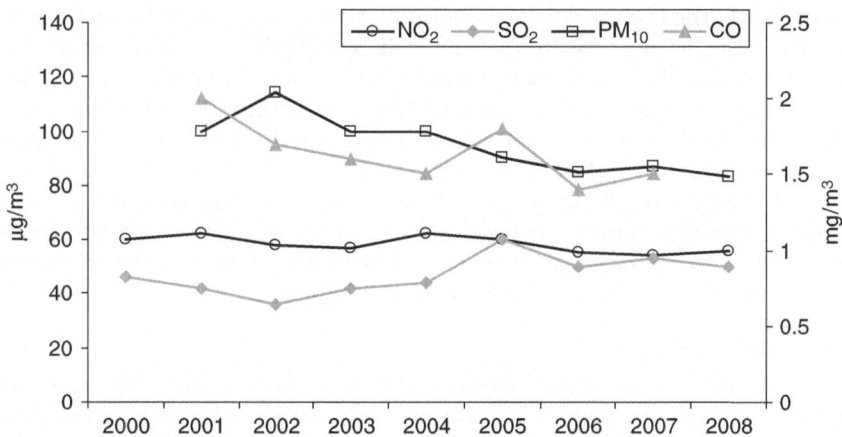

Figure 7.3   Annual mean NO_2, SO_2 and PM_{10} (in µg/m^3; left axis) and CO (in mg/m^3; right axis) concentrations in Shanghai, 2000–8

Source: Shanghai EPB data, UNEP 2009b

PM_{10} and SO_2 levels is necessary. By the end of 2008, emissions of COD had been reduced by 15 per cent from the level of 2000. Compared with the levels in 2000, COD and SO_2 emissions per 10,000 Yuan GDP have been reduced by 71 per cent and 67 per cent respectively. This has not passed unnoticed by the population. According to a Gallup opinion poll, 86 per cent of the citizens believe that more investments have been undertaken in environmental protection, 94 per cent think that the government has done a good job in this regard and 87 per cent experienced a rise in the overall environmental quality of Shanghai.

One important measure for improving air quality in Beijing was to strengthen regional cooperation between the six provinces and Beijing. Following this model, in May 2009, one year away from the Expo, the first Yangtze Delta environmental protection meeting took place in Shanghai, following the *Agreement on Environmental Protection Cooperation of the Yangtze River Delta (2009–2010)*. Tasks were specified for Shanghai, Zhejing and Jiangsu provinces, to be implemented in 2009 and 2010. Shanghai's trans-regional plan can be regarded as an upgraded version of the Beijing regional plan. While Beijing targeted specific air pollutants and relied much on emergency measures, Shanghai aims to establish a new long-term regional environmental governance system based on prevention, coordinated policy measures, institution building and synchronous monitoring-supervision-enforcement mechanisms. Both the Beijing and the Shanghai models have limitations and should not be copied directly to other regions; however they provide interesting examples of regional environmental cooperation in China.

However, the city of Shanghai has not taken extreme measures, such as demanding the temporary closure of polluting companies in Shanghai and nearby cities, to improve air quality for the sake of the 2010 World Expo. On the one hand, air quality in Shanghai is better than in Beijing, in part due to its favourable geographical and climate conditions. On the other hand, the city's Party chief Yu Zhengsheng ruled out the possibility of a large-scale shutting down of polluting enterprises on the grounds of costs and a potential negative impact on employment and government revenue. 'The short-term improvement of environment index will not tackle the problems, let alone the fact that the Expo will last six to seven months. We won't take such measures' (*China Daily* 19 April 2009).

The exhibition area

After several rounds of discussions and comparative studies, the Expo site was designated alongside the Huangpu River, covering a land area of 5.28 km^2. Although many citizens and organizations worried about the costs of and difficulties in cleaning up this polluted piece of land, the government was determined to take the opportunity to turn this former industrial zone into a green expo site. After decades of industrial development, the site was crammed with shabby dwellings, old factories, docks and warehouses. The 272 factories in the area, mostly outdated and heavily polluting, were a mosaic of power plants, steel refineries, chemical industries, mechanical workshops and shipping manufacturers, among them 21 major industrial pollution sources (SEPB 2000). All these pollution sources have now been shut down or removed. About one-sixth of the old warehouses, workshops and other factory buildings, with a total area of 370,000 m², have been preserved and reused for the Expo, turning the site construction into a combination of new projects and renovated old urban areas (Bureau of Shanghai World Expo Coordination/SEPB 2009b). Post-event use has been included in the planning (Zhi & Liu 2008). For instance, the post-Expo utilization of venues and facilities has been taken into account to minimize the environmental impacts of the event.

Environmental considerations have been part of all Expo processes and related activities: the selection of the Expo site, the design and construction of facilities (Shi & Gong 2008), refurbishment, marketing, procurement, logistics, accommodation, the planning of transport, energy provision, office work, publicity and operational affairs. Prior to the bidding process, Environmental Impact Assessments were also conducted on the blueprint of the Expo site. The preparation process of Shanghai Expo itself was a process of exploring and practising the theme of the Expo: 'Better City, Better Life'. Different pavilions showcase new concepts and latest technologies for sustainable urban development, through both the exhibitions and through their architecture and design. Environmental management has been established throughout the Expo, defining the roles and responsibilities of various departments across the entire life-cycle of the Expo (Shi & Gong 2008), so as to ensure the attainment of environmental objectives.

The on-site infrastructure and buildings were planned as demonstrations of environmentally sound technologies, such as shading systems, temperature control materials, natural wind planning, tunnel wind and water cooling systems. Since many Expo buildings will be removed after the event, it was thought to be important to choose building materials that are renewable or can be reused or recycled. Energy-saving technologies, including new insulation materials, ice storage air-conditioning, LED and solar photovoltaic devices (of 10 MW, five times the existing PV capacity of Shanghai), have been applied widely, to reduce on-site energy consumption by 90 per cent. One thousand vehicles powered with new energies (electricity, hybrids and hydrogen) serve the Expo area and contribute to making the Expo a zero-emissions zone.

With assistance of UNEP, the Bureau of Shanghai World Expo Coordination published the *Green Guidelines for Expo 2010 Shanghai China* in May 2009, to guide the hosting agencies, participants and visitors in greening all kinds of practices: from design, construction, transportation and accommodation, to visitors' behaviour (*China Environment News* 18 June 2009). Figure 7.4 shows the structure of the guidelines. The Shanghai World Expo Coordination Bureau (2007) also published *Participants Guidelines,* requiring participants to follow relevant Chinese laws and regulations, among which those on water protection, air quality, waste management and radiation, as well as advocating green procurement, green offices and re-use and recycling.

Both the private sector and public organizations have been involved in these efforts. The Guidelines have 'forced' multi-national investors and sponsors such as Siemens, OTIS, Tetra Pak, Coca-Cola and General Motors to further integrate environmental conditions into their activities. But citizens were also called to participate. Shanghai EPB initiated the 'Shanghai Environmental Protection Map', the first of its kind in China. This map shows, among others, environmental infrastructure, green space, natural reserves, green hotels and restaurants, green schools (#668), green communities (#480), environmental education bases, environmentally friendly towns and eco-villages (www.news.cn 19 February 2009). It may help to kick off green consumerism. And to make the Expo low-carbon, the Shanghai Environment & Energy Exchange established an Internet-based platform to allow the public to buy carbon emissions on a voluntary basis (*China Climate Change Info-Net* 27 August 2009). It is estimated that 9 million tons of carbon emissions have been generated during the Expo period, of which the Expo has reduced 1.5 million tons through the application of energy saving and emissions reduction technologies and measures, and the remaining 7.5 million tons have been for sale to the public. Though similar to the Cleaner

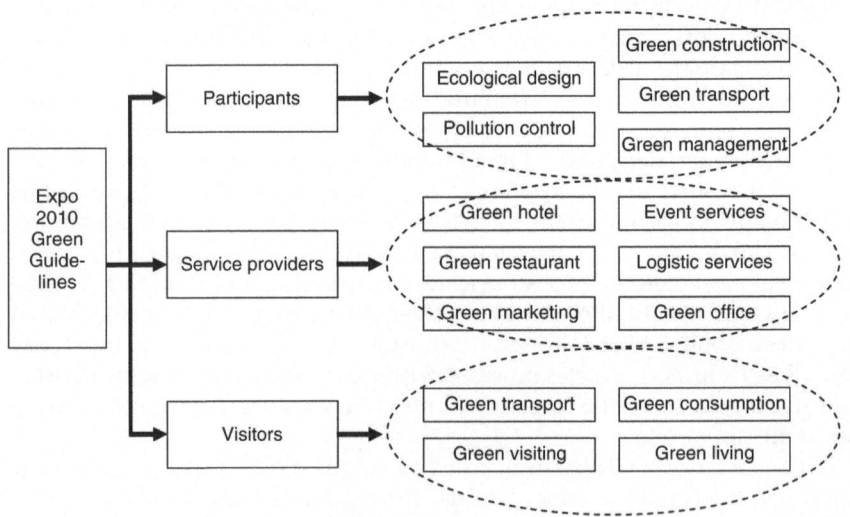

Figure 7.4   Green Guide for Expo 2010 Shanghai, China
Source: Shanghai EPB data

Development Mechanism which mainly operates between businesses and states, this programme aimed to involve the general public.

Criticism and shortcomings

In contrast to Beijing, Shanghai has a barely active environmental NGO community (Xie 2009). During the preparations for the World Expo, environmental NGOs were relatively uninvolved in planning, in designing the exhibition area, and in communicating plans to the wider Shanghai community. Only WWF Shanghai and the US-based Environmental Defense Fund have been involved in the World Expo mega-event. While many public activities were organized in the run-up to the World Expo, they were generally top-down, government directed. The absence of an active civil society community prevented major civil society criticism during the various phases of planning and implementing environmental measures, and this did not change during the World Expo itself in 2010. At the same time, Shanghai handled over 40,000 environmental complaints from its citizens in 2007, showing widespread environmental concerns. UNEP (2009b) and some of the international environmental NGOs called upon the Shanghai authorities and the World Expo organizers to increase the participation of environmental NGOs in the World Expo, with only limited results.

Even in the absence of critical watchdogs, a number of environmental shortcomings have been observed within and outside China. Irrespective of the interesting renewable energy experiments in the Expo area, Shanghai still has a high dependency on coal (over 50 per cent of primary energy use in 2008 still came from coal; UNEP 2009b), and poor performance on renewables. The introduction of LPG taxis and buses has seemed to stagnate. The safe disposal of waste and the still poor quality of its surface waters can also be mentioned as additional serious shortcomings. Shanghai has reduced SO_2 and COD emissions by almost half during the past decade, and CO and PM_{10} levels have reduced notably. But several indicators are not monitored, such as O_3, mercury and $PM_{2.5}$, or did not improve (e.g. NO_2 and SO_2 concentrations). And there remains a lack of clarity as to whether Shanghai has made the six-month World Expo climate neutral.

Patterning mega-events and beyond

This overview of environmental reforms related to the 2008 Olympics and the 2010 World Expo gives of course no final conclusions about the successfulness of a green Olympics in Beijing and a green Expo in Shanghai. A much more extensive analysis would be needed, and (at the time of writing) we also still cannot give a full assessment of the sustainability performance of the World Expo. But it does allow us to draw a number of conclusions with respect to the centrality of environmental sustainability in the two Chinese mega-events.

Certainly compared to the former Olympic Games in Athens, Beijing was widely applauded by international governmental and non-governmental organizations to have lived up to most of its green Olympics promises (see UNEP 2007, 2009a, Zhang 2008; the Beijing Municipal Environmental Protection Bureau and BOCOG were among the winners of the 2009 IOC Award for Sport and the Environment). Few of the environmental promises and plans in the OAP have not been kept. As such, the Olympics have accelerated environmental reform in Beijing, although there remains much to be improved on the environment in the capital. In greening Beijing, local and national resources and networks have been put to work, as has become clear above: on reforming the transport system, on greening Olympic facilities, on water saving and quality, on industrial relocation and clean-up, on air pollution mitigation, on expanding green space and renewable energy and so on. But transnational flows and networks have been equally crucial to the greening of this mega-event. International information flows and media networks articulated the environment strongly through their constant reporting, from the moment Beijing was elected as host city to the end of the Paralympics. Foreign environmental technology, service, design, and consulting industries and networks obtained major contracts for several of the Olympic projects and venues, and considerably contributed to advanced and up-to-date renewable energy use, environmentally sound construction, water-saving systems and the planning of green space, among others. By institutionalizing the environment strongly in their bidding, decision-making, implementation, and monitoring processes, the global IOC network was a major factor in articulating environmental sustainability in all phases of the Beijing Olympics process, as reported above. Most of the main international official sponsors and suppliers of the Olympics (all major multinationals) included the environment in their local and global advertisements, products and activities. These official sponsors (including Coca-Cola, Kodak, Visa, Panasonic, John Hancock, Samsung, MacDonald's, Swatch, Sema, General Electric) formed a non-permanent environmental working body for environmental cooperation with BOCOG (Beijing 2008 2007b; Zhang 2008).

And finally, international organizations such as UNEP, IUCN and WWF developed close links with BOCOG, constantly assessing and offering advice on the greening of the Olympic Games, putting continuous 'pressure' (also via the international media) on the organizing committee to live up to its promise of a green Olympics. Sustainability was a key factor in these global networks and flows. In that sense, local *and* global networks and flows around this mega-event did articulate – in different degrees and ways – sustainability, and as such directed the patterns of this mega-event towards greening. Thus, sustainability can be interpreted as an attractor in both the route towards and in the final Olympic Games, articulated in and redirecting global and local flows and networks, and materialized in discursive processes *and* socio-material systems, including transport, venue construction and use, green space, energy and water.

Environmental sustainability is obviously an aspirational selling point of the Shanghai World Expo, reflected in its theme, the city preparations towards the event, the construction of the Expo site and the practices during the Expo. Many of the lessons and measures from the Beijing Olympics have been taken into consideration by the Expo organizers, for example, active cooperation with UNDP (United Nations Development Programme) and UNEP, measures to improve air and water quality, public transportation investments, higher energy efficiency, green marketing of global Expo partners, and regional cooperation. Some however were notably different, such as the involvement of civil society. Although environmental sustainability is not (yet) formally institutionalized in the global bidding and implementation of World Expos, Shanghai's green efforts are likely to be followed by its successors. The lower levels of (global) media attention, and a less direct relationship between environmental quality and mega-event performance, mean that a 'sustainability attractor' is less visible at the Shanghai Expo than at the Olympics. Around the Expo, public debates, global media attention and international organizations are less visible in directing the Expo towards sustainability. But results do not seem to be less green.

Notwithstanding these conclusions, it would be a misunderstanding to attribute this all to the 2008 Olympics and the 2010 World Expo. Already before – and to a significant extent independent from – these mega-events, China had identified the environment and natural resources as one of its major future problems/constraints for further development. At least from the mid-1990s onwards, we can witness increasing efforts and resources of all kinds being spent on setting up and implementing a more effective system of environmental protection and efficient natural resource use: more environmental investment, more environmental staff, more environmental R&D, more environmental laws and regulations, more (freedom for) environmental NGOs, more eco-labelling, more ISO 14001 certified Chinese firms, more environmental complaints and deforestation turning into reforestation (see Mol 2006, Carter & Mol 2007, Gang 2009). And many of these developments were not merely endogenous, internal, domestic affairs, but were as much related to the economic, political and cultural opening of China: to the WTO and increasing global (inflow and outflow) trade and investments that articulate, for instance, food safety, environmental standards and green labelling; to the increasing activities of China around international environmental agreements and international environmental organizations; and to global norms and values of sustainability and quality of life that have entered China via multiple flows and networks. In that sense we should not overestimate the role of the Olympics and the World Expo. Both further strengthened and articulated existing sustainable developments tendencies in China; and the Olympics and World Expo could successfully include a green discourse and performance because this fitted with existing sustainability attractors in various national and international networks and flows.

One of the main reasons why China aspired to organize these mega-events was to strengthen its position globally, to become (again) one of the world superpowers (Broudehoux 2007). And today, acquiring, safeguarding and strengthening this position is not just a matter of economic power or military strength. It has also to do with becoming a legitimate and respected member of the international community. And environmental sustainability is fully part of such a legitimate position. The global and domestic interests and legitimacy of the CCP and the state are to a major extent fortified by a better environmental record and more environmental transparency. As such it should not surprise us that we were able to witness, in August/ September 2009, a sudden major change in the international discourse on China and the environment: from China as global polluter towards China as leading nation in global environmental investments.

Consequently, we should not be too worried about the lasting effect and legacy of these green mega-events. As the various environmental projects of the two mega-events fit closely with the wider sustainability efforts of the two cities and China, and of international networks and 'projects' in which China is engaged, these green mega-events become easily 'materialized' and institutionalized and gain permanency. This is definitely true for the material infrastructure, such as the extended public transport system, the new taxis and buses, the facilities, the upgraded industries and power plants, and the wastewater treatment expansion, to name but a few. They remain there after the events, and also set new standards for future materialities. By the same token, this is also true for the institutional and procedural changes that took place for environmental protection in Beijing and Shanghai. New standards (e.g. on car emissions), new development planning systems, stringent enforcement practices, more and better air quality policy cooperation between the city and its neighbouring provinces, and better EIA practices will not easily be turned back after the mega-events are over.

This all might be valid for Beijing and Shanghai, but not directly for other parts of China. Although these green mega-events will have some environmental benefits for the provinces neighbouring these two cities – due to their geographical closeness and the direct relation with air quality – this is not necessarily the case for other provinces and urban centres. Most of the environmental innovations and accelerations taken in the course of the mega-events are specifically related to Beijing and Shanghai, and are not so much national ones. That counts especially for innovations related to air quality, such as the car emission standards, the transition of coal-fired power plants into gas-powered energy plants, the extension of green space, the public transport network and the industrial relocations. It is not that environmental innovations and improvements do not happen in other urban centres in China, but that they happen with much less determination, speed, results and resources. In that sense the Olympics and Expo have a rather place-based environmental effect. It remains to be seen – and it is a key question for further study – to what extent and with what speed these environmental innovations in Beijing and Shanghai diffuse (through domestic and global political, economic and civil society networks and flows) to the wider China.

Epilogue: sustainability as global norm

Our analysis of the environmental sustainability of (the route towards) the Beijing Olympics and the Shanghai World Expo can be understood in terms of a 'sustainability attractor' that structures and patterns local and global networks and flows around these mega-events. This 'sustainability attractor' should also be identifiable at the Guangzhou-based 16th Asian Games, in 2010. It can be hypothesized that the reconfigured networks and flows following such mega-events will have some permanency, as in China a 'sustainability norm' is not restricted to such events. The environmental redirection of networks and flows in China (and beyond) takes place in a much wider context than just the Olympics and the World Expo. In that sense, mega-events are time-space compressions where global norms become easily visible and identifiable; however such norms are not specifically related to mega-events. We should expect to find similar pockets of sustainability ordering beyond the mega-events in China, albeit with less dramatic effects in such short time intervals.

However, with the idea of sustainability as a global norm we do not claim that environmental sustainability is becoming the global standard that will be implemented everywhere, nor that we are on an evolutionary path to full sustainability around the globe. Look outside the window and the world is still full of unsustainabilities of various kinds. And that certainly counts for contemporary China and its mega-cities: most levels of pollution, resource use and even environmental standards are still well adrift of what environmental NGOs, international organizations and even Chinese environmental authorities would consider sustainable. Draconian, one-time measures had to be taken to get air quality acceptable during the weeks of Beijing's Olympic Games, and Shanghai was called by UNEP to further improve its environmental conditions for the World Expo. In addition, there remain different interpretations, definitions and uses of the notion of sustainability according to time, place and interests. There is – and continues to be – debate and conflict on what sustainability means, what dimensions should be included and which not, which time line should be chosen, and how it should be assessed spatially. For instance, with respect to the Chinese mega-cities: can we have sustainability with the rigorous demolition of entire city neighbourhoods for the Olympic City; with the large inflow of natural resources from other places, such as Africa; or with the strict media control – which also limits environmental reporting (see Broudehoux 2007, Mol 2009, Gang 2009)? And even if we limit sustainability to environmental sustainability, as we did in this chapter, large differences in practices and definitions of (un)sustainability remain throughout the world.

But, and this is the strength of the global norm metaphor, global pockets of ordering move towards environmental sustainability. In a complex world, quite diverse developments and dynamics in global and local networks and flows show some order by moving in a particular direction: they restructure socio-material systems towards a global norm of sustainability. Such directionality towards sustainability can no longer be denied; neither by us social scientists, nor by the practitioners of mega-events. But of course debate remains on the interpretation of sustainability, and whether the speed and degree of change are sufficient.

8
Olympic Games as an Opportunity for the Ecological Modernization of the Host Nation: The Cases of Sydney 2000 and Athens 2004

John Karamichas

The greening of sports mega-events, and the hosting of Olympic Games in particular, is now reasonably well established. The potential international disgrace that a host nation might experience due to construction delays gives added impetus for their completion on time; the same appears to be the case for projects relating to environmental commitments made by the host nation. This temporal pressure of the event has the potential to accelerate much-needed environmental reform on the one hand, and to bring ENGOs as consultants into the planning of key Olympic projects on the other. Yet evidence from the first decade of environmentally conscious Olympics points to diverging patterns of achievement in the operationalization of the IOC's 'third pillar'. As is now common knowledge, for example, Sydney 2000 was the first 'green Olympics' in the history of the Games; yet four years later, Athens provided a stark contrast, and was the subject of highly critical assessment reports by both the WWF (WWF-Greece 2004) and Greenpeace (2003, 2004a, 2004b).

In this context, this chapter seeks to offer some preliminary thoughts on two questions directly raised by these experiences: first, to what extent has hosting the Games been translated into medium- and long-term improvements in environmental consciousness and performance in host nations? And second; given the universal prescriptions of the IOC bidding files and the globalized nature of both the event and its governance regime, how can we account for the fact that – from the evidence of 2000 and 2004 – some nations are apparently more successful than others in the organization of 'green' Games?

For both Sydney and Athens, the analysis we develop here examines these questions in relation to each of the phases of the event (Hiller 2000: 192), from the pre-event IOC bidding applications, to the fulfilment of the event commitments made by the hosts, and the extent to which they signified a post-event commitment to environmental sustainability (ES). As quintessentially modern events, the Olympic Games and other sports mega-events encapsulate the essence of the normative claims made by the ecological modernization perspective (EM): namely that market-oriented regimes of capital accumulation have the capacity to engage with the environmental dynamic and provide sustainable solutions. Our analysis is therefore broadly influenced by Andersen's (2002) influential discussion of the effects of Europeanization on Central and Eastern Europe. Following the rationale of his analysis, we thus evaluate the effects of hosting the Olympic Games on the ecological modernization of the host nations; like Andersen (2002: 1396), we have identified two contrasting hypotheses. The first hypothesis is that, in the wake of their respective Games (which were after all awarded to them, at least in part, on the basis of a range of green claims), 'one should be able to identify marked signs of environmental improvement' in the host nations (ibid.). Our second hypothesis is more cautious in its expectations: to achieve environmental transformation, the effect of hosting the Olympic Games 'depends more on the supportiveness of domestic political processes'. For Andersen 'the degree of ecological modernisation in a particular country depends on its capacity for environmental reform as fostered and supported by the character of the political and socio-economic reform process' (ibid.). We will test these contrasting hypotheses here.

Our analysis thus starts with an overview of the EM perspective, before discussing the environmental commitments made by the two candidate cities at local and national levels. After a brief appraisal of the degree of implementation of these plans, we discuss the post-event environmental performance of the two hosts by examining six key variables. These variables range from the use of standard practices for assessing the environmental impact of development projects to the ratification of international environmental agreements. We conclude by arguing that analysis of the data suggests that the EM capacity of Olympic host nations depends less on any putative post-Games 'Olympic effect' than on the state of political and ideological competition, allied to the culturally specific structural conditions of the respective polities.

Ecological modernization

By asking whether the Olympic Games can be seen as an opportunity for the ecological modernization of the host nation, we are immediately placing our analysis within discourses of 'reflexive modernization' (RM), a concept which entered the social sciences lexicon through Ulrich Beck's seminal *Risk Society* (1992) and has found great resonance in the work of Anthony Giddens (1989, 1991). The Risk Society perspective was supplemented with the development of what became known collectively as EM, which maintained two core premises of RM: (1) individual and institutional choices are not simple reflections of the dynamics resulting from the dominant structures of capitalism and industrialization and so forth; (2) the resolution of environmental problems would not emanate from the 'de-modernization' or 'anti-modernization' advocated by the most radical sectors of environmentalism but from a gradual modernization of societies (see Buttel 2000a, 2000b, 2003).

The EM perspective owes its emergence to the developments taking place in a number of advanced democracies in Western Europe, which from the early 1980s began to show signs of factoring the environmental problematic into their economic and developmental policies. In particular, in the Netherlands and West Germany there was an evident increase in environmental legislation and a de facto placement of environmental issues at the top of the political agenda. At the same time, one could observe a tendency for the delinking of material from economic flows, thus challenging the prevalent assumption of an intimate relationship between economic growth and environmental deterioration. In contrast, EM discourses posit that environmental reform has played a crucial role in the reduction of polluting emissions and the use of natural resources irrespective of economic or material growth (Mol 1995, 1996, Sonnenfeld 1998). It is perhaps not a coincidence that the initial conceptualization of EM stems from the aforementioned countries, which have a strong environmental movement with significant procedural capacity (such as access to decision-making bodies; see Van der Heijden 1999). EM is, in many respects, characterized by ambiguity of outlook; for instance, it is not clear whether it is prescriptive or analytical in its intentions. Nevertheless, it has found great affinity with sustainable development programmes as promoted through LA21 in the 1992 UN Summit on Environment and Development in Rio de Janeiro and has become one of the dominant discourses in political debates over environmental issues (Hajer 1995: 4), particularly with regards to the institutionalization of environmental transformation. Underlying EM is the assumption that, as Hajer put it in an often quoted phrase, the 'hardware can be kept but the software should be changed' (1996: 252): in other words, existing institutions are able to learn and to adapt, to integrate ecological rationality into social decision-making. There is thus no need for systemic or structural political or economic change.

The Netherlands and Germany still continue to dominate the discussion on EM to such an extent that a well-known environmental sociologist has argued 'that little is said about the social and political barriers that are likely to be faced in trying to implement [EM] strategies' (Hannigan 2006: 26) in countries where the environment is not a major priority. Indeed, the successful delinking of economic growth and environmental protection outside this context can only be appraised through the examination of multiple variables, such as the eight essential social institutions singled out by Mol (1995) for the study of ecological restructuring in Western Europe and adopted by Frijns et al. (2000: 258) to examine its potential in non-Western European countries. Among those institutional characteristics we find:

1. a legitimate and interventionist state with an advanced and differentiated socio-environmental infrastructure;
2. widespread environmental consciousness and well-organized NGOs;
3. experience with and tradition in negotiated policy-making and regulatory negotiations.

Furthermore, according to Cohen (1998: 149, 2000) the enactment of policies consistent with EM is 'likely to depend on, among other factors, cultural capabilities pertaining to societal commitment to science and environmental consciousness'. This cultural dimension partly fits with the second hypothesis put forward by Andersen (2002) and will play a key in our examination. First, however, we summarize the challenges that may militate against attempts to stage an environmentally conscious Games, before investigating the place of environmental themes in the bidding process.

The environment in the bidding process

Eight years before an Olympiad, the IOC publishes a manual for candidate cities (MCC) to inform their bids for hosting the Games. The MCC dedicates a section to environmental matters, outlining the commitment to environmental protection by the IOC and guiding the candidate cities on the policies they have to employ to achieve a positive bid evaluation. The environment section of the MCC is very compact and does not contain more than three pages; but in terms of the guarantees that it requests from a prospective host, it is fairly demanding. The MCCs consulted by the two host cities examined here, as well as those that guided Beijing 2008 and London 2012, differ slightly in their wording, ranging from the fairly brief environmental section in the 1992 MCC for the 2000 Games to the more developed environment and meteorology section in the 2004 MCC for the 2012 Games. For instance, in the 1992 MCC, the introduction of the environmental protection theme starts as follows: 'The Olympic Movement endeavours to contribute to the protection of the environment and the IOC is anxious that the Games should be an exemplary event in this connection' (IOC 1992: 28), while in subsequent MCCs (IOC 1996, 2002, 2004) this opening statement has changed to: 'The Olympic Movement is fully committed to sustainable development and endeavours to contribute to the protection of the natural environment' (IOC 1996: 44). This change of wording is a direct result of international developments, and the entry of sustainable development in environmental protection discourse after the 1992 Rio Summit in particular. Of course, the adoption of this new terminology by the IOC should be seen as a continuation of the IOC strategy of integrating environmental protection into financially sound Olympics, rather than as the expression of something groundbreaking and novel. After all, its initial commitment to environmental protection was never meant as a subscription to 'dark' green concerns (see Dobson 1995). However, sustainable development and earlier environmental standards set by IOC are clearly substantial precedents for the transformation of host country planning along EM lines.

Indeed, all candidate cities are required to

1. map local environmental and natural resource systems used by relevant authorities with emphasis on their interaction with OCOGs;
2. provide 'an official guarantee from the competent authorities, stating that all work necessary for the organisation of the Games will comply with local, regional, national regulations and acts and international agreements and protocols regarding planning and construction and the protection of the environment' (IOC 1996: 45; but repeated, with small differences, in all other MCCs);
3. carry out EIAs (Environmental Impact Assessments) for all venues;
4. describe the OCOG's planned environmental management system (including, possible collaboration with ENGOs and/or their reaction to the Games);
5. describe the application of environmentally friendly technology relating to the Games;
6. draw up state plans for minimizing the environmental impact of infra-structural projects relating to the Games (e.g. road expansion);
7. demonstrate how plans for waste management (including sewage treat-ment) are expected to 'influence the city and region in the future' (IOC 1996: 46; but repeated with slightly different wording in all other MCCs);
8. explain how 'will the OCOG integrate its environmental approach into contracts with suppliers and sponsors, for example, with respect to pro-curement of recyclable or compostable goods, in recyclable or composta-ble packaging' (IOC 2004: 88; the most explicit statement on this issue by the IOC when compared to earlier MCCs); and
9. outline plans for raising environmental awareness.

We can clearly see here the IOC's emphasis on the environmental impact and legacy of the Games in terms of the principles of the EM perspective. The preparation for the Games by a host nation requires the successful coor-dination of various state bodies, and consultation and collaboration with civic organizations and significant restructuring, what Rutheiser (1996) has called 'imagineering'. For the host nation, one would therefore think that the planning and ambitions for the Games goes beyond the successful prepa-ration of a one-off event and has a long-term impact on the realms of policy, decision-making, organization, scientific consultation and the use of new technology; or in other words, on its capacity for EM (see Weidner 2002). We can see a process akin to *engrenage* here, in similar terms conceived for policy-making in the nascent European Community by Jean Monnet, in that the process of meeting the IOC's environmental standards could both drag with it the host nation's institutional framework and set a precedent that other nations would strive to emulate.

Environmental protection and performance in the pre-event and event phases

Both the Sydney and Athens bids make extensive reference to their existing environmental capacity, as well as to plans for further developments which have the potential to increase the environmental capacity of the host nations through Games planning. Chalkley and Essex (1999b: 301) point out that, for Sydney, 'the Olympic plans give an increased priority to ecological sustainability and issues such as energy conservation, bio-diversity and the need to conserve natural resources. Indeed the Sydney Organising Committee has published a detailed set of 'green' guidelines which are intended to govern the design, layout and the construction of the Olympic facilities.' However, if one compares the bid submitted by SOCOG (1993) to the bid submitted by ATHOC (1996), one is soon confronted by a very different use of language. Whereas in the former case, we are presented with an ambition that seems to have been programmed anyway, in the latter case we are confronted with a rather unqualified statement of intention. For instance the statement put forward in the Sydney bid on the use of new environmentally friendly technologies highlights the existing capacity of Australia on this issue in the following way:

New technologies which help protect the environment are being applied to developments in Sydney for the Olympics.

Many of these techniques have been developed in Australia, which is recognized as a world leader in research on the application of new technology to environmental enhancement.

Within New South Wales, waste and water treatment technologies, such as the use of special membranes, are improving the quality of affluent discharge in various applications.

New ocean downfall systems are significantly improving water quality at urban beaches; water supply treatment technologies are considerably upgrading the quality of municipal water for domestic and commercial use.

Australia has led in many areas of solar energy use, and in designing to improve energy efficiency. A prime example of the latter is the flexibility designed into the air-conditioning system for Sydney International Aquatic Centre. The system can create a comfort zone for competitors and spectators. The air-conditioning can run in particular zones of the building when required, thus saving energy.

Technologies are applied, as discussed above, to significantly reduce vehicle pollution and introduce more energy-efficient fuel sources.

ATHOC, in contrast, ambitiously proclaimed that 'the environment will not only be protected: it will be improved' (ATHOC 1996: 52). In terms of using appropriate technology, the Athens bid promised thus:

The projects planned will be incorporated into, and adapted to, general government policy on the protection of the environment and the remodelling of the area.

Completion of the projects will be carried out with the use of environmentally-friendly technologies and materials. Contractor companies will commit themselves to the use of such technologies and materials, for which they will have to submit detailed schedules and lists.

Uses will be planned for all the permanent installations once the Games are over.

The key environmental drive of the Athens bid is thus less one of existing capacity than one of cultural change, transformation and legacy. According to Kazantzopoulos (the environmental manager of the Athens Games), ATHOC, in conjunction with the relevant state institutions, was adopting a pioneering approach towards the environmental planning of the Olympic projects in the pre-event phase. Following the adoption of a relevant legal framework (Law 2730/99), a strategic Environmental Impact Assessment of all Olympic-related projects was attempted. Furthermore, environmental planning was incorporated into this new legal framework, such that 'notwithstanding a few points which might have been improved, this is an initiative which has not so far been visible in other development programmes' (Kazantzopoulos 2002: 111).

In what can be seen as a de facto commitment to environmental sustainability, Kazantzopoulos continued by setting out the various ways that the environmental dimension had been incorporated into different Olympic construction projects, ranging from the use of environmentally friendly materials to the bioclimatic planning of Olympic facilities. In this, Athens very much appeared to be following the Lillehammer and Sydney examples. 'All these', according to Kazantzopoulos (2002: 112), 'are seen through a bigger objective that [...] was set by ATHOC: the creation of a sustainable environmental heritage'. Of course, one may retrospectively see this as more of an attempt to reassure the IOC, which had already warned Greece back in 2000, after a number of delays had became apparent, that 'they might lose the Games if action was not forthcoming' (Gold 2007: 271), than as a realistic forecasting of what lay ahead.

Kazantzopoulos continued: 'it is certain that up to the time of the Games [...], for the construction projects as much as for various aspects of the organisation and preparation of the Games, the environmental character of all of them will be maximized. The key question is to what extent this initiative will continue, or whether we will experience a significant shortfall' (Kazantzopoulos 2002: 112). Indeed, the final plan for the Games appeared to follow the logic of post-Games use. And yet, post-Games analysis has told a different story: 'In reality, however, there was no proper strategic planning for the period after 2004, with the plan containing apparent contradictions. Despite espousing the desire to protect and create open spaces, development focused primarily on greenfield sites and overlooked possible brownfield locations' (Gold 2007: 272). Moreover, development of decontaminated areas is only now (2010) beginning to take some shape in the form of planning for an Athens metropolitan park, instead of the further housing that was promoted by the 2004–9 ND administration and resisted by environmental groups.

It is not only Athens that received substantial post-Games criticism on its environmental record, social exclusion and gentrification. In Sydney, '[t]he promotion of a Green Games came not without controversy, as local activists denounced what they saw as an opportunistic top-down strategy that failed fundamental environmental standards but succeeded in creating a perception of success thanks to the "questionable" or bought-in support of media partners' (McManus as cited in Garcia 2007: 254; see also Hall & Hodges 1996, Hall 2001), whilst a number of commentators, even before the Games, contrasted the vast sums spent with the failure of the Australian government to commit to the Kyoto process. For a leading Friends of the Earth figure, 'the huge sums of money involved could be used, perhaps, not on a two-week athletics festival but on dealing with really pressing environmental priorities, such as a national programme to reduce CO_2 emissions or the dumping of sewage to coastal waters' (in Chalkley & Essex 1999b: 306).

As has been already indicated, the environmental performance of the Athens Olympics was slight in comparison to the high standards set by Sydney 2000. The good intentions, as seen in the candidature file, were signalled from the start and ATHOC appeared to be drawing lessons from the experience of Sydney. However, those good intentions pale in comparison to their lack of materialization. Characteristically, in a list of eight lessons learned from Sydney 2000, Greenpeace scored Greece poorly in all of them. In relation to Greece's environmental promises in its candidacy file, the outcome is predominantly poor, albeit with some minimal but noticeable areas of long overdue improvements in the urban environment (the extension of the metro network, the construction of tramway and suburban rail lines etc.) and the use of EIAs in all related projects (see Greenpeace 2004a, 2004b). In line with these highly critical Greenpeace reports, the WWF also issued an assessment report on the Athens Olympics based on the benchmarks established by Sydney 2000 (WWF-Greece 2004). On a scale of 0–4, Athens was rated at 0.77 by WWF. As with Greenpeace's evaluation, the highest scores were for developments in public transport, a range of infrastructural improvements and the promotion of environmental awareness.

Overall, for both ENGOs, the environmental plans for Athens 2004 closely followed those of Sydney. As such, although ATHOC did not take advantage of the critical points raised over Sydney 2000 by ENGOs to complement its own environmental action plans, they were still good plans. For both ENGOs, the implementation of these plans was inhibited by the political process.

Ecological modernization in the post-event phase

From our reading of the key EM literature (e.g. Mol & Spaargaren 2000; Buttel 2003), we have selected six indicators for identifying and testing changes in environmental-event phases. These are:

(1) the average annual level of CO_2 emissions;
(2) the level of environmental consciousness;
(3) the ratification of international agreements;
(4) the designation of sites for protection;
(5) the implementation of EIA procedures; and
(6) ENGO participation in public decision-making processes.

Given the overwhelmingly negative environmental assessment of Athens 2004, it may be reasonably argued that our original question on the EM capacity of an Olympic host nation should be abandoned in the Greek case. After all, the expectation of identifying 'marked signs of environmental improvement', as formulated in our first hypothesis, cannot be reasonably attributed to the successful application of environmental sustainability principles in hosting the Games. Nevertheless, it could be argued that the promotion of our fifth indicator, environmental awareness (one of the few environmental success stories of Athens 2004), has been paramount in facilitating environmental improvements in the post-event phase, and for that reason we have cautiously proceeded in our examination of the Greek case. We also remain cautious – despite the typical claims of event organizers – in deducing causality, between event hosting and developing EM capacity. A range of key intervening variables, such as economic growth, the environmental policies of the political party in government and possible cultural inhibitors to modernization, also need to be accounted for.

(1) Average annual CO_2 emissions

Australia's CO_2 emissions are among the highest in the world. In 2006, total emissions were 26.9 per cent higher than 1990 levels, the baseline year for the Kyoto protocol, and 13.05 per cent higher than 2000 levels (see Figure 8.1); per capita emissions stood at 19.00t CO_2e in 2006, an increase from 17.36t in 2002, and the tenth highest in the world according to UN calculations (ABS 2006a; UNSD 2009). Among the factors accounting for these high emissions levels are the high use of coal in electricity generation, the high dependence on car travel for urban transport and the aluminium smelting sector (ABS 2006b: 582). Coal is mined in every Australian state and territory, providing 85 per cent of the country's electricity production, and making Australia the world's leading coal exporter – a factor continuously raised during John Howard's Liberal-National coalition governments (1996–2007), which consistently rejected calls to reduce the mining and burning of coal for energy use as an attack on the country's economic interests (Christoff 2005, Curran 2009, Doyle 2010). This intransigent attitude was to some extent responsible for Howard's defeat by Labour (Australian Labour Party; ALP) in November 2007 (see Rootes 2008), though Kevin Rudd's subsequent administration became mired in internal battles over the introduction of carbon pricing and trading. With Ross Garnaut, the Australian government's climate change adviser, referring to the problem of climate change as 'a diabolical challenge' and stipulating that 'nations have a collective imperative to adopt their own policies to cut carbon emissions' (Garnaut as cited in Pietsch & McAllister 2010: 217) it seems that Australia remains committed to adopting 'a stronger version of EM, but resistance has continued with industry pressures on the government [...] to weaken its commitment to climate change policy' (Pietsch & McAllister 2010: 220).

In the Greek case, according to UNSD data, in 2006 Greece's per capita CO_2 emissions stood at 9.90t, with overall emissions having increased by 33.1 per cent on 1990 levels (UNSD 2009). According to European Environment Agency (EEA) data, Greece recorded one of the largest absolute increases in GHG emissions between 1990 and 2006 in EU15 of 24.4 per cent, set against

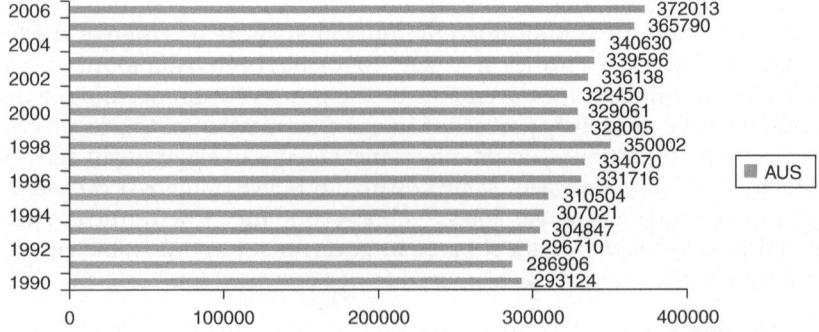

Figure 8.1 Australia's CO_2 emissions (kt)
Source: UNSD (2009)

both an overall emissions decrease of 2.7 per cent by EU15 over this period, and an overall EU-27 target of a 20 per cent reduction in GHG emissions by 2020 on 1990 levels set by the United Nations Framework Convention on Climate Change, and an aspiration to reduce emissions by 30 per cent on 1990 levels. Per capita emissions (tCO_2e) increased by 15.8 per cent over the same period (as against an EU15 performance of –8.4 per cent). Clearly, this is long-term data, with emissions fuelled by increased transport activity and energy demand, more recently partly counter-balanced by the use of natural gas and hydropower, and is furthermore within Greece's Kyoto allowance of a 25 per cent increase in emissions (EEA 2008: 134–5). Intriguingly, however, there was a rather unexpected 1.19 per cent decrease in CO_2 emissions two years after the Games (see Figure 8.2). This decrease is too small for us to draw any conclusions, especially given the continued dominance of the main contributors to CO_2 emissions in Greece. Moreover, Greece was suspended from the UN's carbon trading scheme for three months in April 2008 as punishment for the inaccurate reporting of carbon emissions. Characteristically, the body responsible for atmospheric measurements, at the time, 'overestimated the emissions for 2004 by 37,000 tonnes' (Psaropoulos 2008).

Like Australia, Greece is highly dependent on lignite (or brown coal), which accounts for 63 per cent of the country's energy generation (Kavouridis et al. 2005: 1); Greece ranks second in Europe and fifth globally as a lignite producer country (Kavouridis et al. 2005: 9). As such, coal production in Greece is also a politically sensitive issue for any government: plans for its reduction would – as in Australia – inevitably lead to accusations of betrayal of the national interest, indifference to resulting job losses and increased pressure on household budgets. Nevertheless, in Greece the new Panhellenic

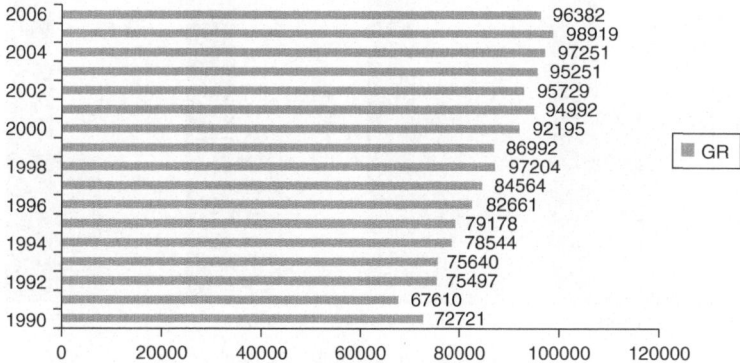

Figure 8.2   Greece's CO_2 emissions (kt)
Source: UNSD (2009)

Socialist Movement (PASOK) government has started a process of closing down extremely polluting lignite power stations with the intention of supplementing energy provision with solar, wind and natural gas.

Overall, it is clear from the available data that CO_2 emissions in both countries have not only not decreased but continued to increase in the post-Games phase, and that Games hosting is at best a relatively minor factor in long-term trends to emissions reductions.

(2) Environmental consciousness levels

According to the ABS survey data, concern for the environment among Australians reached its lowest point four years after the Sydney Olympics, in 2004 (ABS 2006a). Nevertheless, when asked about the importance of the environment as an electoral issue, the Australian public has continuously rated the environment as, at least, a quite important issue. Characteristically, in the election year of 2007, 95.3 per cent rated the environment as a quite or extremely important issue (see Figure 8.3). Furthermore, in survey data linking climate change to the result of the 2007 general election, more than 70 per cent considered climate change to be a 'very serious' problem and over 60 per cent expressed dissatisfaction with the Howard government's

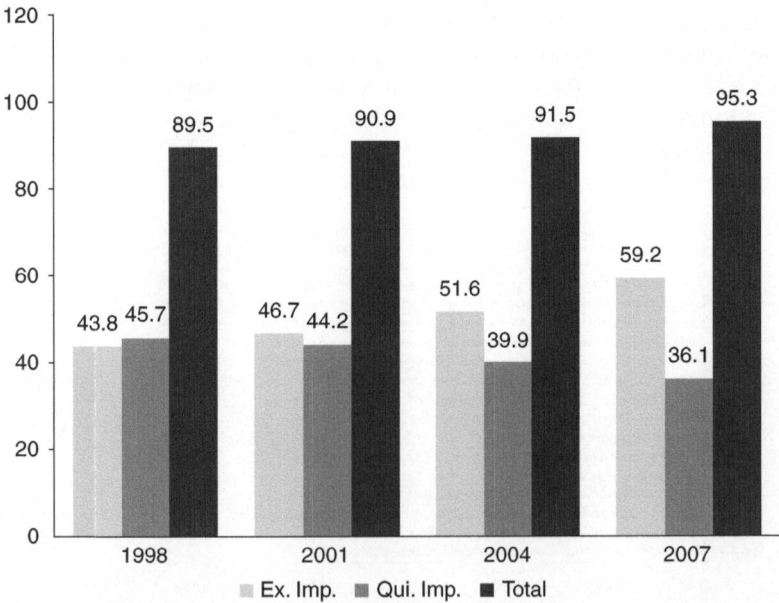

Figure 8.3   Importance of the environment in the 2007 Australian General Election
Source: ASSDA (2009)

response to the issue (Rootes 2008: 473–4). Nevertheless, as Rootes (2008: 479–80) argues, whilst climate change was indeed a very important issue, there is no evidence that it was the decisive issue explaining the election result. Rudd's subsequent loss of public support over inaction on carbon trading is borne out by more recent Newspoll opinion polling data, which demonstrates the increase in the acceptance of the climate change message by the general public, as 'a significant proportion (over 60 per cent) [supported] the introduction of an emissions trading scheme (EMT) by 2010, regardless of how the remainder of the international community, including developing countries, decide to proceed' (Curran 2009: 211). In a poll conducted by the Australian National University (ANU) in September 2008, 58 per cent of Australians were clearly in favour of an ETS and evidence gathered by Pietsch and McAllister (2010: 225) 'suggest that the public is reasonably well informed about the ETS, and able to form opinions accordingly'.

A range of factors explain the environmental concern exhibited by the Australian public before and soon after the 2007 elections as well as their acceptance of the ETS, ranging from the unprecedented high temperatures and droughts that preceded the elections to the (at that time) successful engagement of the public by Rudd on climate change (see Rootes 2008, Curran 2009, Pietsch & McAllister 2010). Again, it is difficult to detect in this issue any direct causal link with Games hosting.

According to successive Eurobarometer surveys, the Greeks continuously exhibited the highest professed levels of environmental concern in the EU for most of the 1990s (see Karamichas 2003), and have continued to do so during the first decade of the 2000s. For instance, the 2009 Eurobarometer (72.1) on European attitudes towards climate change has shown that climate change is the highest issue of concern for the Greeks with 71 per cent seeing it as the 'most serious issue currently facing the world as a whole' after 'poverty, lack of food and drinking water' (European Commission 2009). However, we have some reason to doubt the sincerity of the high levels of professed environmental concern exhibited by the Greek public (along with their southern European counterparts) during the 1990s, based on their continuous perpetration of environmentally harmful practices and their noted lack of knowledge over the cause and effect behind the global issues they were professing their concern for (see Karamichas 2003, 2007).

Nevertheless, there is good reason to believe that the Greek public's concern on climate change is much more sincere and better informed during the first decade of the 2000s than in the 1990s. It is likely that this can be attributed to international factors, such as the promotion of the role that human activity has on climate change since the 2007 Nobel peace prize was shared by the IPCC and former US vice-president Al Gore, and to national factors such as the extremely devastating forest fires of summer 2007, rather than the staging of the 2004 Olympics and the promotion of environmental awareness associated with them.

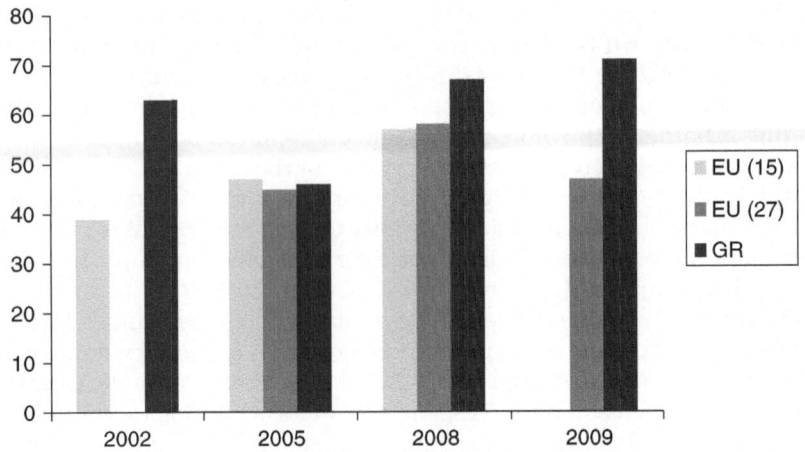

Figure 8.4   Expressed concern about climate change: EU and Greece
Source: European Commission (2009)

(3)   International treaty ratification

From as early as the 1970s, Australia was a signatory to a number of international conventions related to the environment, which had an immense impact on 'the powers of the states to manage and regulate the environments' with the establishment of many departments responsible for the environment (Papadakis 2002: 25). During this period, the willingness of the ALP to promote environmental concerns clearly stood out (Papadakis 2002: 24). Yet, during successive Howard governments, Australia was (alongside the US) a key country not to ratify the Kyoto protocol, despite negotiating an agreement permitting it to raise its GHG emissions by 8 per cent on 1990 levels by 2010. The rationale of Howard administrations was that meeting Kyoto targets was a potential impediment to Australia's capacity to compete in the global market, pointing to the protection of the country's competitive advantage in energy as a net coal exporter, job losses and a reduction of GDP on the one hand, and the small contribution by Australia to global emissions, uncertainty surrounding the human contribution to climate change, and the lack of binding conditions on developing nations on the other (Balkeley 2001, Christoff 2005, Curran 2009). Australia signed the protocol in December 2007.

Greece was similarly slow to ratify the Kyoto protocol, becoming the last EU15 member state to do so in 2002, within 24 hours of a 31 May deadline (Psaropoulos 2008). Its agreement allowing a 25 per cent increase in emissions levels on the 1990 baseline, as Kyoto sought a 12 per cent reduction by 2010, was secured because Greece was seen as a developing country; the country's capacity to meet its targets are heavily circumscribed by a voracious developmental process with many polluting emissions (Elafros 2007). Indeed, students of comparative European environmental policy may well not be surprised by the precarious nature of Greece's ability to meet the requirements of an agreement to which it has been a signatory. Greece, like its south European counterparts, has been notorious for its failure in applying EU environmental directives.

Two perspectives have dominated the discussion on this lack of compliance by southern European EU member states such as Greece: the 'Mediterranean Syndrome' (La Spina & Sciortino 1993) and the 'Southern Problem' (Pridham & Cini 1994). Both approaches explain non-compliance as a function of socio-historical factors, such as (1) the lack of a civic culture, which would enable the development of cooperative behaviour among the citizenry, and mistrust towards state authorities, which inhibits compliance with the law; (2) ineffective and inefficient administrative structures, often characterized by severe lack of technical expertise, which are intimately related to the permeation of patronage and clientelistic tendencies in the recruitment of officials; and (3) a party-dominated reactive legislative process that impedes regulatory continuity. Counter-arguments to these perspectives stress that as relatively late entrants into the European Community, southern European member states have had to import policy and legal frameworks and adapt to the regulatory standards and administrative structures of northern European states, and thus face increasing implementation and compliance costs. Moreover, non-compliance can only be averted if the European Commission opens infringement proceedings – often after campaigning by domestic environmental movement organizations – thus raising the financial and reputational costs of non-compliance beyond the costs of adaptation and implementation (Börzel 2003, Koutalakis 2004).

Greece's record of non-compliance with EU environmental directives continued in the post-Games phase, under George Souflias's environmentally destructive management of YPEHODE (the Ministry for the Environment, Planning and Public Works) during the 2004–9 tenure of the conservative New Democracy government of Kostas Karamanlis. According to Psaropoulos (2008), '[u]nder Souflias [...] Greece has taken spectacular steps backwards, even as Europe has moved forward. Souflias has never attended a European council of environment ministers. He was not only absent from December's round of UN climate talks in Bali, but failed to organise a representation under the deputy minister for the environment.' Indeed, Souflias's evident disdain for the environment exasperated Stavros Dimas, a fellow New Democracy politician, during the latter's term as European Commissioner for the environment (also 2004–9).

In neither of our cases, therefore, did hosting the Games produce a visible effect on governmental willingness to ratify international treaties or agreements, despite the apparent scope to do so.

(4)   Protected natural site designation

Our discussion of the protection of the natural environment is restricted here to measures on forestry protection, as presented by secondary literature and official publications, as an obvious point of comparison between Greece and Australia. Both countries have a significant number of designated areas of natural protection (see Papadakis 2003; Papageorgiou & Vogiatzakis 2006). Australia 'has one of the largest and greatest national park systems in the world, covering over 24 million hectares, with such diversity as lush rain forest to waterless desert' (ANP 2010), whilst Greece has 359 sites incorporated in the Natura 2000 European Ecological Network, and as a signatory to the EU Habitats directive, a number of obligations in relation to the conservation of the habitats and species which occupy these sites (Dimopoulos et al. 2006: 175). Nevertheless, both countries have been beset by the annual occurrence of destructive wildfires, most notably in February 2009 in southern Australia, when 160 people lost their lives, and the immense destruction of the 2007 forest fires in Greece (Karamichas 2007). For the Australian Greens, the fires were a 'sobering reminder for this nation and the whole world to act and put at a priority the need to tackle climate change' (Walsh 2009).

At this point a number of commonalities with the Greek case are evident. The annual occurrence of forest fires and the accompanying practices of illegal conversion of land designated for reforestation to building land with the complicity of the authorities, in Greece, and the building of homes in fire danger zones in southern Australia (Walsh 2009) prohibit us from awarding a high post-Olympic score to either country on this variable. It is worth noting that, at the time of writing, Greece still does not have an official forest and land register, which presumably sustains the aforementioned negative practice in a complex web of political patronage and clientelism (Karamichas 2007). Nevertheless, it is worth noting that one of the first initiatives put in action by the new ministry of Environment, Energy and Climate Change (YPEKA) in December 2009 was a new legal framework for the rejuvenation of the burned forest lands of the Greater Athens area that prohibits any construction in these areas before the composition and validation of a forest land register, finally passed by parliament in January 2010.

If we accept the argument put forward by the Australian Greens that climatic change was the main cause behind the southern Australia fires, could it be argued that if the Howard government had ratified the Kyoto protocol maybe the preparation for such a possibility would have been more effective? Can we possibly claim that the forest fires of 2007 in Greece would have been averted under a different administration, one more receptive to the environmental problematic? Speculation aside, in both countries there were advanced warnings about the possible occurrence of those fires, and fires are an annual event. Olympic hosting does not appear to have had any positive impact in relation to them.

(5) Environmental Impact Assessments

As early as in the 1960s, there was a clear attempt to integrate environmental concerns into economic objectives in Australia with the introduction of the Environmental Impact Statement (EIS), but which evidently fell short of achieving its ascribed objectives (Papadakis 2002: 24). It was much later, during the mid-1990s, that the 'National Approach to Environmental Impact Assessment in Australia' was adopted in an attempt to eliminate the prevalent duplication of EIA procedures among different government jurisdictions (state government and the commonwealth government) and improve consistency (Elliot & Thomas 2009: 123). This approach was very much in action during the planning stages for Sydney 2000. A key aspect of importance here is the identification of 'environmental significance [...] as a key factor in determining the need for EIA' (Elliot & Thomas 2009: 124). Of course, the degree of significance can have many interpretations, and the use of EIAs for the Sydney Olympic projects is a quintessential example of this: a range of Olympic-related projects were authorized without being subjected to the full EIA process, which was considered to impose unnecessary delays on a development of national importance (see Hall 2001, Hall & Hodges 1996).

Whilst a number of studies, published five years or more after Sydney 2000, have identified Australia to be an 'international leader in assessing health impact within EIA' (Harris et al. 2009: 311), studies at state level, such as the qualitative descriptive analysis conducted by Harris et al. in relation to 22 Major Project EIAs in New South Wales (the Olympic state), identified a number of 'deficiencies in practice and legislative provisions across states and territories' (2009: 311), attributed to constraints emanating from pressures for project approval, and land-use planning processes, among others. In effect, little has changed in the post-Olympic phase on this issue, because of the inherent ambiguity of the EIA process.

The Australian case nonetheless remains more positive in this respect than the Greek one. In Greece, we are faced with, in addition to problems inherent in EIAs, the continuation of blatant inefficiencies due to nepotistic practices and a lack of resources. The stringent application of EIAs for Olympic projects had little impact, and no attempt was made to overhaul existing poor administrative practices. A study evaluating the quality of EI (environmental impact) studies produced in Greece between 1993 and 2003 found they had performed rather poorly with respect to most indicators, and that there was little evidence of improvements (Androulidakis & Karakassis 2006). A post-Games study based on findings from nation-wide research reached similar conclusions and pointed to the lack of specialized staff and material shortages in many local authorities that severely compromise the effectiveness of EIAs (Botetzagias 2008).

(6)   ENGO participation

In the mid-1980s, the most dominant and mainstream Australian ENGOs were incorporated into the decision policy-making processes of the Labour government of the time. This process continued in the early 1990s, but was interrupted with the election of the Howard government in 1996, that is three years after the Games were awarded to Sydney (see Doyle 2005, 2010). According to Doyle (2005: 72), this period saw the state 'increasingly remove itself from its role as environmental legislator, monitor, and regulator and again [like the 1960s to mid-1980s period] set itself up in active opposition to environmental concerns'. Post-Howard, there has been an evident reincorporation of ENGOs into policy consultation, but the ALP's apparent dependence on certain economic interests (mining, for example) has already caused significant strain on this relationship (Curran 2009). Moreover, if the legacy of ENGO participation in the preparation for Sydney 2000 is to go by, there is no reason to dispute that this process might also be characterized by immense selectivity in its inclusiveness/exclusiveness of participant organizations.

In Greece, there seemed be, at least initially, an opening to the environmental movement by the 2000–4 PASOK government through the appointment of a number of environmental activists to government and/or advisory positions. As a case in point, the former director of Greenpeace-Greece Elias Efthymiopoulos was appointed as junior YPEHODE minister; Efthymiopoulos projected the concept of EM when questioned about the compatibility of environmentalism with the modernizing project that the government was putting into action (see Karamichas 2008). In hindsight, it is obvious that that collaboration did not produce the expected results on the environmental front, and the proclaimed modernization project of the 2000–4 administration produced only limited results, if any. Under the tenure of the ND government (2004–9), ENGOs had a partially institutionalized status through participation in steering committees, though this was 'largely undefined and based on informal interaction' (Papadopoulos & Liarikos 2007: 302). However, as we have already highlighted, the environment was notoriously further downgraded during that period (see earlier discussion on Souflias's running of YPEHODE). Under the current PASOK government, with its highly ambitious green plans, it appears that there is a de facto openness of YPEKA in collaborating with ENGOs and accepting their input into policy-making.

Discussion

Pace Andersen, we started with two contrasting hypotheses: first, that in the post-event phase, we should be able to identify environmental improvement in host nations by testing a number of clearly identified indicators, and second, that the prevalent structural conditions in the host territory will be the central variable for the capacity of the event to achieve cultural change. Our analysis suggests that for the majority of the key EM indicators examined here (CO_2 emissions, ratification of international treaties, protection of the natural environment, EIAs and participation of ENGOs in public decision-making), not only has there been no noticeable post-event 'environmental improvement' in either Australia or Greece, but there has been an evident continuation of practices leading to a progressive worsening of the environmental status of both nations.

The increase in CO_2 emissions in both countries can be partly explained by their dependency on the extraction and burning of lignite to satisfy their electricity generation needs. Throughout our examination, we have encountered the recurring theme of the intransigence of the Howard government in refusing to ratify the Kyoto protocol, and in the Greek case, a continuous lack of compliance with EU environmental directives. In both countries, annual forest fires have had catastrophic consequences. Although EIAs were used in both countries for Olympic-related projects, the inherent ambiguity of the process – its general tendency to privilege project completion – coupled with perennial structural deficiencies in the Greek case suggest that the use of EIAs has been unable to make made an unequivocal contribution to environmental sustainability in either country. In both cases, there was a remarkable exclusion of ENGOs from the policy-making process in the period after the Olympics. The only EM indicator demonstrating an unambiguously positive result was 'environmental consciousness'; however, in both cases, this can be at least partly ascribed to the general global increase in concern about climate change, coupled with the aforementioned poor national responses to the problem's immediate perceived symptoms (droughts, forest fires, etc.). In the Greek case, public environmental consciousness can be seen as a continuation of trends already visible in the 1990s, but in the post-Olympics phase there is also greater understanding and awareness of the problem.

This outcome can largely be explained by the rationale underpinning our second hypothesis, concerning the importance of the 'character of the political and socio-economic reform process'. Indeed, in both cases we have seen how power-holders have resisted the ecologization of economic processes, due either to the adoption of a specific rationale subscribing to dominant perceptions around development and growth as promoted by a range of powerful interests (such as the mining industry in Australia), or to the influence of a range of cultural negativities, which not only are not conducive to the modernizing process but are extremely incongruent to the EM project (Greece). It is interesting to note, nonetheless, that electoral change in both countries has produced a much more favourable political climate to EM. In both cases, the win-win dynamic of EM has been heavily promoted by the resultant left administrations, and departments/ministries specializing in climate change have been established.

Although, the popularization of EM as a theoretical and normative paradigm can be traced back to case studies conducted by Mol (1995) on the Dutch chemical industry and by Sonnenfeld (1998) on Thailand's paper pulp manufacturing, a range of comparative studies have demonstrated that the successful incorporation of environmental factoring in one particular sector of a national economy does not by any means mean that it can be generalized as indicative of the general practices characterizing the whole of that national economy (see York & Rosa 2003). Similarly, a successful bid to host the Olympic Games may affirm that the environmental standards prescribed by the IOC will be met, but it means that neither these standards will be implemented as prescribed nor the successful implementation of environmental standards for the projects associated with the Games will inevitably lead to a general ecologization of the national economy. Through our exploration of the six identified EM indicators, it is fairly obvious that these institutional characteristics were either severely undermined by the existing political administration (the devaluation of the environmental dimension under the Howard and ND administrations in Australia and Greece) and/or were severely underdeveloped and deficient because of the perpetuation of severe structural deficiencies (e.g. the continuation of nepotism and clientelism in Greece has severely undermined the necessary public legitimacy of the state apparatus to adequately develop its EM capacity).

Concluding remarks

By questioning the potential of the Games to facilitate the EM capacity of the host nation, we have essentially put to the test the proclamations of and aspirations to sustainable development made by the IOC and actual or prospective event hosts. To do this, we have carried out a comparative investigation of the cases of Sydney 2000 and Athens 2004, selecting these cases because of the availability of data in the post-event phase and because of the diametrically opposed green legacy they occupy in Olympic historiography, with Sydney 2000 having an established reputation as hosting the first green Olympics and Athens 2004 widely considered a failure in these terms. Our analysis has enabled us to evaluate the role played by national and local structures in the facilitation of post-Olympic EM capacity; our analysis suggests that neither of the two host nations has fared well in these terms, irrespective of the vast differences that characterize them.

In effect, for neither country can we claim that hosting the Olympics has produced unequivocal cultural change, turned them towards advocacy of ecological modernization, or a consequential adoption of strategies aiming at the delinking of material flows from economic flows. As we have seen, for the best part of the decade following the Olympics, Australia continued to trade off industrial interests against environmental ones under the guise of the national interest; and only after the election of an ALP government in 2007 did Australia ratify the Kyoto protocol. And whilst Greece has certainly developed its environmental legislation, we can detect causality here in its obligations as an EU member state. In any case, the main problem that Greece faces here is one of implementation rather than superficially expressed intention. Greece remains dominated by characteristics that are not conducive to modernization, let alone its ecological dimension; in order to achieve this, changes in the 'software' may have to be so encompassing that they have a serious potential to be seen as changes in the 'hardware'. In effect, this means that a 'strong EM' that is more prepared than the more popular 'weak EM' to confront the 'structural impediments that contribute to environmental risk' (Curran 2009: 204).

Our analysis thus points to a delinking of the idea of a 'green Olympics' from wider considerations of environmental rationality, and the reduction of the former to the more particularistic notion of the environmental management of the local infrastructure. There is very limited evidence pointing to a facilitation of the ecological modernization of the host nation. For sure, our intention here has not been to question the environmental measures implemented for each Olympics, still less to challenge the 'first green Olympics' title awarded to Sydney. Clearly, some important infrastructural changes took place in both cities, though in most cases these were long overdue commitments. Rather, we have sought to contextualize the wider value of claims by mega-event organizers, local, national and transnational, that hosting the event justifies the immense public investment because of its long-term transformative capacity, cast as 'legacy'. The question is not academic: ambitious environmental claims have become a staple of Games since Athens, with the bid committee for the Beijing 2008 Games claiming that it would 'leave the greatest Olympic Games environmental legacy ever' (UNEP 2007: 26), while the London 2012 promotes the concept of the 'One Planet Olympics'. It is possible, of course, that the Beijing and London experiences will provide a firmer base for the realization of Olympic environmental ambitions than Athens or even Sydney; future long-term work will be key here.

9
What Happens When Olympic Bids Fail? Sustainable Development and Paris 2012

Graeme Hayes

On 6 July 2005, the International Olympic Committee (IOC), meeting at the Raffles Hotel in Singapore, announced which of the five shortlisted cities would host the 30th Olympic and 14th Paralympic Summer Games in 2012. Moscow, New York and Madrid had already been eliminated in the first three rounds of voting, leaving the final choice between Paris and London. The Paris delegation – led by Bertrand Delanoë, the city's Socialist mayor; Jean-Paul Huchon, the Socialist president of the Ile de France region; and Arnaud Lagardère, chief executive of the Lagardère media and aeronautics group, and chief representative of the Club des Entreprises Paris 2012 business consortium – appeared confident, if nervous. Previous recent bids by Paris to host the Games for the first time since 1924 had floundered: in 1986, the city had lost out to Barcelona for the 1992 Games, and in 2001 had come a dispiriting third behind Beijing and Toronto for the 2008 Games. This time, however, would be different; Paris was, by consensus, the favourite to get the nomination. Whilst London emphasized its Games in physical, emotional and economic terms (stressing the role of the Olympics in creating urban regeneration, and sporting and cultural legacies), Paris had run a campaign based on the supposed intersection of French and Olympic values: romance, solidarity, popular engagement, ethics, ecology. Much more compact than its chief rival, the Paris bid had enjoyed comprehensive domestic support, with France's entire political, economic and media class firmly behind it; where he had been 'unavailable' to present the previous Paris bid in 2001, this time French president Jacques Chirac had flown to Singapore, on his way to the G8 summit in Gleneagles, to work the room and share centre stage with Delanoë and Olympic legend Jean-Claude Killy. Rated better than its rivals, recognized by the Olympic evaluation committee as technically excellent, the Paris bid would surely not falter now.

It came as a body blow therefore when IOC president Jacques Rogge opened the envelope and announced that the 2012 Games would be held in...London. As slightly disbelieving but nonetheless vigorous celebrations began in Trafalgar Square, in Paris the rain that had threatened throughout the morning started to fall. The 10,000-strong crowd that had gathered in front of the Hôtel de Ville quickly melted away. In the aftermath, the French press picked over the bones of the bid, struggling to find a reason for its failure. For the conservative *Le Figaro*, the Catholic *La Croix* and the communist *L'Humanité* alike, the result was *une gifle*, a slap in the face for France. Many sought to cast the choice as the fruit of a deeper crisis. For *L'Humanité* (7 July 2005), the reason lay in the pressures of capitalism: at a time of the increasing marketization of human activity and of economic and media concentration, a number of IOC delegates had clearly preferred the unconditional European ally of North American neo-liberalism, reinforcing the unipolar shape of the world. Lagardère, from a rather different political perspective, produced a strikingly similar analysis: London won because of the Anglo-Saxon world's much greater adaptation to the new global economy (*Le Monde* 8 July 2005). The immediate reaction of the bid team itself was somewhat darker. For Delanoë and Huchon, it was highlighting their rival's aggressive promotional campaign and particularly its final lobbying drive, where London had variously abandoned fair play, waged an unethical campaign, privileged commercial imperatives over sporting ones and over-stepped the boundary of acceptable behaviour (*La Croix* 12 July 2005).

The reactions of the Paris bid team reveal much about the importance of cultural specificities as well as of global capital flows to the image strategies of host coalitions. In this context, the goal of this chapter is to turn our attention back to the content of the bid, and its fate in urban, political and, especially, environmental terms. Of all the global sporting mega-events, the Olympic Games is the biggest, the extent of its political importance and influence amply demonstrated by the presence in Singapore of not just Chirac, but Tony Blair, José Luis Zapatero, Hillary Clinton and Mikhaïl Fradkov, presenting or lobbying on behalf of their candidate cities. Not only is the Games a significant international media event, but bidding for the Games has also became a global event in itself, as since Seoul and, particularly, Barcelona, the Games have become synonymous with the international promotion of the host city as a global brand, or what David Harvey (2002) refers to as the accumulation of collective symbolic capital as a powerful differentiation tool in the politics of globally competitive urban entrepreneurialism. Typically, bids are designed to convince the IOC and domestic publics alike of the necessity of hosting the Games for the delivery of desirable urban policy goals, conceived in terms of the creation of post-Games legacies, constructed through urban regeneration schemes and the provision of new infrastructures (transport, housing, cultural democratization). In the Paris case especially, these projected legacies took the form of the development of transport infrastructures, promises of sustainable development and environmental best practice, and new social housing.

These goals are thus clearly consistent with the environmentalization of the IOC since the early 1990s. Yet we should perhaps treat these claims with kid gloves. As Kasimati (2003: 442) points out in her discussion of ex-ante economic studies for Games between 1984 and 2012, bid predictions, including the claims of increased long-term economic growth, tourism and employment, have not been confirmed by ex-post analyses, and should be considered 'potentially biased' given that they were without exception commissioned by bid teams themselves. This chapter therefore aims to address two fundamental questions. First, what kind of development is on offer here? What do claims to sustainability, to environmental staging, mean in concretely? Of course, as this volume has already discussed in depth, due to their corporate nature and their potential environmental impacts, Games are highly problematic, whilst bidding and staging offer a number of political and meta-political opportunities, and pose a series of strategic and ideological questions for the Green movement in particular. Second, what do the Olympics themselves bring to domestic policy development and norm diffusion? Prior to the Singapore vote, political and economic elites were falling over themselves to convince the French and Parisian publics that Paris desperately needed the Games in order to achieve desirable policy goals. So what happens to social and environmental legacies when bids fail?

The Paris bid and its environmental credentials

The winner in Singapore was one of the more complex and expensive of the five shortlisted proposals. London had an initial costing of $15.8bn in total, of which $3.8bn (£2.4bn) was dedicated to specifically Olympic infrastructure such as road and rail links, stadium construction and renovation, the Olympic village and so on (IOC 2005). Like its main rival, the Paris bid also envisaged new sporting and transport infrastructures, including a Dôme for judo and badminton at Roland-Garros in the Bois de Boulogne, plus a 'Superdôme' for gymnastics, and new aquatic centre, velodrome, canoe slalom course and shooting centre, in various locations close to the capital. The centrepiece was the conversion of the disused Cardinet railfreight yards into an Olympic Village and a new urban park at Batignolles in the northern section of the city's 17th *arrondissement* (district). La Rochelle, hosting the sailing, would also have a small ancillary Olympic Village, and would benefit from new road and high-speed rail (TGV) connections. Nonetheless, the Paris bid was foremost remarkable for its *lack* of grandeur. Its non-operational budget of USD 6.2bn comprised a USD 2.2bn budget for direct Games-related capital investment, in addition to a further accelerated programme of planned investment for general infrastructure at the cost of USD 4bn. Though considerably more expensive than Madrid's budget (USD 1.64bn), this nonetheless augured for a much more compact and less financially onerous Games than that promised by either Moscow (USD 10bn) or New York (USD 7.6bn), as well as London.

According to Delanoë, presenting in Singapore, the Paris Games would not be about 'waste or excess, but efficiency and a sense of values'. Of the 32 sites designated to host events, 12 already existed, including the 71,000-seat Stade de France, which would become the Olympic stadium. Thirteen more were conceived as temporary installations only. Only seven stadiums therefore would be permanent and new, of which three – the aquatic centre, the velodrome and, more controversially, the Roland-Garros Dôme – have been the subject of long-running demands from the national swimming, cycling and tennis federations respectively. The vast majority of events were to be concentrated in two Parisian 'clusters', one on the northern and the other on the western outskirts of the city. The northern cluster, at Saint-Denis, Aubervilliers and the Porte de la Chapelle, would comprise nine venues and provide a focus for the post-Games urban planning legacy; the western cluster, at the Bois de Boulogne, would symbolically connect with the first Paris Games, in 1900, and would comprise eight venues. The Olympic Village was half-way between the two clusters. Over three quarters of the medals, it was envisaged, would be awarded within a ten-minute radius of the athletes' residences.

On the face of it, therefore, the Paris bid appears to live up to the 'ethical and ecological' strapline. This is perhaps unsurprising. Delanoë was elected the city's mayor at the 2001 municipal elections, leading a coalition of Socialists, Greens and Communists on the basis of a programmatic agreement (*contrat de mandature*). Key posts in the subsequent administration were filled by Greens, such as Denis Baupin (transport) and Yves Contassot (environment, including public parks). Rather than outsiders, Greens were thus junior coalition partners in the Paris administration. As one might expect, the details of the bid reflect Greens' concerns, and particularly the dispositions of the *contrat de mandature* covering transport and green spaces. These focused particularly on two aspects of the Olympic legacy. The first was the plan to construct the Olympic Village in Batignolles. The site, to include a new 10 hectare (ha) urban park, would be redeveloped as a sustainable development showcase, an *éco-quartier* (eco-neighbourhood): the site would be a key element in delivering a 'CO_2 neutral Games', generating its own power through a biomass plant, 2ha of photovoltaic solar panels and a small wind turbine dedicated to driving a rainwater recovery system. All new infrastructure would be built to stringent HQE (Haute Qualité Environnementale) environmental quality standards. Following the Games, the athletes' accommodation quarters would be converted into a mixed housing development, 50 per cent of which would be reserved for social and student housing.

The second legacy centred on the plan to extend and improve the social mobility of the city's public transport system, with the bid dossier bringing together within the 2012 deadline a series of network restructuring decisions that had already been taken, but also adding a number of new extensions. In particular, this featured the extension of the T3 tramway (known as the *Tramway des Maréchaux*, as it runs along the nineteenth-century boulevards ringing Paris, named after Napoleonic military commanders). The extension would connect the northern and western Olympic clusters (clockwise), providing the city with its first new tramway since the last part of the original network was removed in 1938, and – in the face of the metro's overwhelmingly radial model – its first new concentrically designed connections since lines 2 and 6 of the metro were completed a century earlier, linking the city's outer districts directly (rather than via the centre). The bid would also address the service and saturation problems of the metro system in the north-east of the capital. Outside Paris, the final section of the suburban tram line from the Stade de France to the Porte d'Aubervilliers, would be completed. All surface public transport on the capital's lines would be made accessible for passengers with reduced mobility.

In many respects therefore, the Paris bid appears a model of civic and environmental methodology. The T3 extension, for example, has long been demanded by Green politicians and associations in the capital; the choice to build it on the Boulevard des Maréchaux, rather than on the *petite ceinture* (a now disused double-track railway circling Paris inside the line of the Boulevard des Maréchaux) is widely seen by Greens as crucial step towards creating better connections between Paris and its outer suburbs, improving the connectivity of the existing public transport system, and reducing car use. It has also, despite much resistance, enabled work to start on the transformation of the *petite ceinture* into an urban green corridor. Further, where the Athens, Sydney and Vancouver Games each featured a withdrawal of participatory democratic and civic oversight procedures (Searle & Bounds 1999: 171, Hall 2001: 172–3; Karamichas 2005; Whitson in this volume), this was not the case for the Paris bid projects. The Batignolles development featured seven local public meetings and exhibitions between June 2004 and May 2005, and the May 2005 decision of the National Public Debate Council (CNDP) to submit the T3 extension to an independent (if non-regulatory) public consultation procedure in the first semester of 2006 provided 13 public meetings for the discussion of the project and its alternatives, accompanied by a comprehensive public information campaign. The very fact of the CNDP is itself significant: one of the last measures enacted by the Jospin plural left government before losing office in spring 2002, the CNDP was designed as a response to long-term civil society demands for greater clarity and public participation in major infrastructural project implementation. One of the key reasons given by the CNDP for submitting the T3 extension to this process was, precisely, the significance of the project in the Games bid dossier (CNDP 2006: 3); the council had, notably, previously declined to study tram network restructuring projects of similar scale, but not similar national significance, in Montpellier and Marseille. The T3 extension thus became the first Parisian project, as well as France's first tramway project, to be subjected to CNDP scrutiny.

Environmental movements, sustainable development and mega-events

The apparent greening of the Games creates somewhat of an existential crisis for environmentalists, who are presented with a fundamental choice between two contrasting visions of political action, between a constructive politics of engagement and a critical politics of resistance. Of course, these choices are not simply dependent on ideological alignment, but may also depend on the precise political, institutional and cultural contexts of staging, as a function of different standards of civic tolerance, the objective nature of environmental threat, the possibilities for political accommodation afforded by national and local institutional arrangements and so on.

In Olympic terms, the positioning of environmental movements is conditioned by a number of factors, from domestic institutional and cultural arrangements to, in particular, public responses to sustainable development agendas. In France, the situation with respect to the furtherance of the sustainable development agenda is perhaps complicated. On the one hand, the environmental movement has since the late 1980s become increasingly institutionalized: success at various sub- and supra-national elections has facilitated the integration of Green Parties into institutional power-sharing arrangements such as in Paris, whilst Sylvie Ollitrault (2008) underlines the long-term trend to what she refers to as the emergence of a dominant, ideal type model of environmentalism based on professionalism, expertise, legitimization and institutionalization (and with it the evolution of collective action repertoires). On the other hand, though the closure and centralization of policy-making in France has often been over-stressed, the environment ministry is traditionally weak, and national policy-making has typically been resistant to the integration of environmental actors into sectoral policy arrangements (Hayes 2002). Despite a wide and increasing focus since the Brundtland report on allying economic growth to sustainable resource use, particularly visible in the proposals of environmental movement organizations and the regulatory diffusion of norms by international governance regimes, the culture of public policy decision-making in France has remained relatively resistant to the mainstreaming of environmental considerations (Szarka 2004, Lascoumes 2008). For Olivier Godard (2008: 38), in fact, President Sarkozy's arresting and (potentially) transformative *Grenelle de l'Environnement* – a series of environmental policy commissions held in autumn 2007, bringing together state representatives, producer groups, civil society organizations, leading to new legislation, regulations and methodologies in the fields of fiscality, energy efficiency, biotechnology, agriculture, resource and waste management, participatory democracy – should properly be seen as remedial action making up for lost time in implementing measures that have long been proposed by NGOs and even official bodies.

The Paris bid encapsulates these conflicting societal, economic and policy pressures. For the Paris bid, of course, Greens are already institutionally legitimized internal actors, playing fundamental roles of project construction, management and implementation. Indeed, the putative Paris Games studiously sought to avoid white elephant construction, and brought together a series of policy enactments central to the *contrat de mandature*. There was little external opposition. *Le Monde* was able to report ahead of the IOC's visit to Paris in early March 2005 that, since the bid was launched in 2003, 'not a single political voice has been clearly raised in opposition to the event', with two polls finding public support for a French Games to be running at 85 per cent and 87 per cent respectively (*Le Monde* 8 March 2005). Moreover, France's labour unions, despite timing a one-day strike against government industrial policy for the middle of the IOC visit, also took pains to demonstrate their backing for the bid and its promises of job creation and tourist development. In June, the main unions signed a 12-point 'Olympic social charter' with the City Hall, covering job creation, the preferential recruitment of the unemployed, and the integration of social and ethical criteria into Games tenders; Delanoë promised that all contractors would have to meet International Labour Organization standards. The main union leaders appeared in Luc Besson's promotional film for the Paris bid, pledging their support for the Games and promising 'social peace' during the event; a union delegation accompanied the bid team to Singapore.

Nonetheless, the outward display of political consensus was not reflected within Les Verts. Nationally, the Greens' annual General Assembly voted in November 2003 not to support the Paris Games bid. In the Paris City Council, though the Green group decided to back the city's bid, the motion was only narrowly carried. Even whilst endorsing the bid in October 2004 and expressing his group's support for the 'ecological, ethical and fraternal' Games and its promises of new public transport infrastructures and higher regulatory standards, Alain Riou (the leader of the Greens on the city council, since deceased) acknowledged the strong hostility towards the bid amongst many Greens (Riou 2004). Two Green city councillors, Sylvain Garel and Charlotte Nenner, launched a dissenting campaign via an online blog, joining – along with various anti-advertising and counter-globalization groups in particular – an umbrella anti-Games collective (Collectif anti-jeux olympiques, CAJO).[1] Opposition to the Games ranged from criticisms of specific infrastructural projects (with environmental and local inhabitants associations mobilizing against both the Superdôme and the Roland-Garros Dôme), to more fundamental arguments holding the modern Olympics to be antithetical to Green values.

Of course, development, expansion and growth are symbolically central to the Olympic ideal of citius, altius, fortius (faster, higher, stronger), and find a more prosaic economic translation in the big 'more' that a winning bid can derive from hosting the event (as one pro-bid writer put it in *Le Figaro* (4 July 2005), 'more involvement in sport, more tourism, more jobs, more image, more money'). Chief amongst anti-bid arguments were therefore criticisms of the Games as a motor for 'unfettered capitalism', with public investment bearing the risk and private corporations deriving the profit;[2] in a piece published in *Le Monde* during the week of the IOC visit to Paris, Garel, Nenner and Bernard Maris constantly emphasize the 'ultra-liberal economic model' of sporting mega-events, characterized by excessive commercialization and social exploitation (Garel et al. 2005). Specifically, opponents were sceptical about the number, the quality and the security of the projected jobs, about the value and social utility of the returns given the extent of public investment required, and about the ethical record of many of the 'partner businesses' recruited by the bid. Discussing the 2006 Turin Winter Games in this volume, Dansero et al. identify a dialectic clash between the global and the local world of the event, triggered by the apparent disparity between the normative Olympic discourses of peace and human rights and the records of a number of TOP sponsors. Similar concerns are evident in the Paris bid. The hotel group Accor, for example, one of the 20 member companies of Lagardère's Club des Entreprises, each of which pledged the sum of 1M€ to promote the bid, is widely associated by critics with low-pay, low job security and poor working conditions (Brohm et al. 2005); Garel is highly critical of a list featuring construction companies and arms manufacturers. Les Amis de la Terre (the French Friends of the Earth) declined the invitation to be a signatory to a 2012 'Sustainable Development Charter' on the grounds that it was to be non-binding on the private companies involved, some of whom – such as the utilities conglomerate Suez – the association has been campaigning against for years. Under pressure from Greens and Communists, Delanoë was forced to abandon a planned sponsorship and sales agreement with Coca-Cola. Above all, concentrating investment in and around the Games risks diverting funds and attention from the real Green priorities. For Garel, 'We are a left-wing council. Our objective shouldn't be the Paris Olympic bid, but combating social inequality, pollution, and poor housing' (*Les Inrockuptibles* 23 March 2005).

The contours of this debate are neatly exemplified over what became perhaps the most controversial part of the bid. In 2002, the French Tennis Federation (FFT) put together a proposal for a new 12,000 seat court at Roland-Garros in the Bois de Boulogne. Opposed by a coalition of environmental and residents' associations, the proposal was rejected by City Hall on the basis of its impact on the forest, though the administration took pains to express its support for the FFT's wish to expand the stadium

(Roland-Garros is the smallest of the four Grand Slam tennis stadiums, in London, Melbourne and New York). The Olympic bid provided an opportunity for both Federation and City Hall to re-work the proposal in order to minimize its environmental impact and, particularly, provide the extra funding for necessary road access improvements. The resultant project – designed in the context of a sustainable development charter for the Bois, concluded by City Hall in November 2003 with a number of surrounding local authorities – would not only create a new prestige court, but also enclose a section of the capital's circular *périphérique* motorway and modify sliproad access, thus compensating for the loss of green space. For Ile de France Environnement (IDFE, the regional association of the national France Nature Environment federation, itself federating 350 associations and 35,000 members), however, this promise was insufficient: developing the Bois, classified since 1930, would set a dangerous precedent, and construction so close to the *périphérique* would subject spectators to unacceptably high levels of atmospheric pollution (IDFE 2004).

H. J. Lenskyj argues that we should pay particular attention to citizens' mobilizations against the Games precisely because of the power of IOC and host cities to 'suppress local dissent and to promote the illusion of unequivocal support' amongst host populations (2004: 152). Key elements of this 'manufacture of consent' typically include the strangling of opposition through the reform of planning rules, but also the creation of 'illusory' unanimity through the elite interest politics of cross-partisan consensus, media representation, official celebrations and official endorsement. But in this case, opposition remained largely symbolic, and criticism of negative environmental impacts was highly localized, conjunctural and aimed at protecting the wealthiest parts of the capital. Though the IDFE's criticisms of the Dôme were recognized by the IOC – four members of the 13-strong evaluation committee met the IDFE in March 2005 – it is characteristic that these criticisms were 'friendly', seeking to 'save the Paris bid' by 'correcting its weak points' (IDFE 2004). The Green group on the city council couched its opposition to the Dôme in almost identical terms: Les Verts's criticism of the Roland-Garros extension, stressed Riou (2004), should absolutely not be taken as an attempt to sink the Paris bid. The position of many Greens was probably best encapsulated in the speech made in October 2004 by Jean Vincent Placé, leader of the Green group on the Ile de France Regional Council, explaining the group's decision to endorse the council's €1bn financial guarantee for the Games. For Placé (2004), the reality of the Games today is nationalism, commercialization, doping and human rights violations; yet the Paris bid, whilst far from perfect, also offered a number of significant, concrete policy advances in transport, housing and the environment. Consistently critical of the democratically problematic character of the IOC, the value-system of the contemporary Games and, particularly, of the commercial conditions imposed by the IOC on host cities during the event, the majority of Greens in the city and regional councils chose rather to attempt to seize the political opportunity the Games represented not just to realize key projects central to their own programme, but to bring about long-term cultural change in institutional procedure and methodology.

'We are the largest left-ecologist City council in Europe', argued Garel, 'it was important to take a stand' (Interview, Paris, May 2007). But corporate growth oriented criticisms, potentially reflecting wider ideological and strategic debates within the French environmental movement, remained largely symbolic. The *décroissance* ('degrowth') movement is increasingly intellectually influential in the Green and counter-globalization movements in France (Baykan 2007), particularly through the work of economist Serge Latouche, who sees sustainable development as a strategy to preserve a model of economic development on the basis of structural inequality (Latouche 2003, see also Quenault 2004). Latouche has done much to popularize a critical examination of the basic assumptions of sustainable development as a concept (or 'contradiction in terms', as he would have it). Indeed, as Françoise Gollain underlines, attitudes to economic development present a point of rupture between a traditional left committed to managing growth and a Green movement which developed out of a fundamental critique of its consequences (Gollain 2006: 121–6). Yet in the context of the 2012 bid – perhaps precisely because of the historic difficulties of placing sustainable development on the national policy agenda in France in the first place – this fundamental opposition, the potential of a civic politics challenging the assumptions of market growth or the top-down, non-democratic nature of Olympic governance, struggled to find transmission from within a Green movement sensing the opportunity to integrate sustainability goals into major public policy decisions.

Testing the claims: the Olympic village

How effective has the wider sustainability strategy proved, post-Singapore? What happens to public policy promises when bids fail? Event-driven infrastructure development is clearly vulnerable to review. Yet the major new sporting venues – the aquatics, nautical and shooting centres, and the covered velodrome – were all maintained in the government's January 2006 National Sports Development Programme, to be financed from the public purse with the goal of enhancing elite athletics performance and facilitating future events bids. On the other hand, the Superdôme would not now be built, and the FFT dropped the Roland-Garros extension. City Hall abandoned its plans to extend the T3 tramway to the Porte d'Auteuil, a development specifically designed for the Games and much contested by residents and businesses in the wealthy western 15th and 16th *arrondissements*. The much longer extension to the east and north was however maintained; following the conclusion of the public debate procedure and a public enquiry in spring 2008, preparatory work started in 2009, with entry in service scheduled for late 2012. Moreover, in September 2010, City Hall announced its intention to extend the T3 line to the Porte d'Asnières, in the 17th district, for entry into service between 2014 and 2016. The possibility of extending the T3 to complete the circuit round Paris is now distinct.

Particular attention should be paid to the Olympic Village. Since the 1960 Rome Games, Olympic Villages have been used to create specific national/ urban narratives and, in regeneration terms, are central to the drive for post-Olympics 'legacy momentum'. Typically under the aegis of a dedicated funding and design body, supported by a dedicated political structure whose authority cuts across overlapping (and often competing) levels of political jurisdiction, Villages offer an ideal discrete space (physically and temporally) for controlled projections of urban identity (Muñoz 2006: 175–81). Given the Singapore vote, it is of course not possible to discuss the Batignolles Village's architectural projection of identity. Yet, in line with Villages since Sydney, a key interest is not simply the architecture but the claims to sustainable design and technology which underlies it. Typically, the Batignolles Village was conceived as a sustainable development showcase: the image of the global city, and of its own status as such, that Paris as host city aimed to project to the world was primarily figured in terms of environmental responsibility. Delanoë, reacting crisply to Garel et al.'s comment piece in *Le Monde*, cited the Batignolles development as the prime reason why 'Paris needs the Games', creating the clear impression that the development would not be achievable without them (*Le Monde* 10 March 2005). This argument was echoed two days before the Singapore vote by Jérôme Dubus, responsible for urban planning in the 17th *arrondissement*'s district hall, who claimed that with the Games, the Batignolles development would take seven years; without them, it would take 15 to 20 (*Le Figaro* 4 July 2005).

In fact, as part of a wider reflection on the utilization and rationalization of the city's rail estate, the Batignolles site had already been identified as a possible location for the Olympic Village in the 2008 Games bid, for which Aubervilliers was eventually selected. Discussions between the city and the state rail companies continued over the redevelopment of the yards, and studies undertaken in 2001 confirmed the adaptability of the site (Pelloux 2005: 110–14); the following year City Hall launched an architectural competition for an urban redevelopment project. Meanwhile, in May 2003, the mayor's office confirmed that Paris would bid for the 2012 Games. The IOC's previous reserves about the logistical suitability of the Aubervilliers site, coupled with the results of the initial architectural studies, convinced the 2012 bid team to choose Batignolles. This new element was integrated into the redevelopment project which would now have three phases: the rapid conversion of 5.5ha; the completion of a 45ha Olympic Village for 2012; the subsequent reconversion of the Village into a public neighbourhood (Mairie de Paris 2006: 2). By the time of the Singapore vote, the city had acquired 10ha of the site; demolition work on the first section was underway; decontamination of the soil was nearly complete; the city council had created a 7.3ha priority development zone, to comprise a school, housing an underground car park, hotel and retail space, and the first 4.3ha of park area. Despite the much-publicized fears over the city's ability to acquire the rest of the land, a convention was concluded in November 2005 and ratified the following July, enabling the creation of a second priority development zone in the north of the sector. Work on the remaining park area, 3500 housing units (to accommodate approximately 10,000 people), crèche, school and office space would begin in 2009, with the entire development – including the relocation of rail freight terminal and rail maintenance sheds to the north of the sector – scheduled to be completed by 2013.

Most significantly, City Hall's 2007 Climate Plan committed the city to turning the Batignolles site into an 'éco-quartier exemplaire', as part of the generalization of new sustainability norms for urbanization projects. These projects henceforth have to conform to a four-point programme, with environmental impact assessments evaluating energy efficiency, use of renewables, the reduction of carbon impact, the limitation of urban sprawl and the privileging of public transport access. Energy and carbon assessments will be required every five years post-development (Mairie de Paris 2007: 27–31). In other words, the Batignolles development retains its essentially pedagogical significance: the demonstration of new principles championed by the Green group, both in terms of park management (recycling, organic gardening, integration of green space into urban development, inclusion of associative communal gardening areas into new green spaces, opening up of park lawns for public use) and building construction. City Hall has since adopted HQE norms for all new construction or development projects of at least €1m, generalized as part of a series of environmental best practice guidelines distributed in late 2005 to all public and private construction operators (Mairie de Paris 2005). These norms are, for instance, operative for the aquatics centre in Aubervilliers and the 110m Sogelym Steiner skyscraper now to be built next to it. These are significant gains for the Greens, though they were far from automatic either during the bid phase or in its aftermath. The appointment by the Paris bid team of Jules van Heer, a former advisor on Sion's bid to host the 2006 Winter Games as a consultant on environmental measures, was crucial to shaping the environmentally sensitive character of the Paris bid, providing Baupin's and Contassot's directorates with a de facto key ally in internal negotiations on bid detail. Post-Singapore, Contassot's directorate in particular was forced to fight a series of internal battles to ensure that the Batignolles development would retain the same high environmental standards as had been envisaged for the Games. Ultimately, therefore, IOC stipulations created leverage for the Greens for the generalization of HQE across development programmes in the capital, made possible once its principle had been accepted for the Village development. For Sylvie Laurent-Bégin, Contassot's *directrice de cabinet* (chief political advisor), the bid enabled real cultural advances in the way City Hall deals with new development (Interview, Paris, July 2007). In a policy setting traditionally resistant to the mainstreaming of the sustainable development agenda therefore, the Games bid, and with it the expectations of best practice standards, created a series of domestic opportunities for environmental actors to achieve concrete policy reforms.

Discussion

What conclusions can we therefore draw from this set of circumstances? As critical observers, we generally have a problem with events that did not happen: there is no balance sheet, no set of definitive results, no measuring of promises against realizations, no before and after. Instead of the certainty of data and the crucible of practice, we have only a continuum of speculation, estimation and prediction.

This does not mean, however, that nothing happened. Indeed, this chapter is one of three in this volume (the others being of Polo and Dauncey) which contribute to what is perhaps a sub-genre of 'non mega-events' (Hill (1994) on Manchester, Lenskyj (1996) on Toronto, Swart & Bob (2004) on Cape Town). Such studies are revealing of the processes underpinning mega-event hosting precisely because they do not lead to events, but demonstrate the evolution of thinking about the requirements of and adaptive capacity of prospective host coalitions to event staging across multiple bid iterations. They also allow us to disentangle what is promised as event-dependent from what is already planned and will in any case be delivered, and thus to discuss the problematic importance of additionality in bid dossiers.

Foremost, however, the failure of the Paris bid was experienced as a significant national political event. For the careers of the politicians most closely associated with it, certainly: whilst immediate designs Delanoë might have harboured within party, city and country were set back, for Chirac the IOC's decision – hard on the heels of defeat in the European Constitution referendum five weeks before – confirmed the waning of the light, announcing internationally a *fin de régime*. On 12 July 1998, Chirac had beamed from the stands of the Stade de France as the multi-ethnic French national team beat Brazil 3–0 to win the FIFA World Cup that France had hosted. As crowds flocked to the Champs Elysées to celebrate and Zinédine Zidane's face was projected onto the Arc de Triomphe, it seemed that sporting success heralded a new, integrated, harmonious France, momentarily replacing the bleu-blanc-rouge of the national flag with the black-blanc-beur (black, white, Arab) of France's multi-racial population (see Marks 1999, Dine 2003, Hare 2003 for further discussion). That this powerful metaphor of contemporary French identity has since appeared increasingly premature is perhaps besides the point; success off the field as well as on, created by hosting a sporting mega-event in front of a global television audience, had projected to the world *la France qui gagne*. Seven years later, the IOC seemed to confirm the subsequent sense of drift, the development of a social, cultural and political malaise, an unhappy transformation into a sort of national depression named as *déclinisme*. In Agen, in June 2006, Nicolas Sarkozy kicked off his presidential campaign by promising *une France nouvelle* and a politics of aspirational individualism.

For Paris, London's victory also signalled a *return* to politics. For the T3 tramway, Batignolles and even the fate of the Roland-Garros tennis stadium, the defeat re-opened political divisions over the purpose, utility, composition, costs and benefits of the projects. As the precise details of the Batignolles development firmed up post-Singapore, the Gaullist controlled district hall became increasingly critical of the mix and location of social housing intended for the project; subsequently, the district council voted against the details of the second ZAC priority development zone in the northern section of the Batignolles site, judging its public transport provisions for the new housing development insufficient. The Gaullists on the city council were also highly critical of the claimed advantages of the tram over alternative investment in the metro system, boycotting the opening of the southern section of the T3 in December 2006, attacking City Hall policy as anti-car and effectively blocking the western extension. Moreover, questions concerning the balance of social housing to office space in new priority development projects produced fundamental divisions within the governing coalition. In June 2006, the Green group abstained in the council chamber on the city's Local Urbanism Plan (PLU), which sets a medium-term framework (10 to 15 years) for the city's construction programme. For the Greens, the PLU afforded too much space to office development within Paris, and not enough in the suburbs; for the Socialist group, highlighting that the PLU set a minimum of 25 per cent social housing for all new construction projects of 800m^2 or more, the Greens' stance ignored pressing job creation needs which the provision of new business space in the capital was designed to meet.

For Roland-Garros, the circumstances are a little different, characterized by an accentuation of the Olympic problematic. Despite the post-bid knock-back, the FFT did not give up on its plans to extend its signature stadium, instead producing an even more ambitious plan: a 60 per cent increase in surface area, a roof on the central court, the demolition of court No. 1, a new 5000-seat covered court, a new 99-year lease at a preferential rate. The total estimated cost is around €250 million, paid out of public funds. This time the project was underpinned by the threat to relocate the tournament outside Paris (Versailles being amongst the alternative sites mooted). Anxious to retain the tournament in the capital, Mayor Delanoë (backed by the Gaullist opposition) was an enthusiastic supporter, promising the channelling of municipal sports subsidies to the project and abandoning a condition to review the terms of the lease after 25 years (*Le Parisien* 10 February 2011). This time, the planned extension involves converting the neighbouring Hébert stadium, currently used for school and municipal sports, to relocate the national tennis development centre; perhaps above all, it requires the space taken up by a number of the listed belle époque plant hothouses in Auteuil. A petition against the project has collected 40,000 signatures; at the time of writing, the legal and political battle – pitting mainstream parties against residents and environmentalists, political, sporting and economic elites against citizens – was just hotting up.

What happens, therefore, when bids fail? The insistence of Delanoë that Paris must get the Games in order for the Batignolles park to be created, when, of all the projects, this was one of the more secure, might seem obtuse. Yet this would be to miss the function of Olympic bids, and specifically their role in the global communication strategies of major cities. The point about the park is not that without the Games, it would not happen; as Shoval points out in his early discussion of the London and New York candidacies, the existing commercial and residential pressures within the city ensure that the urban development projects at the heart of the bids would be implemented in any case, win or lose (Shoval 2002: 595–6). Paris did not need to win the bid so that there could be *an* Olympic Village constructed along sustainable design principles; in 2012, there will be such a Village, but it will be in east London. Neither did Paris need the Games so that an industrial site in the city's 17th *arrondissement* could be converted into an urban park with substantial domestic and business accommodation; the reclamation of the railfreight yards already figured within the city council's commitment to create 30ha of new parks by the end of its mandate. Rather than 'event urbanism', the Batignolles development was part of a long-term reflection on the redevelopment of both the capital's northern districts and its railway infrastructure. Perhaps above all, recent and increasing pressures on the affordability and availability of real estate in the city made it virtually inevitable that such a prime site would be prioritized for redevelopment. The Olympic Park – now the Parc Clichy-Batignolles Martin Luther King – is thus now to be connected to President Sarkozy's future 'cité judiciaire', to host the signature buildings of the new re-organized Paris judicial courts (TGI, Tribunal de Grande Instance), scheduled to open in 2015.

To put this into some perspective: on 1 January 2006, social housing constituted 14.98 per cent of the city's total housing units. The same year, the city received over 109,397 housing applications in total, including requests for re-housing, for a total social housing stock of 171,502 units, managed by housing associations on behalf of the city council (another 62,000 units are also managed by housing associations, but with higher rents). Given the high cost of rents in the private market, turnover in Paris is only 5.5 per cent, against a national average of 10 per cent. Few spaces are suitable for new construction in the city. Consequently, only about 14,000 housing applications can be met each year (APUR 2007). This shortage of affordable accommodation is set against background trends in Paris of increases in population levels (Paris gained 24,000 inhabitants between 1999 and 2005, a total *intra muros* population of 2.15 million (INSEE 2007)), and in real estate prices, with the cost per square metre doubling between 1997 and 2006. It appeared to many to be symptomatic that in three separate incidents in Paris in 2005, fires in very poor quality buildings claimed the lives of 48 people, including 14 children. Nearly all the victims were of African origin, some squatting, some waiting to be rehoused by the council. The problems caused by this critical lack of affordable housing are particularly acute in Paris, but indicative of the wider French context. In December 2000, Lionel Jospin's plural left government attempted to provide a new framework capable of dealing with the problem, passing the Urban Solidarity and Renewal Act (loi SRU), a major land-use planning reform of radical intent which linked planning procedure to the development and implementation of urban policy (Booth 2003). Amongst its dispositions, the new law imposed a minimum level of 20 per cent of social housing (as a proportion of total housing) on all communes of at least 3500 inhabitants. In order to meet this stipulation, Paris City Hall has set itself a target of constructing or converting 57,000 new social housing units over 15 years, to achieve the 20 per cent threshold by 2020. This target is complemented by a series of obligations on the city's wealthiest districts (principally in the centre and west), in order to re-balance housing development and increase social integration. Once identified in 2002, it was always probable that the Batignolles site, set in an *arrondissement* with a level of only 10.1 per cent social housing, would be redeveloped whatever the outcome of the Olympic bid: the conversion of the site plays a fundamental role in addressing current, pressing housing needs.

What rather is at stake therefore is perception and the claim to global leadership. Zygmunt Bauman has identified the critical opposition at the heart of the neo-liberal global economy to be the distinction between the extraterritoriality of power and the tightly circumscribed territoriality of political institutions and action; indeed, the measure of power is the extent of its capacity to disengage, to flow away. The function of politics, he argues, thus becomes the search for local solutions to global contradictions (Bauman 2003: 18–19). From this perspective, the mega-event becomes the point of contact between global flow and social place. For opponents of the Games, as noted above, these points of meeting, of touchdown – what Sassen refers to as the multiple localizations of the global (2007: 118–22) – generate social and political contradictions that create a grounded context for action. For proponents, too, the meeting of the global and the local is fundamental to the creation of meanings not simply of the event, but of the policies and programmes that underpin it. In this sense, Paris did need the Games for the Batignolles project to have a significance beyond the local: Olympic architecture is not simply a question of functionality, but intrinsically one of spectacle; the creation of a showcase is first dependent on the creation of an audience; this is what the Olympics delivers, globally. Recycling rainwater, constructing wind turbines, producing electricity through solar capture are real advances; yet stripped of a global media audience, this becomes a small, highly localized experiment in ideal urban design. In other words, despite the clear pedagogical value that the development has had within Paris in environmental and municipal terms, its use as a global pedagogy – understood not simply as the will to implement sustainable development practices, but as the will of the city to be perceived throughout the world as their progenitor – disappears the moment it becomes clear that no-one will be watching. And in 2012, the audience will be watching London.

In sum, therefore, what does this bid tell us about environmental legacies? What kind of development is on offer here? What do Olympic bids bring to domestic policy development and norm diffusion? There are, of course, two ways of looking at these questions. The first might stress that, if the proposals are enacted irrespective of the bid outcome, then bids themselves provide only a negligible 'added value'. The second might contrastingly argue that only the fact of the bid enabled the political will and social consensus integral to the elaboration of sustainable development policies and methodologies, with the Games thus acting as a catalyst for the dissemination and application of environmental best practice. Delanoë's often-repeated claim that the Olympic bid was an 'accelerator' for the council's key projects would therefore appear validated, in defeat as well as victory. Rather than comprising new policies, the infrastructure projects included in the Olympic bid were already extant in the Delanoë administration's geographic and functional planning priorities, designed to correct some of the problems inherent in the capital's urban development inheritance, particularly the effects caused by transport infrastructure and high density housing construction immediately to the north of the city. The principal effects of the Olympic bid were to bundle these projects into a single coordinating programme, set a hard deadline for their completion and unblock additional sources of capital, primarily from the central state, but also from the private sector. Even where the additional finance and the hard deadline subsequently disappear, and where national and international agreements (such as Kyoto or the domestic Grenelle regulatory frameworks) and market forces also dictate the improvement and generalization of sustainability agendas, it is the case that expectations have been raised, standards set and the programme (mostly) remains.

In this argument, in purely instrumental terms, bidding for the Games benefits even those cities whose offer to host them is declined by the IOC; we might even suggest that failure can be beneficial, as it avoids the material and symbolic impacts that hosting the Games actually brings. Of course, as the continuation of the Roland-Garros extension project also signals, Games bids also concretize expectations for new infrastructures which themselves bring significant material and symbolic impacts (the channelling of public funding to private sporting federations and from collective municipal to elite professional sport, the loss of culturally significant buildings, the loss of urban green space...). Moreover, the removal of the event-driven hard completion timetable also creates potential space for (perhaps) effective counter-mobilization and resistance.

Final remarks

It is of course difficult to draw conclusions from the experience of one bid, just as it is important to draw distinctions between the experiences of different Games in different social, political, and cultural contexts. The experience of recent North American Games points to the exclusion of non-elite interests and concerns in event staging, and an overwhelming focus on mega-events as a springboard for the furthering of narrowly defined, consumption-based economic growth strategies (Andranovich, Burbank & Heying 2001: 127–8). Given the trend towards 'world cities' in bidding and hosting decisions, this is likely to be an ongoing preoccupation. In terms of the Paris bid, as Garel and Nenner argue, there are other cities in France that might be more deserving recipients of Olympic investment, helping to re-balance historical spatial and social inequalities. In Paris, the presence of Greens as elected partners in City Hall is significant, and might point both to a more inclusive, more civic Olympic project, and a more lasting sustainable development legacy. Yet the Paris bid remained dominated by corporate booster consensus. The French labour unions, like the media and vast majority of the French political and economic class, maintained a consistently uncritical view of the projected social and financial benefits that the Games would bring; the only study of projected employment gains was commissioned by the *Club des Entreprises 2012*. Democratic consultation took the ersatz, superficial form of opinion polling, which found overwhelming popular support for the Games.

We might also remain sceptical about how important or how enduring the sustainability legacy is. On the one hand, not having not come to fruition, the Paris Games did not have the chance to reveal the sort of environmental contradictions that have been evident in each of the Summer Games since the IOC formally adopted environmental protection and sustainable development (Sydney, Athens, Beijing). Garel, for instance, maintains that given the time necessary to undertake the studies necessary for HQE implementation, it would have been precisely these aspects which would have been seen as the most expendable by the local coordinating authority should the timetable have begun to slip (Interview, Paris, May 2007). This may or may not be true, and the global financial crisis of 2008 may or may not have had dramatic effects on the contours of the Paris project. The budget and design problems that have dogged London's venues have obviously not had the opportunity to affect Paris, and neither has Paris had to address the detail of its ambitions: the measurement and reduction of carbon in construction and delivery (rather than simply in Games-time operation); the contradiction between carbon neutral venues and increased international transport and tourism; ethical sourcing, the use of PVC, the elimination of HFCs; not to mention the opposition to the Roland-Garros and T3 extensions in the western cluster and so on.

On the other hand, a number of such contradictions have indeed surfaced in Paris's urban development programme since the bid failed, and particularly since Delanoë was re-elected mayor in March 2008, without being dependent on Green support. The City Hall's new plan to redevelop Clichy-Batignolles now features not only the new court buildings and a tower of between 130 and 200m in height, but also a series of 50m high residential blocks. Given the Haussmanian restriction on the height of buildings within Paris to 37m, this is a politically charged issue, but which also effectively hollows out the Climate Plan, the claims to éco-quartier status, and even the meaning of sustainability norm diffusion, given that the planned towers will also have to respect HQE energy efficiency standards. It is clear that with the global spotlight fixed on London, in Paris the construction of a low-rise 'eco-neighbourhood' is no longer as vital an issue as it was. The timetable for completion has also, inevitably, slipped by two to three years. The conclusions to be drawn from the Paris bid and its aftermath, then, is that the sustainable development legacy should be very clearly understood as the top-down diffusion of technical, regulatory norms rather than as the normalization of a bottom-up, civic, participatory process.

Indeed, though key projects in the Paris bid were subjected to collective oversight (the tramway planning process), this was predominantly technical in nature, and is far from suggesting that there was the type of participatory democratic decision-making, either at bid stage or after, that could be identified alongside what Gursoy and Kendall (2006: 603–5) locate as a potentially more inclusive, community-based trend in mega-event planning. The Paris bid was directed and managed from the mayor's office; even the Greens, junior coalition partners, were presented with, rather than consulted on, City Hall's decision to bid (Interview with Sylvie Laurent-Bégin, Paris, July 2007). And though far from negligible, even in its own terms the sustainable development legacy of Paris's bid is thus also partial, incremental, has to be continually parlayed and fought for within institutional bargaining regimes, is dependent on public scrutiny, is subordinated to what is achievable within the accomplishment of previously defined local public and corporate development goals, and is ultimately, always, contingent. Without the Games, the redevelopment of Paris has continued apace, given other, pressing, prior exigencies; without the global media spotlight, and without strong checks and balances from environmental and other non-institutional actors, the character of urban redevelopment has been free to be changed. One prestige project has effectively replaced another, with the TGI taking over the Olympic Village at Batignolles, just as the site identified for the Olympic aquatic centre in the 1992 Paris Games bid was subsequently recuperated for the new national library (Bibliothèque François Mitterrand), at Tolbiac on the left bank of the Seine.

What, finally, does Paris's failure tell us about the IOC's priorities? In Singapore, the IOC rejected two bids, both widely perceived as credible alternatives to London, which had constructed their candidacies around the use of already existing venues and emphasized their legacies in environmental terms. Madrid had promised a car-free, 'humanist, sustainable and environmentally friendly Games', featuring a Strategic Sustainability Plan, new 'green zones', an Olympic Fund for Biodiversity, and a significant attempt to build NGO capacity by funding civil society associations to manage the rehabilitation of key sites (IOC 2005: 84–5, 94). The Paris bid had been deliberately tailored, after two previous rejections, to meet the exigencies of the new Rogge doctrine. Indeed, the IOC evaluation committee recognized that the Paris bid's use of a large number of temporary venues was designed to meet the movement's own recommendations on the use of existing infrastructure and the need for sustainable development (IOC 2005: 10). Whatever the reasons for its choice of London for 2012, and whatever the objective merits of the respective candidacies, the IOC chose the largest, the most costly, the most complex, and the most 'virtual' of the options on offer, preferring the expensive and extensive construction of new transport and sporting infrastructures to the incremental extension or improvement of largely existing ones. Despite a general feeling that, after Beijing, London would be the last massive new infrastructure Olympics for some time, the IOC decision in Copenhagen in October 2009 to give the 2016 Games to Rio simply feels like Groundhog Day.

Notes

1. Garel made the first move against the Olympic torch on its passage though Paris in April 2008; see Renou in this volume.
2. See 'Les JO à Paris 2012? Non merci!', http://jo-paris.blogspot.com/2005/03/les-jo-paris-2012-non-merci.html, posted 6 March 2005.

Part III

Constructing Civic Resistances to Mega-Events

10
Olympic Games, Conflicts and Social Movements: The Case of Torino 2006

Egidio Dansero, Barbara Del Corpo, Alfredo Mela
and Irene Ropolo

Olympic rhetoric usually presents the Games as an opportunity for peace, as demonstrated by the so-called Olympic Truce, inspired by ancient Greek tradition and re-established by the IOC in 1992, with the aim of reducing conflict in sports competition. But conflict doesn't only exist at the international level. In addition to conflicts of interest between the supra-local Olympic system and the local players in any hosting arrangement, the Olympics increase conflict at the local level too: while some use the event for urban renewal objectives, others see it as an invasion of the local community by a supra-local resources-consuming elite, leaving behind often useless and unsustainable facilities and forgetting other social needs (i.e. healthcare and school services).

The IOC has learnt from prior negative experience (especially Denver 1976) that a measure of local consensus behind the event must be secured. As a consequence, a multi-dimensional paradigm of sustainability has been applied to the bid file. In this context, the Torino 2006 Winter Games are an interesting case, because their bipolar spatial organization (a large urban area plus mountain skiing resorts) brought different development strategies and collective actors into confrontation, and because the host territory had already proved highly reactive to top-down project imposition. On the one hand, the organizers of the event paid great attention to the relationship with the local system, implementing a Strategic Environmental Assessment (SEA), creating an Environmental Consultative Assembly (Assemblea Consultiva Ambientale; ACA) and various opportunities for stakeholder dialogue. On the other hand – as has been the case in other host cities – in a context of widespread public scepticism, associations and spontaneous movements criticized the event but without developing into an effective opposition movement.

In this chapter, after introducing the idea of the mega-event as a disputed place, we therefore briefly present the spatial structure of Torino 2006, in both its pre- and post-event phases. Subsequently, we focus on the associations and movements which expressed (in various forms) strong opposition

to the hosting of the Games, adopting positions ranging from radical opposition to constructive criticism. The attention we pay to the role of civil
society actors in their opposition to the event focuses on the discourses and
event representations articulated by these actors, especially with regard to
the spatial and environmental changes generated by the event. Our aim is
to understand whether opposition to Torino 2006 can be qualified as a 'protest movement', and to what extent the hosting of the mega-event created
an opportunity for the development of alternative projects for the city and
the host territory as a whole. We conclude by examining the interpretations
of these oppositional actors of different post-Olympic legacy scenarios.

The Olympics as disputed places

Mega-events are undoubtedly of great interest for understanding the complex
relationships between globalization processes and national and local par-
ticularities. With reference to an increasingly consolidated international literature concerning the definition and classification of mega-events (Ritchie
1984, Roche 2000), it's important to emphasize how these events (and the
Olympic Games in particular), owing to their intrinsic nature as points in
space-time and thus potential moments of discontinuity, effectively reveal
both the dynamics and tensions of globalization processes, and the dialectic relationships they generate with the different host venues (Cashman
& Hughes 1999, Cochrane et al. 1996). Countries, regions and cities seek
to use mega-events to improve their position in international geopolitics
and global economic flows, a phenomenon confirmed by the strength of
the competition seeking to be awarded the latest of the Winter (Chappelet
2002) and Summer (Shoval 2002, Chalkley & Essex 1999b) Olympic Games,
or World Expos. However, besides the global economic players (such as TOP
sponsors), other collective actors also seek to exploit the opportunity for
media (over)exposure provided by the mega-event, as we have seen with the
emergence of the Tibetan question and the heated protests at the passage of
the Beijing 2008 Olympic torch.

In this sense, the Olympics can be interpreted as a *disputed* place. It is also
a complex place: it exists as a typical-ideal reference in Olympic Movement
rhetoric, and periodically acquires a tangible form in a definite location that
must win a competition against other locations in order to host the event,
by preparing a view of the future, of what the location wants to be. This
location, projected into the future, must be considered not only as a functional plan for hosting the event, but also as a complex, multi-scale, multi-
dimensional and actor-related construction, the first objective of which is to
legitimize its own participation in the event. The host location is thus mobilized by the event to bring it within its own logic, but at the same time the
location also mobilizes the event, incorporating it within a strategic plan for
its own transformation: this, at least, appears to be what we can learn from

the cases which the existing international literature consider to have been successful, such as the Barcelona Games of 1992 (Chalkley & Essex 1999b). The competition to host the Olympics is at the same time a competition between scales (the global scale of the Olympic Movement and of the major economic-political actors, the national scale, the regional and local scales), between places (which compete to host the Olympics on an international scale, as well as on the regional and local scale, for the distribution of Olympic events and functions) and in local terms, between the different views of the future for the locations selected (Short 2004). To some extent, as Roche (2000) has underlined, an event like the Olympics is also an opportunity for the formation of a global civil society, as again the protests relating to Beijing 2008 appear to suggest. The display of dissent can remain confined to the local context, or establish inter-local or transnational alliances with other movements and forces which can predominantly be classified as articulated around resistances to globalization.

This aspect contrasts enormously with the rhetoric of the mega-event in general and the Olympics in particular, where events are described as peaceful places, where competition is only exalted at the sporting level and where even the 'Olympic truce' has been reinvented to be upheld at the local and international level. Such rhetoric is extremely questionable and does not consider the significant impact which a mega-event generates at different geopolitical levels. In particular, the event's significance at the local-regional level places it at the centre of a broader, intrinsically conflictual set of territorial and urban changes. In this context, 'civil society' appears to have a role to play, beyond passive spectatorship or the role of active extras assigned to it by, and subject to the strict control of, event organizers. Instead, civil society may adopt an active role which is not regulated by political and economic elite coalitions, whether limited to local frameworks or carrying its voice to other geopolitical levels of activism (see inter alia discussions in Burbank et al. (2000) of movements opposing successive north American Games in Los Angeles, Atlanta and Salt Lake City; of Lillehammer in Leonardsen (2007) and Spilling (1998); of Sydney in Cashman (2006) and Waitt (2003); and of the bids of Cape Town 2004, Berlin 2000, Atlanta 1996 and Barcelona 1992 in Preuss (1998)).

Accordingly, our analysis focuses here on the case of the XX Winter Olympic Games. Undoubtedly, the winter editions of the Olympics have a much smaller media impact than their summer counterparts. However, numerous experts have highlighted the fact that, considering the respective sizes of the two events, the impact of the Winter Games tends in fact to be greater in infrastructure terms (May 1995), largely because they are necessarily located in mountain areas with ecosystems which are more fragile and a man-made environment that is more difficult to manage – for example, in relation to transport (Chalkley & Essex 1999b, Essex & Chalkley 2004). Protests against Winter Games are not necessarily smaller than those directed at Summer Games, and it is perhaps not by chance that the greatest protests so far were against Denver's hosting of the 1976 Winter Olympics, ultimately forcing the city's withdrawal, and replacement by Innsbruck.

Torino 2006: a territory in search of new opportunities

The Torino Winter Olympics were held in February 2006, and are to be seen as an advanced stage in the process of transition of the city of Turin and its metropolitan area away from the one-company town based on the automotive industry (Dansero & Mela 2007), while in the Alpine valleys the Games were used to strengthen pre-existing economic strategies.

Turin, a 'too simple' city (Bagnasco 1986), grown in the shadow of Fiat, revealed the full extent of its frailty when its main industry was thrown into crisis. The crisis of the Fordist production system was aggravated by chronic delay in urban renewal, particularly regarding infrastructures and services (Conti & Giaccaria 2001). So, at the beginning of the 1990s, the city found itself forced to look for effective ways to recover its competitiveness, by increasing economic diversification (Vanolo 2004, Governa et al. 2009). A master plan for deep-seated urban renewal was adopted by the city council in 1993 and approved by the Piedmont Region in 1995. It focused primarily on the Spina Centrale, a broad tree-lined avenue that constitutes a sort of new central axis, running alongside the main railway line and crossing the city from north to south. The most important projects in the plan were the construction of the 'cross rail' system (still ongoing at the time of writing) the recovery of vacant sites (over 2.5 million square metres, owned by the municipality and partly by major industrial groups stretching along the railway line (Saccomani 1994); new prestigious and highly symbolic buildings such as the new offices of the Region and the Polytechnic (still ongoing at the time of writing, the Law Courts and the Environmental Park (a technology park dedicated to environmental services, eco-compatible technologies, etc., co-funded by the EU)).

The master plan was followed by the first Strategic Plan for the city of Turin (making Turin the first city in Italy to adopt such a plan), published in February 2000. The Strategic Plan defined the strategies for the development of the city and its metropolitan area to 2010. It aimed to create relationships and synergies between institutions, the public administration, local businesses and the public – setting up less hierarchical and more inclusive governance processes than those typical of the city's Fordist heyday – as well as to promote the city's growth abroad and to attract external resources to be integrated with local potential (Pinson 2002, Governa, Rossignolo & Saccomani 2009). In this context, major events constitute an opportunity for Turin to show an international audience the achievements made possible by urban renewal and diversification, communicating a new image of the city.

However, the decision to bid for the 2006 Olympics was taken separately from the Strategic Plan, although the Games were integrated (almost at the last moment) in the Plan under the fifth strategy: 'Promoting Turin as a city of culture, tourism, trade and sports'. As a consequence, we can say that in the case of Turin none of the most important urban projects (such as the *Metropolitana di Torino* metro system, or the new high speed/high capacity rail line to Milan) originated from the Olympics (Crivello et al. 2006). Nevertheless, the Games gave a decisive impetus to those projects – making them now truly necessary for ensuring a successful event – and fixed a deadline for their completion.

From an analysis of the territory in which the event took place – what Kovac (2003) refers to as the 'Olympic scene' – it emerges that Torino 2006 presented a number of distinguishing characteristics. Following the model of multi-location urbanism, typical of the Winter Games of the period 1992–2006, the 'Olympic region' of Torino 2006 was clustered around two kinds of spaces (Figure 10.1):

- an urban space, centred on the city of Turin itself, a medium-sized metropolis (of approximately 900,000 inhabitants), capital of the Piedmont Region;
- a mountain space, with various sites located in the upper Susa, Chisone and Pellice valleys, converging on the plain surrounding Turin, some 80 km away from the city.[1]

The 'Olympic places', covering a significant portion of the Province of Turin, were a network of nine different localities (the urban areas – Turin, Grugliasco and Pinerolo – and the mountain areas – Torre Pellice, Pragelato, Bardonecchia, Sauze d'Oulx, Claviere, Cesana-San Sicario and Sestriere) where all the 'Olympic functions' (Olympic Villages and sports facilities) were distributed. All the snow disciplines were hosted in the ski district of the so-called Via Lattea (Milky Way), whereas Turin hosted the ice competitions, together with the Media Villages, the Main Press Centre and the International Broadcasting Centre for television and radio services. Spatially, the 'heart of the event' in Turin was Lingotto, formerly a Fiat manufacturing plant, and now the Fiat SpA Headquarters and a tertiary and quaternary multi-functional centre; in the mountain areas the heart was Sestriere, a winter resort created by the founder of Fiat (Dansero et al. 2003).

This innovative 'Olympic region' thus displayed the problems and opportunities typical of Summer Games located in urban areas (though to a lesser degree), along with those typical of Winter Games, located in mountain skiing localities. With this time-space concentrated territorialization, sustainability constituted a big issue: in the mountain areas, some of which are relatively unspoilt zones of considerable natural interest; in the city, where the Olympics multiplied the construction yards promising to improve the quality of urban spaces, and has left numerous sports and service facilities to be reconverted in the post-Olympic period (Dansero et al. 2006), so as not to become 'white elephants'. The impacts on the environment were heavy,

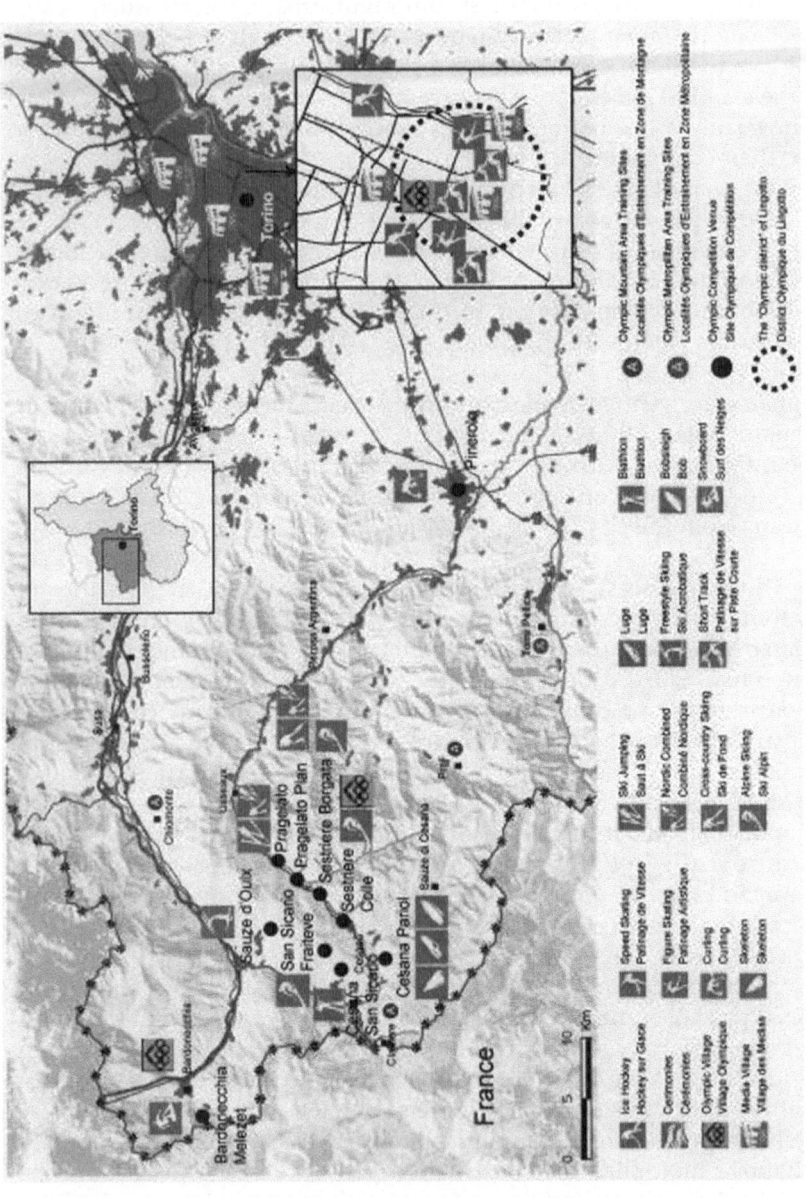

Figure 10.1 The Winter Olympic scene
Source: LARTU-DITER, based on TOROC data

both during the preparation of the event and afterwards, and required an overall strategy to manage them. Inextricably, debate and conflict over the environmental impacts of the Games were linked with other physical transformations affecting those same areas, chiefly the Turin-Lyon TAV high-speed rail line, whose route passes through the Susa Valley.

Strategies and conflicts in a disputed place

Torino 2006 brought together numerous actors, led by different logics and situated at different geopolitical scales (see Figure 10.2):

- Sports organizations, including the IOC, the Italian National Olympic Committee (CONI); the international and national sports federations have been attached, as we have already anticipated, especially because the world of the event was busy exploiting the local resources for creating the necessary preconditions for the sports event.
- Political and institutional actors, primarily the city council, the Province of Turin, and the Piedmont Region; central government; the Olympic municipalities and the 'Midland' territories located between the Olympic sites. These actors all sought to exploit the exposure offered by the event, aware of its potential to integrate the territory in global networks attracting international investment and new production and service activities, chiefly tourism.

Figure 10.2 The world of Torino 2006

- Corporate actors, among which were the local organizing committee (TOROC), Olympic sponsors and suppliers, the Chamber of Commerce, the Employers' Federation, the Torino 2006 Agency (responsible for establishing the sports venues and facilities), construction companies and tourist boards.
- Environmental and cultural associations, together with the education and research sector, who have been involved in an important series of reflections over the future of a widely transformed city, and over the effects on the quality of life of its inhabitants.

We can broadly distinguish here between 'project territory' actors (those who made the mega-event possible, in particular TOROC), and 'context territory' actors (those who were not directly involved in event organization, but who were nevertheless influential actors in Torino 2006 networks). A number of actors (in particular, the municipal authorities) straddle the line between these two categories, carrying out measures targeting both the material aspects of spatial planning and urban imaging (Dansero & Mela 2007). Their work was significant in that it involved civil society actors, above all through communication strategies designed to create legitimization and consensus, to overcome the elitism of the Olympic bid, which was a product of, as Andranovich et al. have typically identified, 'influential members of the city's business elite, with the endorsement of elected officials' (2001: 120). However, as was the case in other editions of the Games, critics and opponents were progressively marginalized.

Our analysis here focuses on this relationship between civil society actors and event organizers. A number of questions arise. To what extent were the concerns of local actors about the long-term effects of the Olympic Games, especially over environmental impacts, taken into consideration? To what extent did they lead to material, policy or organizational changes by Games organizers? To what extent was there conflict between the 'project territory' and the 'context territory'?

The first point to make is that, considering the risks and negative impacts on host localities driven by the implementation of the Olympic Programme, the Games cannot be considered a sustainable event. During the past decade not only the host regions but also the IOC have started to adopt longer-term perspectives, with the aim of building positive legacies for the populations involved (Cashman 2003). Among the tools provided by the IOC to candidate cities to improve event sustainability are the manual for candidate cities (MCC) – which, among other things, demands post-Olympic use of sports facilities – and the OGGI (Olympic Games Global Impact) set of indicators to measure the economic, social and environmental impact of the Games on the host region from the period preceding the selection of the city to two years after the event itself (Hug et al. 2002).

Turin and TOROC adopted a proactive approach to the impact of the Games (Dansero et al. 2006). Following the 'Green Card' included in the bid file and the 'charter of intents' (TOROC 1998a, 1998b, 2002), the Strategic Environmental Assessment (SEA), under Law 285/2000 concerning 'Interventions for the Winter Olympics Torino 2006', represents a very important tool. It was applied – for the first time in Italy, and for the first time in the field of mega-events – to the entire Olympic Programme (rather than to single actions, in contrast to Environmental Impact Assessments, which were used for the bobsleigh and ski jumping facilities), from the ex-ante project to implementation and the ex-post management of the Olympic infrastructures, in order to monitor, avoid or minimize the negative effects of the Programme, and to reinforce positive impacts (on the distinction between EIA and SEA, see the chapter by Caratti and Ferraguto in this volume).

The subsequent important step was the creation, within TOROC, of an Environment Department, composed of engineers and environmental experts (see Table 10.1).

TOROC obtained EMAS registration at the planning and construction stages for the 29 sites hosting the Games, including the training sites and Turin Olympic Village, and adopted the ISO 14001 management system. In the valleys, eight municipalities registered with EMAS and 12 accommodation structures were granted the Eco-label. Finally, TOROC worked in collaboration with other local agencies and public institutions to achieve sustainable procurement policies (choosing, for example, products that carried the Eco-label, or natural gas-powered vehicles) and to reduce GHG emissions to zero. This project was known as HECTOR (HEritage Climate TORino). Following the 2002 World Summit on Sustainable Development of Johannesburg and the 2003 Milan Conference on Climate Change, Torino 2006 was the third

Table 10.1   TOROC's environment department

TOROC's Environment Department		
Objective	Following SEA requirements, its task was to prepare the plans for preventing environmental damage and monitoring the quality of the environment:	
Plans	• General plan for	inert refuse
		sustainable mobility
		employee and community safety
		the prevention of natural risks
		water management
	• Environmental landscape plans	
	• Guidelines for the sustainability of the project, in the construction and operation of the Olympic and Multimedia Villages	
	• Environmental Monitoring Plan, which evaluates the quality of the environmental components from the building-yard phase to immediately after the event.	

global event to qualify as 'carbon neutral' (see the Introduction to this volume for further discussion on the problematic use of this term); overall emissions from transport, waste management, accommodation and sports facilities were estimated at 103,516 tons of CO_2e. In order to remediate emissions, TOROC identified several investments in reforestation, energy efficiency and energy production from renewable sources. Given TOROC's budget constraints, the Piedmont Region decided to co-finance a number of energy efficiency and renewable energy promotion projects in Piedmont; however, no data has been produced on the amount of CO_2 compensated.

The most prominent example of TOROC's efforts to set up dialogue and consultation over the Olympic Programme is represented by the ACA. It was a body provided for by TOROC's Charter in article 8, 'a place for consensus building' (Del Corpo 2006), where the projects were discussed prior to their approval by the Torino 2006 Agency. It was composed of representatives from 13 environmental associations (among which were the four associations from the Environmental Observatory 'Torino 2006') and ten local governmental institutions, among which were the Turin City Council, the Province of Turin, the Piedmont Region, and the Mountain Communities of Val Pellice, Val Chisone e Germanasca, Alta Val di Susa and Pinerolese Pedemontano.

From the environmentalists' perspective, the assembly was a way to ensure that the environmental impacts of Olympic infrastructure development were taken into consideration from the initial stages of the decision-making process, thus broadly following Sydney's model where Green Games Watch (in part established on the initiative of the Australian and New South Wales governments to coordinate proposals from environmentalists) was an important element in Olympic Programme decision-making and monitoring. In Turin, the ACA was an effective tool for identifying the extent and problems of Olympic projects, and for the dissemination of information about the Olympic Programme, by allowing associations and local governments to act as an intermediary towards host communities. For example, the environmental associations which participated in the ACA contributed to the EIA, leading to modifications of some projects, and reducing the number of snow-making systems. However, having only a consultative role, its main constraint was that only minor adjustments to the projects could be made, and even where this did happen, projects were not resubmitted to the ACA but directly changed and approved by the Torino 2006 Agency.[2]

Concerning the environmental governance and impact of the Torino 2006 Winter Olympics, a number of issues have emerged from surveys conducted before, during and after the Games, clarifying the viewpoints of the various local actors involved to different degrees in the event (Scamuzzi 2006, Guala & Crivello 2006, Bondonio 2006). In particular, Caratti and Lanzetta (2006) asked representatives from the Piedmont Regional Administration, TOROC's Environment Department, the Polytechnic of Turin, ARPA

Piedmont (the Regional Agency for the Protection of the Environment) and Comitato Nolimpiadi! (the main opponent of the 2006 Games) their opinion about the quality and effectiveness of the Torino 2006 environmental programme, with specific attention to the SEA procedure. The analysis underlined some peculiarities and limitations of the decision-making process. For example, the fact that environmental assessment was only introduced after the Olympic Programme had been established, and money had already been spent on specific structures, reduced the possibility of choosing whether to build or to locate a facility in a specific place.

Dansero and Mela carried out a series of interviews in two different phases (2003 and 2007) with the aim of monitoring the opinion of mayors and presidents of Mountain Communities, as well as business leaders, social actors and experts, from the Turin metropolitan area or the Olympic valleys. Whereas in 2003 the concerns about the environmental impact of the Games from initial preparation to the post-Olympic period were prevalent, attention is now mostly focused on the capacity to exploit the visibility offered by the event (Dansero & Mela 2007).

The 'voices of dissent'

We used qualitative analysis to investigate the associations, groups and spontaneous movements which mobilized against the project of hosting the XX Winter Olympics in Turin and the Alpine valleys, from the bid onwards. We paid special attention to those associations which maintained their action over a number of years (conducting semistructured interviews in 2003, 2006 and 2008, with representatives of each association in Table 10.2), in order to identify changes in opposition motivations and strategies.

In general, the mobilization against the Turin Winter Olympics comprised a large number of counter-globalization and environmental movements, both of which are well covered by the existing international literature (Williams & Matheny 1995, Routledge 2003, Juris 2004, Featherstone 2005, Lewicki et al. 2003). In the 1980s and 1990s in particular, many studies saw neo-liberalism as responsible for an epochal transformation which constrained the organization of broad-based, socially transformative movements, facilitating instead fragmentary opposition to specific and precise targets (Thériault 1987, Melucci 1989, Hamel 2000). However, more recent analysis has highlighted the rise of movement coalitions which, though retaining a wide variety of objectives, values and models of organization, have been able to achieve some ideological convergence and shared forms of action (Della Porta & Tarrow 2005, Ruggiero & Montagna 2008). For some, this convergence is the sign of a new 'class struggle' (Harvey 2006); others emphasize the opportunities offered by international events, such as the UN Rio (1992) and Johannesburg (2002) conferences, G8 meetings and sports

Table 10.2 Organizations/associations which mobilized against the project of hosting the XX Winter Olympics in Turin and the Alpine valleys

Organization/Association		Environment sphere	Ethics sphere
International Associations (Italian Branches)	Amnesty International		– Collaborated in drawing up the Charter of Intent for Social Responsibility (Carta di Intenti in tema di responsabilità sociale) – Was involved in the critique of one of the Olympic sponsors
	Cipra	– Focused on monitoring the event – Disseminated information and data abroad – Supported the local governments of the Olympic valleys in the EMAS process – Concerned about the impact of major events in Alpine areas	
	Greenpeace	– Advocated use of the already existing bobsleigh facilities in Albertville – Concerned about heavy environmental impact	
	Mountain Wilderness	– Launched a petition to stop the project to floodlighting Mount Monviso for ten days during the Games. The project was withdrawn thanks to the intervention of the Piedmont regional administration	
National Associations	Environmental Observatory	– Sent TOROC comments on SEA and on the EIA of the Olympic projects, contributing to the improvement of some of them, such as artificial snow and freestyle – Concerned about the environmental impact of the facilities built in the Alpine valleys (in particular the bobsleigh, ski-jumping and artificial snow) and the contrast between the needs of the event and the welfare of the host communities	

	Greens Group	– Piedmont delegation: Underlined the spiralling increase in organization costs, as well as the episodes of corruption which emerged in the course of secondary construction work – Greens national branch: expressed its concerns over attempts to replace Italian doping laws with the IOC's less stringent regulations, and over the lack of consensus over projects to be carried out in Circonscrizione 2 in particular
	Lilliput Network	– Jointly with other associations has promoted the web site www.giocapulito2006.org (no longer active) highlighting the inconsistency between sponsors who sell weapons and the Olympic Truce initiative. Thanks to Lilliput's lobbying, in June 2002 TOROC published the Carta d'Intenti – Focused on the use of public funds for the Olympics and security in the construction sites
Local Associations	Nolimpiadi Committee	– Opened a website (www.nolimpiadi.8m.com) to question and disseminate information about Torino 2006 – Initiated a series of civil procedures by the European Union and the T.A.R. (Regional Administrative Court) – Concerned about: The presence of 'amoral' sponsor companies, and the environmental impact of the event.
	Turin Social Forum	– Set up a working group to study the Turin Strategic Plan, to develop alternative proposals and inform the public, and implemented actions aimed at reducing the environmental and social damage caused by the event – Criticized the waste of energy – Criticized unethical sponsors, the exploitation of illegal workers and the many accidents which occurred in construction, as well as public order issues – Criticized the Mayor's decisions to forbid rallies and strikes and to adopt reserved road lanes

events. In these analyses, emphasis is placed on the temporary connection between different cultures and 'repertoires of action' (Della Porta & Mosca 2003, Della Porta 2006) in opposition to a common adversary (in this case, neo-liberal capitalism), different features of which are foregrounded by each movement, such as the dominant position of multi-national corporations (Juris 2008), the exclusion of minorities (Ong 2006) and the destruction of the environment (Heynen & Robbins 2005).

This latter approach is highly relevant to our investigation of the Turin Games protest, which marked a momentary convergence of groups and organizations of different ideological orientations and movement structures, that is global justice and environmental groups, often affiliated to national and international organizations. The former have a significant representation in Turin, and also in the Susa Valley, where they are a basic element of the opposition to the construction of the Turin-Lyon high-speed railway line (known locally as the No TAV movement) (see Bobbio & Dansero 2007). The latter – which are also part of the No TAV movement – have taken up the legacy of a tradition of Italian environmentalists, active for at least the past three decades (Della Seta 2000, Della Porta & Diani 2004).

The various association/organizations considered are presented in Table 10.2, maintaining the distinction between international/national/local associations, with a description of the principal features of each action (for further information, see Del Corpo 2006). The international organizations concerned – Amnesty International, Cipra (International Commission for the Protection of the Alps), Greenpeace and Mountain Wilderness – mobilized and intervened through their national branches; both the Lilliput Network (Rete Lilliput – not an association per se but a network coordinating the initiatives of the various associations, groups or individuals who signed up to its manifesto) and the Italian Green Party (Verdi) were active through their Piedmont delegations, the national environmental associations united in the Environmental Observatory 'Torino 2006'. This was established in 1998 by the regional delegations of the principal Italian environmental associations – Italia Nostra, Legambiente, Pro Natura and WWF – with the aim of jointly evaluating the bid file presented by the Torino promotional organization.

The following year, these four associations, together with Comitato Nolimpiadi! (established in late 1997 as a reaction to the destruction of the Alpine environment caused by that year's Alpine skiing World Championships in the purpose-built resort of Sestriere), Greenpeace, Mountain Wilderness Italia and other environmental and peace movements, together with representatives from Italian cultural and political associations, created the Coordinamento Controlimpiadi. Following Turin's selection as host city, the Environmental Observatory adopted a constructive approach, with the aim of influencing the decisions of the organizing committee. In fact, the Observatory was not at any point opposed to the Turin Olympics per se, considering that the Games would represent an opportunity for Turin's urban renewal. The Observatory also joined the Assemblea Consultiva Ambientale, which it had championed from its initial stages.

A further set of civil society actors came together in the Turin Social Forum (TSF), a network of 44 associations, parties and trades unions, established in 2001 at the time of the mass rallies against the Nice European Council meeting, and, finally, at the local level, two noteworthy ad hoc movements: the Comitato Spontaneo di Quartiere della Circoscrizione 2 (Spontaneous Committee of Borough 2, which despite its name is not a 'spontaneous' movement proper, but rather a body established 30 years ago to monitor the city administrative authority) and the Comitato per la Difesa della Piscina Olimpica (Olympic Swimming Pool Defence Committee). Both these groups are of local significance in that they sought to protect the rights of users of services and public space in Turin's Borough 2, which was affected by the construction of the ice-hockey stadium, a major Olympic venue. However, in the discussion that follows we do not analyse these two movements in detail since our attention is focused on the associations concerned with the Torino 2006 Winter Olympic event (see Table 10.3).

We argue that in the case of Torino 2006, it is barely possible to contend that there was a real 'opposition movement' for two reasons. First, because opposition failed to gain widespread popular support in the locality. The majority of residents in the city and in the valleys didn't take part in the discussion, the critique and the protests concerning either the candidacy or the subsequent Olympic Programme. Rather, the protest was limited to a small number of organizations (though with a large membership, as in the case of the environmental associations), without developing into a truly collective effort (indeed, in this context it is significant that only two spontaneous movements formed). Moreover, this failure to mobilize the local population should be set against increasing polarization between the event's supporters and its critics. As the authorities became increasingly confident in the event, and the energies of local political and economic actors became increasingly stretched in the pursuit of urban transformations and event organization, dialogue with and response to critical voices became increasingly sidelined. At the same time, our analysis demonstrates that popular approval of the event also increased: in a series of public surveys, the median score given to the overall experience of the Winter Games (from 1 to 10, where 1 is the lowest score) was 6.9 in 2004, 7.3 in 2005, 9.1 in 2006 and 8.1 in 2007 (for further discussion, see Guala 2007).

Second, because few of the critical voices were opposed in principle to the staging of the event. Analysing the 'opposition front' in greater detail, we can distinguish between 'maximalists' – who articulated a comprehensive critique of the event (Comitato Nolimpiadi! and TSF) – and 'co-operators' – who cooperated with the organizers (Environmental Observatory, Cipra and Amnesty) or tried to negotiate with TOROC and the Agency (Rete

Table 10.3 The opponents and their principal characteristics

Classification according to the scale of action		Appearance and involvement (highlighted parts) in the Torino 2006 life-span (from candidacy to the post-Olympic period)										Sphere of action		
		1998	1999	2000	2001	2002	2003	2004	2005	2006	2007	Environmental sphere	Ethics sphere	Civic sphere
International associations (Italian branches)	Amnesty International					▓	▓	▓	▓	▓			✧	
	Cipra			▓	▓	▓	▓	▓	▓			✧		
	Greenpeace		▓									✧		
	Mountain Wilderness	▓	▓				▓	▓				✧		
National associations	Osservatorio Ambientalista				▓	▓	▓	▓	▓	▓		✧		
	Rete Lilliput					▓	▓	▓	▓	▓			✧	
	Gruppo Verdi							▓	▓	▓		✧		

Local associations																			
Comitato Nolimpiadi!		❖		❖															
Turin Social Forum		❖		❖															
Comitato Circoscriz 2																		❖	
Comitato per la piscina																		❖	

Lilliput and the Comitato per la Difesa della Piscina Olimpica) or with Turin City Council (Rete Lilliput and the Comitato Spontaneo di Quartiere della Circoscrizione 2). However, even maximalists didn't resort to direct action (illegal, non-violent or violent) (see Table 10.4), but pursued opposition through legal and conventional channels. Legal challenges were ineffective. In brief, except for the Comitato Nolimpiadi! and TSF, despite initial opposition to the idea of organizing the 2006 Games in Turin and in the Alpine valleys, oppositional actors adopted a prevailingly collaborative, positive and proactive stance towards the event.

Only during the candidacy and in the immediate wake of Turin's selection as host city was there evidence of a coalition and collaboration between the main associations engaged in detailed study into and debate over the impacts of the mega-event; this was the Coordinamento Controlimpiadi, formed in 1999. After Torino won the Games, Greenpeace (considering this 'victoria' to be an own goal) left the scene while the remaining actors became increasingly 'specialized', choosing different topics and strategies. The most important fact here is the Environmental Observatory's decision to co-operate with TOROC, joining the ACA, whereas the others adopted increasingly fundamental opposition to the Games. In the meantime, Rete Lilliput and TSF entered the debate, introducing the issues of the social responsibility of Games organizers through the 'Biancaneve' (Snow White) campaign. Biancaneve aimed to pressurize TOROC into defining clear rules concerning human and environmental rights, governing the selection of suppliers, contractors and sponsors.

Thanks to the Biancaneve Campaign, in June 2002 TOROC published the Carta di Intenti in tema di responsabilità sociale, or Charter of Intent for Social Responsibility, committing itself to making the Games 'an occasion for peace, tolerance, justice, freedom, solidarity and equality among peoples and individuals' (TOROC 2002: 2). This, in fact, is the first time that a local organizing committee had adopted a charter based on international standards in this area. A third current of opposition to the 2006 Games was thus opened up, alongside the environment (represented by the Environmental Observatory and the ACA) and the 'sharp opposition front' (which didn't engage in any form of dialogue). However, despite the success of the Biancaneve campaign, we suggest that the overall position amongst opposing groups is one of division rather than productive diversity. Apart from Biancaneve and the Giocapulito ('play right') website through which Rete Lilliput and the Environmental Observatory tried to repeat the Coordinamento Controlimpiadi experience, and apart from the diffusion of the information about each association's initiatives, we were unable to observe significant episodes of unitary action. Later, the emergence of the spontaneous committees caused another division in the 'opposition group': between those aiming to preserve common interests and those (the committees) wanting to safeguard very specific interests.

Table 10.4 Classification according to tactics

		Media				Actions					
		Press	Internet	Rallies	Public debates	Seminars and working groups	Participation in conferences and hearings	Law suits	Complaints against TOROC and the Agency	Comments on ESA and single projects	Petitions
International associations	Amnesty International		X								
	Cipra Italia		X							X	
	Greenpeace	X	X	X			X				
	Mountain Wilderness	X	X								X
National associations	Environmental Observatory	X	X		X	X	X		X	X	
	Rete Lilliput		X	X		X	X				X
	Greens			X				X			
Local associations	Comitato Nolimpiadi!		X		X			X		X	
	Turin Social Forum	X				X					
	Committee of Borough 2				X		X			X	
	Swimming Pool Committee	X						X			X

Even though such complexity of position and argument may in principle be considered to be very positive, in this specific case it nonetheless did not increase the bargaining power of dissenting groups, which consequently staked out separate positions from which to confront their 'enemy'.

Assessing the opponents: two years after the Olympics

Two years after the Torino 2006 Winter Olympics, the representatives of the associations and organizations which had opposed the event were again interviewed in order to verify their opinions (see Table 10.5). In general, the attitude that emerges from all interviews is of generalized disappointment: everyone agrees that the successes of the opponents were rather marginal and had a very limited impact on the final balance of the Olympics. This is confirmed by the fact that, according to almost all the interviewees, the negative consequences initially foreseen actually occurred. However, this general disappointment varies according to the initial positions of the associations in relation to the Olympics. On the one hand, those associations that did not adopt a fundamental oppositional stance towards the Games now acknowledge that even when their constructive proposals were accepted, they produced few tangible results. On the other hand, those associations which radically rejected the Olympics emphasize the difficulty experienced in obtaining wide media coverage for their positions, and denounce the limited attention paid by the justice system to their complaints.

Furthermore, in a number of interviews, the disappointment is associated with aspects of self-criticism. Here, the inescapable conclusion is that the actions taken by the associations failed to generate a public environmental consciousness: this result is attributed by the actors interviewed both to the indifference of institutional stakeholders and to the insufficiency of the actions of the environmental movements themselves. In essence, the general climate of consent perceived in relation to the event, the restricted timeframes and the limited resources available to the associations influenced the action of the opponents – especially of those who, thanks to their deep-rooted position in the territory, perhaps could have taken more incisive action to try to increase public awareness of the problems of the sustainability of the Games development model in the Alpine context.

During the interviews, special attention was devoted to two specific topics: the Olympic legacy and development scenarios. Concerning the positive and negative aspects of the Olympic legacy, the interviews highlight in particular the negative effects; however, it is interesting to observe that a fundamental distinction emerges between the perceived effects in the Alpine Olympic territory and in Turin (see Table 10.5). As regards Turin, only the most radical opponents (Comitato Nolimpiadi!) do not identify any positive legacy of the Olympics; the others maintain that Turin has succeeded

Table 10.5 The Olympic legacy in opponents' opinions

	Olympic valleys	Turin
Positive effects	– Renewal of sports facilities – Slight increase in tourism, in particular foreign tourism	– Improved image of the city – Improved road system – Slight increase in tourism – New mental attitude of Turin inhabitants – Reconversion of industrial areas – Energy-saving solutions
Negative effects	– Damage to the environment and to the landscape – New buildings with the subsequent use of the land – Under-utilized sports facilities – Failure to develop a new model of tourism – Weak occupational returns – Exclusion of the 'half-way land'	– Financial shortfall – Unused facilities – Failure to attract investors – Failure to apply sustainability criteria – Failure to create a relationship with the mountains – Increased building volume – Minor increase in tourism

in improving its image abroad, and some also highlight a contribution to changing the mental attitude of the inhabitants of Turin (such as greater attention to the international tourist dimension by traders, a feeling among citizens that it is a city with renewed dynamism and potential and so on). However, some interviewees emphasize that there is no reasonable trade-off between costs and benefits: the city's renewed image could also have been achieved for a much lower financial commitment (some events and secondary initiatives could have been avoided). Three main critical aspects were highlighted in the interviews:

• The existence of a financial debt that could result in reduced social services and the difficult reconversion of the infrastructures built for the Olympics. At the end of the Games (March 2006), the TOROC budget was running a deficit of 33 million USD. In TOROC's final year of activity (TOROC was formally wound up in March 2007), this figure was reduced to 4 million USD (at constant 2000 prices), corresponding to 0.3 per cent of TOROC revenue, thanks to contributions and a rebate ('riconoscimenti') from the IOC and the Italian NOC and the sale of Olympic Village buildings (Bondonio & Campaniello 2006);

• The positive effects invoked as a justification for the significant financial commitment for the Games, namely the predicted increased tourism returns and the attraction of potential investors, have not materialized. In

2007, Turin saw a sizeable drop in tourist arrivals and stays, which shrank by 10.85 per cent and 26.05 per cent respectively on the previous year (Dansero & Puttilli 2010). This downturn, however, was offset by the increases in the metropolitan area as a whole, where arrivals rose 20.67 per cent and stays by 17.74 per cent. Recently, differing opinions as to how these figures should be interpreted have sparked a lively debate in the local media as well as in the world of politics. Some (for the most part tour operators and local researchers) see these figures as cause for alarm, pointing to the possibility that the effect of the Games was only temporary, and that the city's attractiveness to tourists is waning even more than was expected before the Olympics. For others (for the most part local and regional authorities, and public tourism agencies), the figures are more likely to indicate that the city is in the process of embedding within the national and international tourist routes, and that the increase in tourist flows to the metropolitan area is an encouraging trend. Indeed, they say, it underscores the advisability of a unified approach to promoting the metropolitan area as a whole, identifying synergies between the Turin area and its surroundings rather than seeing the city as ending at its municipal boundaries, unconnected to the resources that lie beyond (Dansero & Puttilli 2010);

* The missed opportunity of using sustainable techniques for the Olympic Village and sports facilities.

In contrast, the balance outlined by interviewees is much more negative as regards the Alpine area, on which their attention is focused more. The positive legacies are only cited by a few and refer exclusively to the renewed sports facilities and to the slight increase in tourism. Vice versa, the negative effects are fully illustrated by everyone and concern two aspects in particular:

* The failure to transform the tourist model: the Olympics have confirmed the role of the upper valleys as locations for winter sports and have not favoured any form of diversification. In addition, the returns of the Games are modest as regards new employment and mainly concern foreign seasonal workers.
* The negative impact on the environment: the negative change of the landscape associated with the construction of the sports facilities and the lack of real attention to environmental issues by local administrators and TOROC (such as the possibility of re-using the Olympic sites, the impact on the landscape of the bobsleigh track in particular, etc.).

As regards the opponents' analyses of the current situation and the possible future scenarios for the Olympic territory, it should be stressed that, except for some individuals more accustomed to dealing with broader topics, interviewees focused their attention only on specific criticisms, especially associated with environmental features, and were interested in particular in the mountain areas. The answers emphasized that, despite the temporary cooperation between Turin and the valleys during the Olympic cycle, the two contexts remain substantially far apart and the valleys take second place in relation to the city.

For Turin, the Olympics represented an incentive to renew its image even though the city displayed growing indifference to the surrounding context and its tradition (Bondonio et al. 2008). In contrast, for the valleys the Games represented a missed opportunity for recovering their identity. In fact, the valleys' identity has been compromised by a long industrial season which obscured local culture and subordinated it to the central city, undermining the capacity of local institutions to open up to different perspectives and to develop an authentic welcoming culture.

In relation to possible future positive scenarios, only a few interviewees focused on Turin, stressing the need to solve traffic congestion and develop new forms of tourism which do not distort urban identity. With reference to the valleys, besides the more radical positions targeting the destruction of everything built for the Olympics, the majority of the recommendations refer to a possible alternative socio-economic model including the following features:

- The enhancement of the territory's resources: landscape and cultural assets, local products and handicraft
- A tourism able to break away from 'snow-related monoculture'
- Attention paid to activities with low environmental impact
- Encouragement to less advantaged users, such as low-income families, the elderly, young people
- The capacity to reconvert unused historical assets

In spatial terms, in general, the perceived priority is the restoration of environmental systems, reforestation and the reduction of mobility impacts by increasing public rail transport services. These themes are shared by the majority of interviewees. Conversely, on the relationship between Turin and the Alps, two different positions emerge:

1. Scepticism over the possible strengthening of the relationship between the city and the mountain since the two territorial systems are seen as being too different to be able to develop mutually beneficial relations. In this perspective to speak of Turin as the capital of the Alps is meaningless.
2. The need to ensure more structured links between the city and the mountains to promote the reconversion of the valleys and a relationship based on the awareness of the specific nature of the Alpine area and of its 'charm'.

Conclusions

The opponents of the Turin Games formed a heterogeneous front characterized by different positions; however they tried to focus public attention on some critical aspects of the Games, especially environmental problems. They attempted to create their own counter-information structures, such as the Comitato Nolimpiadi! and the 'Environmental Observatory', which significantly remained separate at a local level and did not manage to create a solid and effective coordination with the national leadership of the Italian environmental associations or with an international no-Olympics and counter-globalization movement. The protests also concerned the excessive concentration of financial flows on the physical infrastructures needed for the Olympics, to the detriment of other social uses, the excessive use of energy, the overexposure of Turin abroad which ended up by reducing the visibility of the mountains. Other protests concerned the under-consideration of the territory's cultural resources, the limited attention paid to ethical aspects in international supply relations, the subordination of local interests to the interests of the multinationals sponsoring the event and the exclusion of the lower valleys from the Olympic relations system.

However, opponents failed to become an effective social movement, in contrast with the opposition movement to the other major infrastructural changes which involved and are currently involving the entire Olympic territory, for example, the new high-speed TAV rail link between Italy and France. It is not by chance that the attempt to block the Olympic Torch on its arrival in the Olympic mountains of the Susa Valley was organized by the protesters of the No TAV movement, which intended to exploit the overexposure provided by the event, and was not organized by the associations that challenged the mega-event as such. However, various other counter-information and mobilization initiatives were crushed by the overwhelming promotion of the mega-event and failed to achieve counter-information outside a limited circle of local environmental activists.

This failure doesn't imply that the concerns of the opponents were meaningless. Actually, in many cases their criticisms of the environmental aspects of the Games' organization proved to be based on realistic analyses, and are now more widely shared by public opinion. However, especially in the period closer to the Olympic event, they underestimated the attraction created by the Olympic place, a supernova sparkling brightly throughout the 15 days of the Games, successfully developed in particular in Turin, where the inhabitants experienced the event with great popular involvement and great pride.

Notes

1. This distance is the longest so far between the Olympic venues of any Winter Games. Turin is the largest of all European and American Winter Games host cities (Guala & Crivello 2006).
2. Better results could probably have been achieved had the assembly intervened not at the planning stage but during project implementation, in order to monitor effects and to counterbalance the absence of environmental experts within the agency.

11
Vancouver 2010:
The Saga of Eagleridge Bluffs

David Whitson

From the outset, Vancouver's successful bid to host the 2010 Winter Olympics was a 'corporate-civic project'. The agendas of the local growth coalition – city and provincial politicians, as well as major players in the local business community – involved showcasing Vancouver as a destination for global investment, and revitalizing the position of Whistler in the intensely competitive global tourism market. In this, of course, British Columbia (BC) political and business leaders were following a now familiar script in which Olympic Games and other mega-events are understood as opportunities to demonstrate the attractions of a city/region to global visitors and investors. Indeed, pursuing mega-events and promoting them as catalysts for the competitive repositioning of a city is a strategy that has been tried before in Canada, in Montreal and Calgary (Whitson 2004) and in other countries, too (see, for example, Bennett 1991, Whitelegg 2000, Hall 2006, Horne & Manzenreiter 2006). In BC, the provincial government has sought to capitalize on Vancouver 2010 by improving the transportation infrastructure serving Whistler (and other ski resorts, too), and it has viewed this as an investment in the growth of the BC tourism industry (British Columbia 2004).

Certainly, one of the widely perceived 'weaknesses' of Whistler in recent years has been that the Sea-to-Sky Highway (Hwy. 99), built to connect Vancouver and Whistler when the resort opened in the 1970s, has become inadequate to handle increasing traffic flows, and the 2010 Olympic bid included a plan to bring Whistler 'into the 21st century' with a multi-lane motorway connecting the city to the resort. With no space to widen the existing highway in places where the coastal mountains push right up against the Pacific, the main options were to tunnel through the mountains, or to drive a new route over the mountains at a place called Eagleridge Bluffs.

The government of British Columbia chose the latter option, believing it to be the simpler engineering challenge, as well as significantly cheaper.[1] However, the Eagleridge Bluffs route was strenuously opposed, both by environmentalists and by local residents whose views, recreational options and property values all stood to be damaged by the construction of a motorway through this very beautiful part of West Vancouver. These groups tried the mechanisms normally available to citizens opposing development proposals – demanding environmental impact studies and public hearings, and ultimately seeking the intervention of the courts. However, the province was determined to get the new highway completed well before the Olympics, and construction began in the spring of 2006. Opponents of the highway resorted to civil disobedience, placing themselves in the path of construction equipment. The BC government responded by taking out civil injunctions against the protesters, several of whom went to jail. One of these, a 70-year-old First Nations woman, contracted pneumonia while in prison; she was admitted to hospital on her release and died shortly afterwards.

One important dimension of this narrative, obviously, is the destruction of the environment of Eagleridge Bluffs itself, despite promises that Vancouver 2010 would be the 'greenest Olympics ever'. The Eagleridge Bluffs rose steeply above the old Hwy. 99 for 2–3 kilometres (Google 'Eagleridge Bluffs, West Vancouver'.), and were the site of a 500-year-old dry arbutus forest, the last old growth arbutus forest on the North Shore. This coastal forest has provided nesting sites, every spring, for bald eagles and over 20 other species of protected migratory birds, as well as for a number of rare native plants. The adjacent Larsen Creek watershed/wetlands also provided critical habitat for several additional species of endangered flora and fauna.

Another critical part of the story, of course, will be how the government of British Columbia rode roughshod over citizen opposition in its efforts to ensure that Olympic-related infrastructure was completed on time. We will see that the BC Ministry of Transport appeared determined to press ahead with a highway that destroyed the ecological integrity of Eagleridge Bluffs, and did not seriously explore options suggested by civil society groups concerned about this destruction. We will also see that public agencies with a mandate to monitor and protect wildlife populations (notably the Canadian Wildlife Service) did not do their jobs in a timely fashion; nor did they support the efforts of local environmentalists to draw attention to the effects of habitat destruction. It will be argued that Canadian laws protecting wildlife are ignored when they conflict with commercial contracts, or with the development agendas of provincial governments, and that this privileging of economic rights and obligations also extends to the Canadian courts, adding to the challenges faced by civil society opposition.

A final dimension of the story of Eagleridge Bluffs, though, concerns why the protest failed, and why those who fought to save the Bluffs did not attract wider public support in their confrontation with the BC government. Here, it may be important to understand that West Vancouver is the most affluent municipality in the Vancouver metropolitan region (indeed, in the entire province), and that homes developed over the post-war decades by British Pacific Properties in the hills overlooking Eagleridge Bluffs are among the most beautifully located – and expensive – homes in Canada. It will be proposed that even though the environment of Eagleridge Bluffs was recognized to be of great ecological significance, and even though the government's treatment of the protesters aroused some public unease, support for their cause was undercut because it could be portrayed as a self-interested campaign by homeowners whose million-dollar views stood to be spoiled by the new highway and its attendant traffic. To critics in the Vancouver media, and to some environmentalists who had been active in other protests that divided BC in the 1990s, affluent professionals from West Vancouver were seen as acting more out of 'NIMBY' interests than a genuine concern for the environment when they opposed the highway. And indeed, although the new road has destroyed a rare and significant piece of habitat, property development has been eroding wildlife habitat in West Vancouver for over three decades, and the suburban communities that some of the protesters live in, and the roads that access them, are part of this history of destruction.

Much of the literature on mega-events has focused on the impacts that global sporting events and World's Fairs have had on the poorest and most marginalized in the host communities, and for very good reasons. All too often, preparing the city/region to host a mega-event has meant the sort of urban redevelopment that clears poor people out of downtown cores (Reasons 1984, Olds 1989, Whitelegg 2000). In other instances, the public spending devoted to facilities and infrastructure for the mega-event has been accompanied by cutbacks in spending on public services in the city/state/province – cutbacks that typically hurt those poorer people whose lives are most dependent on public services and facilities (Ley & Olds 1988, Wilson 1996, Whitson 2004). These kinds of impacts clearly matter, and each can be found in Vancouver and BC as the province prepares for Vancouver 2010 (Shaw 2008). This chapter, however, focuses on a different struggle in which citizen groups opposed the government's plans for an Olympic-related road improvement, on environmental grounds, and it raises the counter-intuitive question of whether opposition to mega-events that is led by socially privileged protestors can be undermined by this very fact.

Eagleridge Bluffs: in the wrong place at the wrong time

The events summarized here followed from Vancouver/Whistler's successful bid to host the 2010 Winter Olympic Games. Vancouver business leaders and British Columbia's political leadership were always strongly behind the Olympic bid. However, Vancouver's new mayor as of November 2002, Larry Campbell, a populist former coroner, had at first taken a sceptical stance; indeed he had campaigned on a promise to hold a referendum on hosting the Olympics. However, during the subsequent referendum campaign held in early 2003, Campbell, who was the nominal leader of a loose centre-left coalition on Vancouver City Council (the Coalition of Progressive Electors, COPE) came out in favour, urging the citizens of the city to support the Olympic bid, which they did by a margin of 64–36 per cent. It was, in part at least, the vocal support of the 'Yes' side by Campbell and another COPE councillor and community activist Jim Green that helped to convince some voters that Vancouver was committed to staging a 'sustainable Games', that the Olympics could bring infrastructure benefits that would benefit Vancouver's poorer citizens (including new public housing and improved public transportation) and that Olympic-related decisions would be open to public debate and influence.

In practice, not surprisingly, subsequent events have raised serious questions about these promises. Mayor Campbell became increasingly frustrated (and unpopular) as he attempted to mediate between the business interests and IOC interests represented in VANOC (the Vancouver Olympic Organizing Committee), the government of British Columbia (who were providing most of the money, and making many of the important decisions), and some of the more left-leaning COPE councillors (and Olympic critics) and the community interests they represented. Campbell effectively functioned as an independent in the last of his three years in office, and he did not run again in 2005. He was succeeded by the more right-leaning Sam Sullivan, who has likewise failed to build bridges between increasingly polarized groups.[2]

The back story here, or at least part of it, is that in Canada, cities have very little power to make important infrastructure decisions. They are limited in their tax powers; they are further limited by the often competing ambitions of neighbouring municipalities; and they are limited finally by their dependence on provincial governments for the funds for virtually any major capital initiative. This 'normal' powerlessness of Canadian cities is exacerbated when 'mega-events' like an Olympics or a World's Fair are undertaken, because it is provinces that have responsibility for regional transportation infrastructure and for providing the necessary funds. In practice, then, it has been the government of British Columbia, not Vancouver, that has called the shots in regards to major Olympic expenditures, including the improvements to the Sea-to-Sky Highway (see Note 1).

It is important to realize here that the Sea-to-Sky Highway, linking Vancouver and Whistler, is absolutely fundamental to the success of Vancouver 2010. Whistler has a reputation as a 'world-class' skiing destination, and it was host to all the skiing events in 2010. Although Whistler itself has lots of tourist accommodation, it was still necessary for many people – coaches, officials, media and spectators – to travel between Whistler and Vancouver, sometimes repeatedly. The old Hwy. 99, built in the 1970s, was a two-lane road in many places, winding along an often precipitous coast before climbing into the mountains. It is regularly subject to poor driving conditions in winter, and has been the site of many traffic delays and serious accidents over the years. These problems have increased as Whistler's skiing has expanded and more Vancouverites visit the resort for the day, and as the permanent populations of Whistler and the bedroom communities en route (e.g. Squamish) have also increased. A particular bottleneck exists between the West Vancouver suburbs of Caulfeild and Horseshoe Bay, where Whistler-bound traffic is often delayed by traffic backed up for the Vancouver Island ferries that depart from Horseshoe Bay. In its original form, Hwy. 99 was clearly not able to safely carry the volume of traffic needed for the Olympics, and this was noted by several of the IOC members who visited the area before Vancouver was awarded the Games. The Vancouver bid acknowledged these concerns, and included a commitment to improve highway access to Whistler (VANOC 2003: 67).

It was in this context, then, that the BC Ministry of Transport commissioned a study of the costs of different alternatives in late 2003 (previous studies of potential highway upgrades had been done in 2001 and as early as in 1993). The options presented to the public by the ministry in March 2004 were $130 million for a new four-lane bypass going 2.4 km over Eagleridge Bluffs from Caulfeild and routing Whistler traffic around Horseshoe Bay; or almost $200 million for a 1.4 km, four-lane divided tunnel *under* Eagleridge Bluffs that would have left the arbutus forest on the Bluffs and the nearby Larsen Creek wetlands intact. Other options identified, but not costed in a formal way, were to expand the existing highway to three lanes, or to build an entirely new highway from Vancouver to Whistler on the inland side of the Coast Range, thus avoiding the coastal suburbs that comprise West Vancouver altogether. However, the former would have done little to alleviate congestion around Horseshoe Bay, while the latter would have cost far more than improving Hwy. 99. Thus the Ministry of Transport announced its intention to proceed with the overland bypass.

Opponents of the overland route (the Coalition to Save Eagleridge Bluffs, CSEB) pointed out that the ministry had 'rounded up' (to $200 million) the consultant's estimate of $187.4 million for the tunnel, and that even this lower figure had included almost $30 million in auxiliary costs (for overpasses to separate ferry traffic from Whistler traffic) that were not included in the estimates for the overland route, even though both options would have required the construction of similar exits and overpasses. The estimate of $130 million for the overland route, moreover, included only $22 million for lands to be expropriated from the District of West Vancouver and from British Pacific Properties; this ignored the likelihood that compensating British Pacific would involve a much larger settlement. The difference in the projected costs of the two routes, in other words, was estimated by CSEB to be much less than the $70 million claimed by the ministry, and perhaps as little as $10 million (CSEB 2006a). Finally, the Coalition noted that the government presented only 'ballpark' figures for the option of adding a third lane to Hwy. 99 (claiming that this, too, would cost close to $200 million), when other opinions – from contractors, as well as from engineers consulted by the Coalition – suggested figures between $40 million and $100 million. What is clear, in any event, is that estimates of the costs for each of the highway options differed wildly. What is also clear, however, is that the BC Ministry of Transport was committed to the overland route and did not want to see work delayed by any further study of the environmental impacts of building a highway over Eagleridge Bluffs, or by a serious examination of the costs of alternatives.

At issue, of course, and the main reason there were such strenuous objections to the overland route was that it involved the levelling of a stand of dry arbutus forest on the coastal bluffs that is unique in Canada, and that provides important nesting sites for eagles and several rare species of migratory birds (including *Vireo solitarius*). The ecosystem also supports a threatened species of snake (*Charina bottae bottae*) as well as lichens, wildflowers and grasses that occur nowhere else on Canada's west coast. The adjacent Larsen Creek wetland provides habitat for one endangered species of frog (the Northern Red-legged frog), for more than a dozen migratory birds that Canada is committed to protecting as a signatory to the Migratory Birds Conservation Act (1994), and for other wildlife whose habitat has been radically diminished by the progressive suburbanization of Vancouver over the past three decades. However, the province was able to secure an Environmental Assessment Certificate (EAC) approving highway construction from the federal government in 2004. This was issued on the basis of a 2003 assessment by Environment Canada which found that the proposed highway would not have environmentally significant impacts.

In late 2005, West Vancouver District Council voted unanimously to reject the overland route and took the province to federal court, challenging the validity of this EAC on the basis that the 2003 assessment had been cursory. In particular, it had not included wildlife and plant studies in all seasons, and had thereby failed to note the presence of migratory birds, or of other fauna and flora species that are threatened by the destruction of these ecosystems. The federal court upheld the EAC nonetheless; but it admonished the province that obligations itemized in its Table of Commitments must be honoured – in particular, those requiring the filing of Environment Management Plans before the commencement of any construction work, and requiring the protection of bird life. This ruling in the government's favour severely disappointed many West Vancouver residents, as well as a group of BC biologists who formed the Eagleridge Environmental Stewardship Alliance (EESA). The latter alleged that the federal staff (from the Canadian Wildlife Service) who had performed the environmental assessment and issued the EAC had been derelict in their duty (or had been over-ruled, politically), and that the provincial Ministry of Transport was ignoring both Canadian and British Columbia laws – the

Migratory Birds Conservation Act (1994), the BC Wildlife Act (1996), the Species at Risk Act, Canada (2004) – as well as local development procedures requiring approvals and permits in its rush to proceed with construction. Meetings and protests were organized, but the ministry responded to this antagonism by cancelling several public meetings that were supposed to allow for public input.

In March 2006, logging commenced near Eagleridge Bluffs to clear forest for the overland route, but opponents quickly drew it to public attention that the Table of Commitments in the EAC clearly states that 'No clearing of vegetation is permitted during the general bird breeding time period of March 15 to July 31, unless pre-approved by the Canadian Wildlife Service' (CSEB 2006b). Logging was suspended on 16 March; however, it resumed in mid-April, with the ministry and the contractor now claiming to be operating on the basis of a 'phased' Environmental Management Plan, covering only that small section of the route (Eagleridge Bluffs itself) where construction of the new highway was to begin. The province's own Environmental Best Management Practices (approved in April 2005) call for complete environmental assessments (including avifauna surveys) prior to the start of any construction went unheeded, so the province was clearly ignoring its own guidelines here, as well as the terms of the EAC. The Coalition responded by filing a petition in the BC Supreme Court seeking a ruling that the EAC being used to authorize construction did not comply with the legal requirements of federal and provincial environmental legislation. They also sought an injunction prohibiting the contractor, Peter Kiewit & Sons, from doing any further work at the site until the required wildlife studies were done and a legitimate EAC issued. However, none of these legal remedies was granted by the court.

Civil disobedience

A situation had thus developed where a number of scientists and other professional people (including, but not limited to, well-educated homeowners in West Vancouver) believed that the courts were collaborating with a government determined to push the highway through, in defiance of both federal and provincial laws mandating environmental protection. For most of these upper middle-class professionals, civil disobedience was not something they had ever contemplated before; however, it now appeared to be the only avenue left open. In April, with logging poised to begin again, CSEB sent its supporters a carefully worded appeal, asking them to consider camping at the site. A small group responded, erecting a tent city at Eagleridge Bluffs and physically blocking the access of Peter Kiewit's workers and equipment to the site.[3] Kiewit responded by seeking a court injunction that would prohibit protesters from trespassing or otherwise blocking access to the worksite, on the grounds that their contracts included financial penalties if the work were not completed on time, and this was granted. Lawyers for the CSEB sought a stay of this injunction. However, this was rejected (some legal opinion suggests that the time-related penalties in such a contract were unusual, and perhaps a masterful pre-emptive move that anticipated protest activity) and on Wednesday, 25 May 2006, BC Supreme Court Judge William Grist signed an order authorizing enforcement of the injunction he had granted the previous week.

The following morning, police moved in and arrested 23 protestors, charging them with violating a court order (to vacate the site). Those arrested included Dennis Perry, the official spokesperson for Coalition, and his 19-year-old daughter; Betty Krawczyk, a 77-year-old veteran of many of BC's anti-logging protests; and 70-year-old Harriet Nahanee, an elder of the Squamish First Nation whose land Hwy. 99 passes through further along the route to Whistler. They also included a number of middle-class residents of West Vancouver. The unusual characteristics of those arrested – two grand-mothers, along with middle-aged, middle-class professional men – helped to make the arrests a major news story for several days. A lawyer for the protestors attacked the government and the contractor for having used injunctions the way they had (i.e. for having the protestors charged with contempt of court, rather than simply with trespass, thereby substantially escalating the potential penalties). However, Minister of Transport Kevin Falcon attacked the protesters as failing to respect the law, claiming that the court's repeated decisions with respect to the injunctions sought by both sides demonstrated that there was no merit to Coalition claims that officials had intervened to bias the EAC or that the environmental studies had been inadequate. 'That has been reinforced now in Court three separate times,' Falcon asserted, going on to claim that 'they have failed in the courts, they have failed in the election, and they have failed in the public relations campaign' (Grabowski 2006).

It is worth noting that the Sea-to-Sky Highway upgrade, and Vancouver 2010 more generally, has divided the Squamish people and indeed other First Nations people in British Columbia. As early as 2003, even before Vancouver was awarded the Games, the BC government and VANOC began negotiating with elected leaders of the Squamish and Mount Currie First Nations (both with lands adjoining Whistler) a number of ways that their people could participate in the economic development generated by the Olympics. In 2004, an agreement was announced with the Four Host First Nations Society (including also the Musqueam and Tseil-Watuth, other Vancouver-area First Nations) in which VANOC provided funding for a Native cultural and craft centre in Whistler, and signed a contract for garbage removal that has pro-vided employment for members of the Squamish and Mount Currie bands, as well as continuing income for their respective band councils. Chief Gibby Jacobs of the Squamish was also appointed a board member of VANOC, and for him and the other elected Native leaders, all of this was evidence that

First Nations could 'take advantage of all opportunities including economic, and establish a clear First Nations presence in the Games' (No2010 2007b).

There are critics, however, who viewed such accommodations as selling out, both individually and collectively. Critics included traditionalists such as Nahanee, who voiced pride that she had never been co-opted or assimilated, as well as younger 'warriors' associated with the Native Youth Movement (NYM). For these critics, collaborating in Vancouver 2010 is collaborating in the ongoing exploitation of lands that rightfully belong to indigenous peoples (No2010 2007b). Asserting their First Nations identities, and fighting for the historical territories in which those identities could be actively lived, is for them far more important than promises of jobs and economic development, which they see as leading inexorably to further development, to further environmental destruction and to the assimilation of more of their people. It should be emphasized that most First Nations in British Columbia (and indeed Canada) have consistently elected 'moderate' leaderships, whose primary focus has been economic development, and improving standards of living in First Nations communities. The No2010 arguments, though, echo some of the more radical views expressed in recent years by prominent First Nations intellectuals at BC universities (e.g. Alfred 2005, Coulthard 2007).

It is worth reiterating that most of the protestors whom Falcon sought to depict as 'radicals' and as people who refused to respect the law were people – with the notable exceptions of the two 'grandmas' who were each protest veterans – who were new to civil disobedience of any sort. The CSEB and the EESA, the groups who between them had spearheaded the opposition to the overland route, were composed of middle-class and in some cases affluent professionals, some of whom would likely have voted for the government. They were homeowners in West Vancouver and North Vancouver, they were retirees and birdwatchers, and/or they were scientists (in the EESA) at one of the Vancouver area universities.[4] They were people accustomed to presenting reasoned alternatives, and they believed they had presented conclusive and evidence-based data, both with respect to wildlife presence in the Bluffs and in the Larsen Wetlands, and with respect to the costs of the tunnel alternative. This was not, as one reporter noted, a protest involving 'Svend Robinson and four aboriginals blocking a logging truck' (Boyd 2006).[5]

We shall consider below whether the failure of the CSEB and EESA to generate more support owed something to the popular perception that these were privileged people whose self-interest in protecting their own neighbourhood did not correspond with a wider public interest. For the moment, though, what is noteworthy is that the contempt of court charges would indeed incur much more serious penalties than trespass charges ever would have. In early 2007, when the charges were heard, protesters were offered the opportunity to apologize for their contempt of court and to pay modest fines, and all but the two grandmothers chose this option. Harriet Nahanee, however, refused to apologize and was sentenced by Justice Brenda Brown to 14 days in jail. In poor health at the time, she fell ill during her incarceration in the Surrey Pretrial Centre (a prison for men, and without its own hospital facilities) and died in hospital of pneumonia shortly after her release. On 5 March 2007 (i.e. *after* Harriet Nahanee's death), Betty Krawczyk was sentenced to ten months' imprisonment for her non-violent act of civil disobedience in disobeying the court's order to leave the Kiewit construction site. A familiar figure in BC's logging protests, she was treated by the judge as a repeat offender. All this in a country that claims to honour the memory of Mahatma Ghandi (Mair 2007).

Issues

Despite the best efforts of Nahanee, Krawczyk and others who associated themselves in the Spring of 2006 (and earlier) with the work of the EESA and the Coalition, the construction of the overland route went on, and the new Sea-to-Sky Highway was completed in the summer of 2009. The events described above raise a number of general issues that bear thinking about, though, in the context of the needs/demands of mega-events, and relationships between the governments that stage them and are committed to their success, and civil society groups concerned about the various impacts that event-related infrastructure may have on their communities. I do not devote time here to the debates about the need for improvements to Hwy. 99 (though I believe it did need to be improved, and by more than one lane), or about the costs of the alternative routes (here I believe that the Coalition raised valid points that were simply brushed aside). I focus the discussion in the remainder of the chapter on three issues: what it means to have a 'Green' or 'sustainable' mega-event; how governments respond to civil society groups who challenge decisions related to the staging of the mega-event, and what responses are available to citizens when normal legal requirements and processes are short-circuited by governments (and courts) anxious to get infrastructure built on time; and, finally, why the fight to save Eagleridge Bluffs did not gain wider traction.

   1. In the case of the 'Green Games' issue, the point is that much official rhetoric was devoted, in the Vancouver bid-book and subsequent VANOC public relations documents, to proud claims that Vancouver would stage a 'sustainable' Olympic Games, even the 'Greenest Games ever'. Such claims were heard again during the referendum campaign, as part of VANOC and government attempts to answer concerns that Olympic infrastructure would have harmful environmental impacts, or leave behind some kind of damaging environmental legacy. It needs to be asked whether this was ever anything more than empty talk, offered up in a country where the public now demands that governments sound environmentally friendly, but is less

inclined to support environmental legislation that would require any real sacrifices. It's also worth asking what a 'sustainable' or 'green' Games might mean or look like, and whether this means anything more than green technologies in new facilities built for the Games (solar panels in the Athletes' Village, low-flush toilets in the washrooms at the competition sites).

In the case of Vancouver 2010, it may be fair to claim (as VANOC do, at every opportunity) that the green technologies were incorporated into Games facilities to an unprecedented degree. Olympic officials will also point, and with justified pride, to the Richmond Oval, the speedskating venue which has been awarded LEED certification for its energy efficiency and its re-use of rainwater. Critics such as Shaw (2008) reject this as 'green-washing', however. They counter that the carbon emissions generated by Olympic-related construction will be staggering, and that the environmental destruction authorized in order to build the Sea-to-Sky Highway, as well as new Nordic skiing facilities in the Callaghan Valley, can never be repaired. What the events surrounding Eagleridge Bluffs also demonstrate, I propose, is that questions of governance should matter much more than the use of green technologies, to any judgement of whether or not Games deserve to be labelled 'environmentally friendly'.

Perhaps the most important of these questions is whether a government is prepared to enforce its existing environmental legislation, and whether that legislation specifies the circumstances in which communities or civil society groups are able to request environmental impact studies, and expect a science-based result. Further to this, it is essential that civil society groups are able turn to the courts (as the 'last court of appeal'), with every expectation that courts will require that governments and their industrial partners abide by the laws of the land. The importance of these factors (and what happens when they are over-ridden) is illustrated in Lowes' account (2004) of how the state government of Victoria guaranteed extraordinary powers to the organizers of the Australian Grand Prix – including exemptions from planning and environmental laws – and suspended the rights of citizens to challenge plans for development in public places (a popular urban park), in its efforts to bring this Formula One event to Melbourne.

In the events described here, the governments and courts of British Columbia and Canada quite clearly failed to meet this test. Even if we suspend judgement, for now, on how seriously the federal agencies concerned initially surveyed the wildlife populations that make Eagleridge Bluffs and the Larsen Creek wetlands their homes for some part of the year, it took only a few days for biologists associated with the EESA and with Vancouver area universities to confirm the presence of a significant number of species that are nominally 'protected by law' in Canada. The laws in question are the federal Migratory Birds Conservation Act of 1994 (based on a 1916 International Convention for the Protection of Migratory Birds), and the British Columbia Wildlife Act. Each of these explicitly prohibits the disturbance of nesting migratory birds, or their nests or eggs, during the spring/summer migratory season, and provides for penalties of up to $50,000 against those who wilfully do so. However, when biologists associated with EESA documented nesting sites and compiled photographic evidence of these nests and of the effects of logging activity by Peter Kiewit & Sons, and took this evidence to the regional offices of the Canadian Wildlife Service, they were told to make an appointment and come back two weeks later (i.e. after more logging was done). It can be said in defence of CWS that they are seriously understaffed, and have been for some years. However this itself reflects the priorities of a federal government that came to office hostile to environmental enforcement, and has since announced plans to limit the situations in which federal Environmental Impact Assessments are required by law (Kwasniak 2009).

British Columbia's courts, for their part, were not prepared to grant an injunction to halt logging activity at the highway construction site so that comprehensive wildlife surveys could be done and environmental impacts officially reassessed. They were prepared, conversely, to grant the injunction that Peter Kiewit applied for to halt the protesters' activities so that work could proceed. In the eyes of the law, apparently, the needs of business and the potential financial losses to the contractor were of greater import than the nesting sites (or the lives) of the Solitary Vireo and other 'protected' wildlife. Although the decisions reported here may accurately represent Canadian legal precedent, what this confirms is simply that when laws that unequivocally protect bird habitat come into conflict with laws protecting the contractual rights and obligations of corporations, the latter can be expected to prevail, and further that the resources of government will be mobilized on behalf of the corporations. Indeed, the BC government's active support of the contractor in this story, and the BC Attorney General's decision to prosecute on contempt of court charges, can be said to exemplify what US political scientist Charles Lindblom described more than a quarter century ago as the tendency of liberal democratic states to privilege the interests of business, whenever these conflict with other social or environmental priorities (Lindblom 1982).[6]

2. This leads directly into the second issue I want to discuss, which pertains to the relations between governments such as British Columbia's and civil society groups such as the Coalition to Save Eagleridge Bluffs, groups that seek to use the legal architecture of their country – environmental impact assessments (which are mandated by law), public hearings (which are supposed to give a fair hearing to public views about development applications), laws (including those described above, protecting migratory birds), and courts – in order to challenge the legitimacy of projects undertaken by governments that citizens believe to be in clear violation of these legal obligations. For Vancouver writer and radio host Rafe Mair, what the saga of Eagleridge Bluffs demonstrates most clearly is that when governments and

their industrial partners want a project to go ahead, there is little chance of a fair or serious environmental assessment process. 'There is a process, the outcomes of which average about 98 per cent in favour of the government. That's not surprising, since the assessor is employed by the government. There are public meetings, which are designed to inform the public'. But these are not genuinely intended to solicit public input or listen to new evidence, since projects like the highway over Eagleridge Bluffs 'are done deals when they're announced', and the minister 'had no intention of changing his mind no matter what the evidence' (Mair 2007).

What is highlighted here is the Government of British Columbia's commitment to stage an impressive Olympics and its determination to fulfil its commitments to its Olympic partners – to the IOC, to VANOC *and* to the government's major industrial partners and supporters. Of particular import, moreover, from a civil society point of view, was British Columbia's use of injunctions to punish those protesting officially authorized environmental destruction. 'This results in more severe sentencing [...] as the judiciary responds to a perceived attack on the rule of law, rather than to the protesters intent of holding their government accountable [...] and [it] has the effect of criminalizing public dissent.'[7] As Mair sees it, 'We all agree that court orders must be obeyed if the rule of law is to prevail. But what if the court allows itself to be used as the strong arm of government and big business? What if [the latter] can operate irrespective of public opinion and without having to fairly obtain legitimacy for their projects, while those who protest go into the slammer?' (Mair 2007). For Mair, this recalls an earlier era of labour-management relations when labour leaders were regularly jailed for contempt of court orders that were unfair in the sense that all the laws of the day favoured management.

Must Canadian environmental groups, then, simply accept that the laws of our day, with respect to major development projects, will be biased in favour of 'governments and their industrial partners' (whether the partnership is formalized, as it was in the case of Eagleridge Bluffs, or in the more general sense alluded to by Lindblom, Mair 2007)? Or, are there circumstances in which we can imagine sufficient public pressure being exerted that Canadian governments will alter the legal framework in which environmental impacts are assessed and decisions about development are made? Here, Mair proposes that Canadian governments are likely to abide by their own environmental legislation only if they come to understand that doing otherwise is going to spark intense protest, *and not only from those directly affected* by a development, but also from people across the country. Canadians, he argues, need to become willing to protest environmental destruction even when it is *not* in their own backyard.

3. However, one of the sadder lessons of Eagleridge Bluffs is that, quite simply, not enough people outside the immediate area cared. Although the activities of the Coalition attracted public sympathy at first, and considerable

media attention as wildlife surveys gave way to civil disobedience, as time went on the cause was one that attracted less and less public sympathy. Part of this can be attributed, almost certainly, to the popularity of the Olympics (recall that Vancouver voted in favour of the Olympics by a margin of almost 2–1, and in wealthy and conservative West Vancouver, opinion was probably even more pro-Olympic). For the mainstream Vancouver media, who were very pro-Olympic, it was tempting to frame opposition to the highway as a fight against a vital piece of Olympic infrastructure, even though the stated objective of the Coalition was not to block the improvement of Hwy. 99, but to push the government to choose a different route. As the dispute dragged on, moreover, other voices chimed in saying, in effect, that in a democracy sometimes you lose, and that even if you believe that an outcome is wrong, you have to ultimately accept the decisions of public institutions. This line of thinking was reinforced, not surprisingly, by successive court decisions that defined the occupation of the Kiewit worksite as 'illegal'. Most Canadians, it might be suggested, may support civil disobedience in principle but are not comfortable with protestors.

However, there was also a discourse worth mentioning in some parts of the Vancouver media which was dismissive of those who had opposed the destruction of Eagleridge Bluffs on the grounds that many of them were affluent professionals, opposing a development 'only' because it was in their own backyard. They could be positioned as NIMBY protestors, in other words, and dismissed on this basis. However, I want to ask why this should matter? I suggested at the outset that that this contributed to the failure to save Eagleridge Bluffs, and I want to propose now that even though (with the notable exception of Harriet Nahanee, of course, and Betty Krawczyk) the people fighting to save the Bluffs were easy to depict in pro-government media coverage as privileged homeowners fighting to keep a road that was important to the success of the Olympics out of their very attractive neighbourhood, their class position should not be allowed to deflect attention away from the merits of their case. Granted, the professionals and business people who are homeowners in that part of Vancouver live in one of the most beautiful places in Canada, and granted also that their expensive neighbourhoods have been part of the suburban expansion of Vancouver that has made the Larsen Creek wetlands the *last* wetland on that part of the North Shore.

Yet none of this, I submit, changes any of the fundamentals of the events described above. EESA were clearly correct in their claims that both the bluffs and the wetland were habitats that were home to threatened species of bird, amphibian and plant life. The arbutus forest that was the distinctive attribute of Eagleridge Bluffs cannot now be replaced. The nests that were displaced and the young birds killed by logging during the spring nesting season also cannot be salvaged. Furthermore, it is difficult not to agree with the Coalition submissions that the Ministry of Transport was in clear violation of both provincial and federal laws requiring proper environmental impact studies, as well as international conventions with respect to migratory birds that Canada is a signatory to. In encouraging logging to proceed in May 2006, moreover, the ministry (and Peter Kiewit, the contractor) were clearly violating the Table of Commitments that formed part of the Environmental Assessment Certificate that they did have.

As we have noted above, British Columbia courts consistently ruled against the Coalition and in favour of the Government and its industrial partner. However, following Mair, I would submit that this only illustrates, yet again, the biases of the law in Canada, whenever environmental values (or obligations, or rights) are pitted against the economic rights of corporations. This bias is further reinforced by provisions in the North American Free Trade Agreement that permit corporations to sue governments that 'unfairly' deny them access to profitable business opportunities, or penalize them for circumstances beyond their control. Canadians largely believe in the rule of law. However, if our environmental laws are ignored by our governments when it suits their economic agenda, and if our courts punish not those who break these laws but those who protest, it is hard not to be cynical about the real motives of our lawmakers.

Mair, nonetheless, argues against defeatism and for a much more active citizenry, proposing that the struggle at Eagleridge Bluffs was lost because the government was able to divide and conquer, by positioning the protesters as privileged NIMBYs. 'What if the protestors at Eagleridge Bluffs had been supported by the physical presence of hundreds from around the province?' he asks (Mair 2007). And why were they not there, except for Harriet Nahanee and Betty Krawczyk? The affluent homeowners who opposed a highway through their upscale neighbourhood were mostly not people who had been part of other environmental protests, in other places, so others in the environmental movement 'owed' them nothing. But if veterans of these protests – or simply people in other parts of the province – stayed away because of this, Mair asks what is the end result of this? The answer, it should be clear by now, is that destruction of the environment continues apace, in place after place, and we are all ultimately losers. It is necessary, he argues, that instead of accepting our governments' present attitudes to environmental law enforcement, attitudes that could be summed up as 'business first and business as usual', Canadians have to start holding their governments – both provincial and federal – to a higher standard. In this, it must be perfectly legitimate to begin by caring about your own backyard; that is what activates many people to get involved. However, change to environmental enforcement is likely to be achieved only when more of us are prepared to show concern about what is happening in other Canadian communities, too: in other cities and suburbs, in rural places threatened by oil extraction activities or by intensive livestock operations, and in First Nations communities without safe drinking water.

The short answer to the question posed above – why were more people not there? – may simply be that not enough of us cared enough to make it our business to get there.

Conclusion

Mega-events have often been occasions for major transportation infrastructure projects that are said to be 'public goods', positioning the host city/region for a bigger and better future, as well as meeting the more immediate needs of the event itself. In this, such projects serve the growth agendas of ambitious regional governments and their industrial partners, and they are accordingly very difficult to oppose. Indeed, as the example discussed here illustrates, communities and civil society groups that stand in the way of such projects face almost insurmountable difficulties in challenging what the authorities have decided will be done. The 'normal' difficulties of opposing growth are multiplied when the needs of the mega-event (in particular, the need that facilities be ready 'on time') are used by officials to short-circuit legal checks and balances (such as environmental impact assessments, or public hearings) or to override legislation entirely (as happened in Melbourne), and attempts to win public sympathy run up against popular support for the event and widely shared hopes that it will be a 'success'. There are no easy answers here, given that the campaign to stop the highway through Eagleridge Bluffs ultimately failed. Following Mair, though, we need to make our governments aware that not only will we actively defend our own backyards, but we will support the struggles of other communities whose environments are threatened. And we must do a better job of holding governments to account if they continue to pay lip-service to their own environmental legislation.

Notes

1. In Canada, although cities are responsible for the construction and maintenance of local roads, provinces are normally responsible for inter-community highways. In this case, moreover, the government of British Columbia's commitment to the construction of Olympic infrastructure meant that all decisions about the routing and costs of the new Hwy. 99 were made by the province.
2. In the November 2008 election, Sullivan was in turn succeeded by the more left-leaning Gregor Robertson. Robertson's victory over the front-runner, pro-business Alderman Peter Ladner, owed much to public anger at a leaked document revealing that the City was on the hook for a large loan guarantee to Millennium Developments, builders of the Olympic Village (which will become high-end condos after the Games).
3. A photo album giving some sense of the environment of the Bluffs, and the protest, can be seen at (www.flickr.com/photos/holycola/sets/72057594113506788/). The numbers at the site ranged from 20 to 40, and there were 23 when the site was

cleared by the police following the court injunction obtained by the contractor. The 'occupation' lasted about five weeks, from mid-April until 25 May 2006.

4. The numbers who associated themselves with the work of these two organizations, or were sympathetic to their cause, is difficult to reconstruct with any precision. It is estimated that there were between 200 and 300 who turned out to a rally in downtown Vancouver in February 2006, and smaller but still significant numbers who phoned or wrote to West Vancouver councillors, urging them to oppose the overland route. There were also significant numbers who donated money to defray the legal costs incurred by CSEB and the protestors. However there were never more than 40 who camped at the protest site, at any time during the occupation, and only 23 who remained until late May, when the site was cleared by the police.

5. The reference is to a controversial former New Democrat MP from the Vancouver area who was an active supporter of the anti-logging protests of the 1990s.

6. I am indebted to my colleague Ian Urquhart for drawing this to my attention, in his paper 'Prison Break? The Politics of Energy Royalties in Alberta', delivered to the Canadian Political Science Association, Vancouver, 4–6 June 2008.

7. From CSEB petition to the Attorney-General of BC, May 2006 (CSEB, 2006c).

12
Resisting the Torch
Xavier Renou

The Olympic flame relay in Paris ended in farce today when police cut the event short after protests forced officials to repeatedly extinguish the torch.

It was a second day of severe embarrassment for Beijing following similar skirmishes in London yesterday as activists demonstrated against China's recent violent crackdown in Tibet.

The Paris stage of the relay ran into trouble immediately after leaving the Eiffel Tower at lunchtime, even though hundreds of riot police and security officials flanked the torch bearers.

With only 200 metres of the planned 17-mile journey to the Charlety stadium on the edge of the city completed, the scale of the demonstrations meant officials had to extinguish the torch and seek shelter on board a bus.

The torch was relit and handed back to the French athletes carrying it through the streets, but it soon had to be extinguished again.

After this had happened for a fourth time, and with the procession hopelessly behind schedule, police decided not to go ahead with its second section.

Instead, the torch was again loaded onto a bus and driven to the stadium, arriving at around 5.30pm local time (1630 BST).

By the time the relay was abandoned, a planned ceremony to greet the torch outside the French capital's city hall had already been cancelled as members of the Green party hung a giant Tibetan flag from the building.

'Olympic torch relay cut short amid Paris protests',
The Guardian, 7 April 2008

When the IOC announced that Beijing was to host the 2008 Summer Olympic Games, the Tibetan cause was almost dead in France. Traditionally active organizations were quiet, demonstrations weren't attracting protesters;

neither the press nor the public authorities were showed much concern, with the only narrative being the huge economic opportunity represented by the Chinese market, in terms of available labour and potential consumers. But the expected exceptional level of media coverage for the Olympic Games, added to the involvement of a large number of political, economic and cultural actors in the organization of the Games, convinced a handful of pro-Tibet activists that the event could be instrumentalized to both mobilize supporters and raise general public awareness.

The Olympics: a public relations exercise for the Chinese dictatorship

The use of international sports events for political purposes, such as providing legitimacy for an ideology, or for the demonstration of military power, is nothing new. Indeed, such tendentious uses are probably inseparable from the organization of sports events of this size. The examples of the Berlin, Moscow and Los Angeles Olympic Games, as well as of the 1978 FIFA World Cup, organized by the Argentine generals, show clearly that political regimes do not invest in such events for the sole purpose of realizing the potential economic and strictly national political benefits.

It was both widely expected and accepted that the Chinese regime would use the opportunity created by hosting the Games to create a public image consistent with its spectacular expansion as a capitalist power with planetary ambitions. Of course China's internal political, social and environmental tensions would remain unspoken, or at least minimized. Among comparable regimes, China's current record is one of the worst, on a series of indicators: China today holds the record for the number of judicial executions, imprisonment without trial, and sundry other human rights violations. The current Chinese regime lacks basic democratic freedoms, and harshly exploits large sectors of its workforce who are deprived of any essential rights regarding health, work conditions, hygiene or security. Finally, but not least, China is rapidly destroying its national biodiversity as well as causing multiple forms of damage to the global environment because of its exponential production of greenhouse gases, and nuclear and chemical waste.

Moreover, the Chinese regime behaves towards countries on its borders such as Tibet and Outer Mongolia in the same way that European colonialist powers behaved during the previous century, with practices now rejected by publics throughout the world. But the widespread diffusion of information concerning Chinese policies, as exploited by China's competitors, seriously compromises its efforts to enhance its reputation and play a major role on the world stage, and to deepen relationships with developed democratic countries, or attract inward investment from multi-national corporations obligated to justify their links with China to their customers and clients.

In this context of China trying to maximize the political gains attached to the Olympic Games, even deploying impressive technological and economic methods such as pollution reduction and rain prevention, it seemed to many observers that the organization of this event was above all designed as a formidable communication tool for establishing the legitimacy of one of the world's worst dictatorships. All this with the active complicity of the Western powers, seeking to put a positive face on their mercantilist enthusiasm for commercial, industrial and political cooperation with China.

The March 2008 demonstrations by Buddhist monks, followed by those of students, came as very bad news for the Chinese regime, undermining its propaganda effort. The regime's violent repression of the monks and students proved to be a major mistake, showing how difficult it can be for even a dictatorial regime to control its image at a time when the world-wide press, because of the coming Games, was desperate for news stories on China. In spite of the regime's rigorous censorship, news of the repression spread across the world, reminding global public opinion of China's dictatorial nature as well as its colonial annexation and rule of Tibet. It also showed how vulnerable the Chinese regime could be with regard to its image, and the regime's propensity to act without measuring properly the impact of its policies on public opinion. This was a weakness that opponents of the Chinese regime would soon make good use of.

Given the Chinese regime's vulnerability on human rights and related issues, the context seemed right to organize protest against an Olympic Games of money and hypocrisy, and against the sponsor regime of China, whose authoritarian methods and significant economic growth had already created a diffuse sense of threat in the public mind.

The French national context of action

Since the mid-1990s, France and other major industrial polities have seen the re-emergence and multiplication of acts of resistance by minority protest groups who, given their limited size and constituencies, resort to unconventional modes of operation, including acts of civil disobedience and transgression. For instance, after the opening in 1989 of the French branch of the US-based anti-AIDS movement ACT UP, French activists started to organize direct actions and provocative dramatized situations to alert public opinion, blaming the inaction of the public authorities in the face of the AIDS pandemic. At the same time, groups such as the homeless, undocumented immigrants and others started to form action groups. They balanced their low numbers of members by mobilizing expressive modes of action, such as the requisition of empty houses, church and embassy occupations, and the public seizure of food in supermarkets. These non-violent but spectacular actions helped these activists gain public support. In Seattle in November 1999, massive non-violent actions prevented the 'Masters of the World' from meeting at the World Trade Organization conference in that city.

At the same time, if there were indeed fewer strikes in the working environment, more and more individual and collective actions of resistance took place including sabotages and wildcat strikes, allied to a general decline of deference. Finally, the awareness of the coming ecological crisis has given a new momentum to an organization known for its spectacular actions, Greenpeace, which has seen the number of French members multiply fivefold in the past ten years, and which, in the first decade of the new century, has been a central actor in a civil disobedience campaign against GM crops by environmental and agricultural activists. Among progressive activists, there is a palpable sense that 'another world is possible', together with the feeling that each and every individual can make a difference, fighting with creative and innovative means to improve the political balance of forces on behalf of popular movements.

A new form of political organization: *Les Désobéissants*

This favourable environment enabled the emergence and growth of an original and innovative political group, the network of *Les Désobéissants* [www.desobeir.net]. Often described as a collective organization, it might be better termed a tool aimed both at training activists in the organization of non-violent direct action, and at helping them on the ground to pursue their campaigns and to achieve their actions. Created at the end of 2006 from a group of anti-globalization activists involved in the opposition to the recommissioning of French nuclear weapons, *Les Désobéissants* soon attracted activists from a variety of protest sectors. The training sessions, organized every weekend in different parts of the country, provided an opportunity for these people to meet, learn and think about new action strategies. At the time the Olympic flame was about to cross Paris, *Les Désobéissants* mobilized more than 4000 people, including several hundred in the capital city. Contacted by a number of *Désobéissants* involved in Tibet liberation groups, the network decided to commit to the campaign, and mobilized activists from environmental or anarchist-related interests in particular. With their experience of direct action and dramatization, and their strength in planning and organizing action, they planned the systematic sabotage of the torch relay and, as it were, a counter-mediatization of the event. At the same time, their structuring as an informal network composed of autonomous groups made it difficult for the police to protect the torch effectively.

The sociological profile of the *Désobéissants* activists, similar to that of the pro-Tibet activists, facilitates undertaking spectacular and daring actions. Highly educated and middle-class, these activists are not very different from those identified in the 'new social movements' of the late 1960s. They work not only in computer science, graphic design, education, the social and non-profit sectors, but also in marketing, trade, art or fashion design. Many of them enjoy privileged professional and social status, at the middle management level, or are self-employed. A minority of them are retired, and well-off, and therefore have free time for social action. Some are students; a few have made the choice to live on welfare. Their symbolic, economic and social resources are numerous: they are literate, can articulate their demands to the press, live in apartments that are big enough to facilitate meetings or to serve as headquarters for actions. They have some genuine economic resources which enable them to pay for their actions, whose costs, however modest, such as fines, remain. They know their rights, have lawyers amongst their extended families who can help at a minimum price, and most of them have been to the training sessions on non-violent direct action run by the *Désobéissants*. In general, they belong to social networks with strong contacts to journalists, local politicians, artists and so on. Through these social networks, they have relatively easy access to strategic information, and can find out their adversaries' weaknesses, that is, to know where to strike with the maximum impact. They know they can rely on political connections and allies such as the Greens, and to a lesser extent the traditional left and far left, and sometimes but rarely, those of the moderate right. Some of them possess skills that are quite useful in actions, such as climbing (which is very useful when unfurling a banner at the top of a historical monument is planned), or filming (to protect activists during actions, and subsequently promote what has been done, on the Internet and with the press).

Ruining an act of communication is also an act of communication

The production of the Olympic torch relay is traditionally the first event (as both global media show and athletic demonstration) in promoting the coming of the Olympic Games. The flame or torch is presented as the symbol for the unification of the international community around the common objective of peaceful participation in sporting competition. For this reason, this symbolic event could barely fail to attract the interest of human rights and pro-Tibet activists, with both groups willing to exploit the event to protest the Chinese regime. The failure of this symbolic event would also symbolize the failure of the Chinese regime to impose its favoured image of itself, and would remind everyone of the scandal of an Olympic Games organized in the context of the repression of human rights in China itself and of Tibetan democratic demands. It would at the same time revive the energy of pro-Tibet activists. We would, in other words, be blowing out their flame to light up ours.

From a technical perspective, the coming of the flame offered an obvious opportunity to the activists: the Chinese Games were too distant and too expensive to reach, and the Chinese regime too powerful and too brutal to be confronted on its own field, in its own territory. But the coming of the flame to us, in countries in which the police response to protest violates recognized human rights laws to a much lesser extent, where the press is quite free to publish dissident voices, and in which numerous human rights and Tibet support organizations exist, was a chance not to be missed. Because this action became accessible for many activists too, the chances for its success increased. In a sense, acting against the torch was both the sign of our powerlessness to stop the gigantic machinery of the Games, and also the expression of our desire to regain some increase of democracy in political affairs, in particular with regard to the decision to organize the Games in a country suffering from its own government. The flame became the symbol for our belief that the intervention of a few would be enough to shake powers.

Moreover, the democratic nature of European countries made it possible that our attempts to extinguish the flame, or to prevent a public event aimed at promoting the Olympic Games from going ahead, would remain unpunished, as our efforts did not constitute any sort of crime under European laws. Our risk of being sued would remain very small, perhaps at most some petty fines. In other words, the chances of undermining the credibility of our authorities as well as of the Chinese regime were much higher than our risks of experiencing problems in court. Hence the success of our mobilization, and the real pleasure experienced by many of the participants, who enjoyed trying to deceive the police and the Chinese agents in charge of the protection of the torch. For these reasons this action created significant momentum for the continued organization of direct actions afterwards.

In the weeks prior to the torch relay, a spontaneous convergence of interests among pro-Tibet and human rights activists took place across Europe: we would all try to stop the torch, and to extinguish the flame, country after country, in order to build a common momentum. We would take advantage of the actions undertaken in each country, in terms of public support, political attention, press coverage, and activists' experience in action. We would engage everyone in the battle of images, to put pressure not only on the Chinese government, but also and primarily on our own governments, to convince them to stop supporting the Chinese regime, and to start speaking out against human rights violations in China.

The idea circulated on Internet forums dedicated to human rights or to Tibet, as well as in activists' meetings. But it was the French NGO Reporters Sans Frontières (Reporters Without Borders) who opened fire: its spokesperson Robert Ménard tried to catch the torch at the March ceremony in Greece, where it was being lit. This (carefully staged) action was covered by the French media at least, and generated enthusiasm among activists, aiding and accelerating the recruitment for and preparation of our own actions on 7 April. The older and more active Tibet activists know each other and maintain links across Europe in order to exchange information. They are also in touch with the Tibetan community in exile in Europe. Of all European groups, the British and the French exchanged information in greatest numbers, as the torch had to cross London on the 6th, one day before its arrival in Paris. A few British activists and some Tibetan refugees from London decided to join our actions after undertaking their own action the previous day. The *Désobéissants* continued their recruitment drive in the Paris region, and made extensive use of their press contacts to ensure that their actions would not go unnoticed. We fielded more and more calls from journalists as the event neared, confirming our view that we had made the right choice in concentrating our resources on targeting the torch.

Ambushes, skirmishes and drama: a day of fire and flames

The day before the 7th, the best-known pro-Tibet activists left their homes to prevent any pre-emptive dawn arrests. This type of arbitrary arrest is in principle prohibited, but sometimes takes place anyway. It is easy for the police to follow activists when they are leaving their houses, or to organize unexpected identity checks lasting as long as four hours just to ensure that activists arrive too late wherever they have to go, thereby jeopardizing the activists' plans.

At dawn, activists gathered in several locations across Paris for briefing. Principally, activists were organized within affinity groups belonging to activist networks from across the entire political and NGO-related spectrum (in particular, Free Tibet, libertarian and human rights groups), but also including numerous individual citizens who had seen the emails posted on lists by these groups, and were integrated into the affinity groups at the briefing. Some activists from leftist political parties (such as the Greens and the Fédération Anarchiste) came without asking for permission from their organizations, who are not used to – and sometimes are explicitly opposed to – non-violent direct action. All shared a sense of urgency, filled with images of the recent repression in Tibet, and a feeling that this was a crucial historical moment for opening a window on the gross violations of human rights in China. Some *Désobéissants* chose to film or photograph the entire action, in order to provide images to the press.

Once all were present, the various groups received a brief presentation of agreed political points to communicate to the press if interviewed, and the precise scenario worked out for the action. The principle was (and is), however, to give a great deal of autonomy to the activists. The security forces in charge of the event and the Ministry of the Interior would be unable to ignore the many actions being undertaken against the torch, and would be preparing for it, mobilizing hundreds of police. It promised to be hard for us to coordinate it all, as well as to ascertain which group had the best chances of actually reaching the torch, to extinguish it, or to at least stop it. The groups had to get to their arranged places on their own, without aid from the other groups. We identified several strategic places to strike, and spread the groups all along the course to be followed by the procession bearing the flame. The idea was to prevent groups from getting in each others' way. In particular, we focused on a series of nodal points: the point of departure of the torch, at the Eiffel tower; outside the French state television broadcasting centre; and Paris City Hall, where the torch was scheduled to stop for a short break. A crew working for a prime-time television show was following us, and had been filming the preparation of the action for several days. Adhering to principles of civil disobedience, we act publicly, taking full responsibility for the entire action and its potential legal and political consequences.

Reporters sans Frontières activists scaled the Eiffel Tower to the first and largest floor, in order to unfurl their large banner depicting the Olympic rings as handcuffs. Other pro-Tibet activists, spread around the foot of the Eiffel Tower, were ready to launch themselves in front of the relay as it started. But before even these activists could act, and unbeknownst to them, Sylvain Garel (a well-known pro-Tibet activist and Green Paris city councillor) attempted to grab the torch as it was being lit, on the steps of the famous monument. The footage of his action was to be shown across the news networks' coverage of the event.

After the Eiffel Tower, and as the first incidents were breaking out, the torch was expected to follow the line of the River Seine, carried by a disabled Chinese athlete surrounded by police officers, some of them Chinese and also dressed as athletes. Our main ambush was located at the end of a road tunnel, closed to traffic, that the relay was to take under the river. It is an ideal location, situated near the television broadcasting centre. Many journalists were present, unaware of what was about to happen. We scattered ourselves along both sides of the road, hidden by small crowds of onlookers who had come to watch the relay. We blended into the crowds, disguised as small groups of friends, couples, apparent lovers, pretending not to recognize each other. Some of us had received news from the Eiffel Tower, which enabled us to judge more precisely when the torch would arrive. Just before its arrival, dozens of riot police buses parked along the road in an attempt to form a protective wall around the relay. Surveillance of the crowd increased. In vain.

At the agreed moment, the activist coordinating and triggering the action launched himself into the roadway shouting 'Freedom for Tibet!'. At this signal, pairs of activists joined the action, bursting through the riot police's human and mechanical blockade to invade the road. The police responded immediately, of course, but as had been expected, chaotically and inefficiently. Some kicks were received from panicking police officers, and activists were pinned to the ground and then dragged from the road to be held against a nearby wall. Camera crews captured the entire action, and kept filming the events, as we had our hands secured behind our backs. Two of us still managed to free our hands and set off red distress flares, which provoked another short panic among the police, and provided spectacular pictures for the cameras: these images led the television news bulletins that evening. Particularly as we had achieved our tactical objective: the torch convoy was forced to stop inside the tunnel, and Chinese security forces ordered the flame be extinguished, for the first time that day. The procession was then loaded onto a group of large coaches in order to evade us.

We were detained by the police, inside a police bus. Those of us who weren't arrested would try to wait for the torch on the other side of the Seine, and make another attempt to stop it. Those of us under arrest simply thought that that was it for the day. The bus was driving us to what we expected to be a police station, where we would be detained for the rest of the afternoon. But surprisingly, the police decide to free us after only an hour, and allowed us to cross to the other side of the Seine, right on the route of the torch! We immediately decided to try to block the torch again. It was only at this moment that we realized how significant our protest had become, and how much it had succeeded in attracting people from circles far wider than solely those of the committed activists. The action now comprised hundreds of people, including high school students, onlookers and retired people, forming a sort of guard of dishonour for the Olympic procession, whistling at the police and insulting the French athletes taking part in the procession. The faces of the latter reflected their moral discomfort, if not shame. And every now and then, people, especially the young, tried to grab the torch. The police were overwhelmed and responded with some brutality, but without deterring anyone; on the contrary, the police reaction added to the public anger. Some journalists were assaulted by police officers, which became a serious problem for the political authorities as the media would speak out even more loudly against police violence, and the entire organization of the event, in the following days. Other journalists were even removed by police from the official press vehicles, at the orders of Chinese officials concerned that the media was filming the crowd and protesters instead of concentrating on the relay itself. This incident would only add to public disapproval at the apparent participation of French authorities in the censorship of the French media. As one *Désobéissant* said to journalists filming his arrest, in a phrase repeated across the news bulletins: 'Instead of exporting democracy, France is importing the dictatorial methods of the biggest dictatorship in the world.'

The route turned into utter chaos. The authorities, like the athletes, were now despairingly trying to save face, but had lost their momentum. After a final ambush on the main square outside Paris City Hall, and the unfurling of a banner against the flame by Green City councillors from the windows of City Hall, Chinese officials decided to put an end to what now appeared to all as a giant fraud. The flame was extinguished, the torch placed in a car to be rushed to its final destination, for a much abbreviated ceremony. The dishonour was complete.

Tibet: in the headlines, and in the minds of the public

That evening and the following day, the French media showed the images of China's public relations disaster at great length, analysing in detail the unexpected success of the protest. The numerous criticisms made by activists of the Chinese dictatorship were repeated in every report. Journalists also emphasized how the torch relay had been organized and controlled, thus emphasizing the authoritarian nature of the Chinese regime: the role played by Chinese elite police officers, dressed as athletes to protect the torch; the presence of Chinese officials giving orders to French police officers; the eviction of French journalists from the press convoy at the behest of Chinese officials; the fact that Chinese journalists were only filming the torch itself, covering up the protests in an obvious act of censorship. They also stressed the illegal confiscation of Tibetan flags by French police, and thus the denial of basic democratic freedoms in France. The French public authorities were also charged with police brutality against peaceful demonstrators, with assaults against journalists, and were called on to pressure Chinese authorities to improve their human rights record. Because of the protest, President Sarkozy, who had previously declared his intention to attend the opening ceremony in Beijing, now pretended to give it a second thought.

Through this one highly visible episode, the deeply cynical nature of public relations between European and Chinese governments, whose interests in trade seem to overcome any moral, social or environmental scruples, came out of the shadows and became a matter of public debate. European leaders, by nature sensitive to public opinion, were now obliged to find justification for their compromising attendance at the Games. This first result was extremely encouraging, and opened an avenue for pro-Tibet and human rights activists. In the following days, they received dozens of calls from journalists seeking information on new actions on the same subject. There's a new interest in Tibet: the dominant media organizations, reputed for their close alignment with French diplomatic positioning, began to question the relevance of national support for the Games and the Chinese regime. Activists felt energized and supported by both public opinion and the media. In Paris, activists decided to create *Réactions citoyennes* (citizen reactions), a group dedicated to non-violent direct action in support of the Tibetan cause. It marked the beginning of a new cycle of spectacular political actions, including the interruption of live TV shows on the Olympics, occupations of the French National Olympic Committee, and the systematic targeting of the main sponsors of the Games, with blockades of UPS and Air China agencies or of major Adidas stores at peak hours, the occupation of Coca-Cola's corporate headquarters, the staging by activists of fake executions of Tibetan monks on the Champs Elysées and of removing organs from monks in front of McDonald's fast-food restaurants. Between April and July 2008, many activists were inspired by a powerful current of creativity and enthusiasm, which helped them hit the news and maintain pressure on the French government. Their flame, our flame, lights up the sky; their fire, our fire, is burning again.

Conclusion

13
Conclusion. Sports Mega-Events: Disputed Places, Systemic Contradictions and Critical Moments

Graeme Hayes and John Karamichas

Sports mega-events are interesting to us because of the excitement they create: they capture the popular imagination, provide a platform for individual and collective feats of skill and dedication. To treat them as purely sociological or political phenomena is to miss their symbolic significance; it is also to miss the pleasures they generate. But sports mega-events are also interesting because they are peculiar, recurrent, time-space compressions where global norms and the ideological operations that sustain them are made visible and identifiable. To treat them simply as stages of technical excellence, courage, endurance and accomplishment would also be to miss their fundamental meanings. To this effect, the central line of enquiry around which this volume has been structured is the ways in which mega-events, beyond their politico-institutional processes, event management procedures and media projections, are located within and impact upon the publics that are called to support and spectate – and perhaps even participate in – these events. The evaluation of these impacts and relationships is especially important given, on the one hand, the cultural, social, political and economic significance of these events; and on the other, the seemingly ever-increasing disconnect between their top-down, elite, nature and the ostensible redistributive and participatory agendas staked out by their governance regimes. As numerous contributions to this volume have underlined, a key legitimizing discourse in the staging of these events is their claims to social transformation, through an aspirational public framework which establishes collective participation in the production of the event, and the creation of long-term 'legacies'. These legacies – 'a means to redirect and expand the growth of the Olympic Games' (Girginov & Hills 2010: 438) – are primarily conceived in terms of the physical re-organization of the urban environment through the completion of transformative mega-projects. Whilst legacies are typically tied by hosting coalitions to the staging of the mega-event, they respond to longer-term social and political agendas; and – since Sydney hosted the

2000 Summer Games – claims from event promoters that their staging of the 'greenest' or 'most sustainable' event so far will have long-term impacts on social and economic processes and structures. Indeed, the terms 'sustainable development' and 'legacy' were both introduced into the Olympic Charter in the 1990s (Chappelet & Kubler-Mabbott 2008: 180). The aim of this book has thus been to provide a wide-ranging, comparative assessment of the effects of these developments.

At the end of this volume, therefore, we should first return to the definition of mega-events with which we started. According to Roche, mega-events are 'specially constructed and staged large-scale international cultural and sport events. They are short-term events with often significant long-term pre-event and post-event impacts on the host nation across a range of dimensions of national society, particularly cultural but also political and economic dimensions' (Roche 2006: 260); for Spilling, 'a mega-event is an event that generally attracts a large number of people, for instance more than 100,000, involves significant investments and creates a large demand for a range of associated services' (Spilling 1996: 323). Not untypically for evaluations of complex processes, these sorts of definitions – different as they are, from symbolic to managerial – are general enough to encompass the wide range and variation of manifestations of mega-events, whether sporting or otherwise cultural. For our purposes, it is probably not useful to create confusion by proposing a counter-definition; moreover, this volume has dealt with very specific manifestations of mega-events, not just sports mega-events in general, but predominantly (though far from exclusively) Summer Olympic Games. Yet one of the common threads running through this volume has been the identification of mega-events as, in the terms staked out in their chapter by Dansero et al., intrinsically *disputed* places, characterized by competition between scales, places and worldviews.

It is useful therefore to underline how our collected analyses of the political and behavioural sociology of these sports mega-events have enabled us to identify, beyond their sporting and cultural character, a series of common traits which are reproduced across specific event iterations:

1. they are global communicative events with high potential reputational costs and benefits;
2. they promote a specific mode of neo-liberal capital accumulation, and are 'critical moments' of social and cultural change for host cities and nations;
3. they are underpinned by a series of legitimizing discourses which establish their social utility and promote specific sets of universal values and worldviews;
4. they are accordingly characterized by a series of *systemic contradictions*.

By 'systemic', we mean here that these contradictions are fundamental to the symbolic and material operations of the (sports mega-) event, and are reproduced across time and space. These systemic contradictions lie in the seemingly inevitable gap between the aspirations promoted by hosting coalitions and the processes and impacts of the staging of the event. Of course, these impacts may take a variety of forms, be lesser or greater, more or less acute, depending on the nature of the hosting regime: the level of its economic development (though in order to host Games and Cups, states must have the necessary infrastructure capacity), the extent of existent democratic freedoms and so on. One of the most fascinating aspects of mega-events is that though their structural processes remain consistent, their precise contours are ever-changing. Yet the evidence of the last 20 years or so is that these contradictions are common to and visible across different iterations in different national and cultural contexts.

The most prominent of these systemic contradictions are located in the gap between the discourses of collective inclusivity and civic participation on the one hand, and both the shift of public resources from mass to elite professionalized sport, and the imposition of delivery-driven, top-down decision-making structures on the other; between the aspirations to promote social justice and reduce inequalities, and the corporate growth model which directs urban infrastructural transformation and demands the public underwriting of private profit and risk, exacerbating social inequalities; between the promotion of universal norms such as civil liberties and human rights, and the event requirements for controlled (corporate, securitized) space which restrict freedoms of movement and expression (indeed, if the two dominant discursive and strategic dynamics of sports mega-events are the extension of rights on the one hand and of markets on the other, then it is also amply clear that it is the latter that takes precedence); between the promotion of cultural difference and the imposition of cultural standardization to service the requirements of the global communication strategies of the host coalition; between the promotion of environmental responsibility and sustainability on the one hand, and the material environmental impacts of event staging, however mitigated, and, again, the top-down nature of the decisional process, on the other.

In this concluding chapter, we return to a number of these contradictions, before setting out some questions which might guide future research.

Behavioural change

We started from the premise that sports mega-events are not simply sporting events, but have profound political, economic, social and cultural consequences. Equally, our starting premise was that the nature of these consequences has evolved significantly over the past 20 years or so. Whereas we might at one time have restricted our view of them as key moments within nation-building strategies – pure celebrations of collective identity, creators of cultural and social capital, part of the glue that forges the national solidarity of 'imagined communities' (Anderson 1991) – the past two decades have seen a change of scale and of ethos, and a change in the political economy of their purpose. Of course, as the contributions of Jean-François Polo, Hugh Dauncey and Anne-Marie Broudehoux in this volume have demonstrated in detail for Turkey, France and China, capturing sports mega-events remains integral to state power strategies; this is just as much the case for FIFA and UEFA tournaments as it is for Olympic events, given the spatial diffusion of football tournament hosting and the smaller numbers of competitors they attract (there is no requirement, for instance, to build a village for competitors, nor necessarily to focus on one city above others). As Roche (2006: 267) points out, the rise of global consumer culture on the one hand and of regionally integrated economic and institutional actors in the post-Cold war settlement on the other 'provide stimuli and opportunities not only for a maintenance but also for an increase in nation-states' interests in staging mega-events'.

Dauncey describes how, by awarding the 1924 Games to France so soon after the end of the First World War, the IOC endorsed France's status, her political and economic recovery from the war; mega-events have thus perhaps always been understood as imprimaturs of development and of global integration. Today, this is particularly the case for what Black and Van der Westhuizen (2004) refer to as 'semi-peripheral' states: developing nations, and nations whose 'membership' of the club of dominant global states, politically and economically, remains to be settled, and who seek heightened visibility and prestige in the context of globalization. In their different ways, South Africa's hosting of the World Cup and Delhi's hosting of the Commonwealth Games in 2010 testify to the continental reverberations of competent mega-event management, as was the case for Mexico and the 1968 Olympic Games (Bolsmann & Brewster 2009). Likewise, Greece's hosting of the 2004 Games, and Poland and Ukraine's joint hosting of the 2012 UEFA Championships, bring narratives of national cultural identity into sharp focus, again focusing on the political and economic passport of 'competence', of the ability to deliver technical requirements to order. Brazil's hosting of consecutive World Cups and Olympic Games in 2014 and 2016 – the first in Latin America since Mexico hosted these events in 1986 and 1968 respectively – is likely to have similar national and continental projections to Delhi, Beijing and South Africa; and the same is true for Russia's hosting of the 2014 Sochi Winter Games, or for Qatar's hosting of the 2022 FIFA World Cup.

Mega-events therefore remain important in the projection of narratives of state and continental identity; and as John Horne argues, in his chapter in this book, they also are of paramount importance in revealing the directions that nations take in the management of global capital flows. For Horne, mega-events have become a central element of urban modernity, exhibiting many of the contradictions of what Naomi Klein terms 'disaster capitalism': in other words, the peculiar time-space compression of event staging has the effect of creating a shock to the political system, of over-riding the established processes of interest aggregation and collective mediation. The shock is felt on the urban fabric of the city, but also on the democratic process. Indeed, we might also cast sports mega-events as *post-national*, in the sense that, under the impact of this shock, the projection of socio-cultural goals (such as the constitution of collective national identity) now appears secondary to political and economic ones: sporting mega-events are increasingly based on the promotion of highly competitive 'world-class' cities, seeking to attract highly mobile capital and people flows. The key relationships are thus now multi-scalar, situated between the metropolitan and the global, including but also transforming the national and the popular. Indeed, even where the Games continue to act as an instrument of state power – such as in Beijing – transnational flows and networks, in the form of media reporting, event information and technology, institutional auditing and corporate partnership, remain crucial to the conception and realization of the event, as Mol and Zhang neatly underline in their chapter.

How, therefore does this paradigm, the creation of these transformative legacies, impact upon civil societies, upon the domestic populations who share directly in the fruits and bear the costs of the staging of the event? Discussion of mega-event hosting has typically focused on the external projection of idealized narratives of identity, whether by nation-states or, more recently, global cities; Getz points out (2008: 407) that mega- (or 'hallmark') events have long been analysed within tourism studies in terms of their tourist attractiveness and associated image-projection and developmental functions. Since Barcelona in particular, the imaging strategies of cities have also been re-imagining strategies, harnessing the event to bring about urban renewal and regeneration. As Dansero et al. underline in their chapter, hosting coalitions typically seek to legitimize the impacts of event staging (the economic costs, social disruption, political transfer of priorities, etc.) by incorporating it within a strategic plan for its own transformation. But one aspect our collection here demonstrates is that the mega-event is also a vehicle for the cultural transformation of domestic publics themselves, going beyond identity-signalling, 'social beautification', or infrastructural development. The emphasis placed by Horne on the mega-event as a form of 'shock and awe' is apposite: borrowing from Boltanski and Thévenot's analysis of dispute resolution within human relationships (1991), we consider that hosting a mega-event such as the Olympic Games, or the FIFA World Cup Finals, should be seen as a 'critical moment' for the host territory, a point of realization and rupture where the ordinary course of action can no longer be continued, but a new agreement within a general order of principles must be found (1991: 52, 1999: 359–61). For Boltanski and Thévenot, the critical moment thus occurs as conflict, where antagonistic parties 'are subjected to an imperative of justification' which rests on the search for equivalence and the construction of legitimacy.

We can extend this model metaphorically to the systemic requirement of mega-event hosting. Here, the critical moment of event staging also rests on equivalence and legitimacy. The capacity of the host territory to stage the event is inserted into the symbolic order of the event regime through the demonstration of worth, and the host's construction of a legitimizing discourse which justifies such capacity. This justification rests both on its similarity to the event as constructed across multiple iterations, and its singularity, or difference from the set of previous host territories and events to which it claims equivalence. The resolution of the critical moment is thus not simply a question of outward-facing justification, a chance to change global perceptions (of a city, a nation, a continent), but also entails inward-looking action: the application of a shock to the organization of civic life, the instrumentalization of the momentum and civic pride attached to hosting the event, to bring about long-term behavioural transformation, whether cultural or economic.

To this effect, both the chapters by Polo (on Turkey) and Broudehoux (on Beijing) remind us that the televised global spectacle of elite sport is also a window on cultural difference, and that hosting coalitions – specifically, national institutional actors – see the standardization of cultural norms and practices as paramount in order to integrate into global society, to be seen as 'civilized', and to expand market relationships (or 'equivalence'). For Turkey, the stakes of mega-event hosting are part of a strategy of European Union membership. But the 'civilizing' and marketizing dynamic is most obviously the case in the Beijing Games where, as Broudehoux discusses in great detail, the Chinese regime enforced not just urban and environmental transformation, but a wider transformation in daily life to fit modern norms of public behaviour and global expectations of civility. Of course, the authoritarian nature of the Chinese regime, with its available technologies and processes of repression, facilitated this transformation campaign. In liberal democratic regimes, our collection suggests that it is the contemporary drive to 'sustainability' which fulfils the similarly transformative function of the critical moment: the harnessing of the mega-event to demonstrate global legitimacy (externally) and produce collective cultural change (domestically).

Sustainable development

Following Cantelon and Letters (2000), we argued in the introduction to this volume that the development of significant environmental capacity by the IOC was, at least in part, a function of the Olympic Movement's own image-making strategies, designed to rehabilitate the Games from a series of environmental, financial, political, diplomatic and ethical crises. But as well as a global legitimation strategy for the Olympic Movement – and indeed, increasingly, for FIFA – sustainable development has, it seems to us, two further symbolic functions within mega-event regimes.

First, it is integral to the reputational and communicative strategies of host cities, states and regions. In their chapter, Mol and Zhang argue that one of the principal reasons explaining China's aspiration to organize the Games and the World Expo was its desire to strengthen its position globally, that is as a legitimate and respected member of the international community. Environmental sustainability, argue Mol and Zhang, is central to this strategy. There are, of course, considerable risks involved in instrumentalizing mega-events within a strategy of demonstrating global environmental leadership, given the negative environmental impacts of event staging: as the Commission for a Sustainable London (CSL) 2012 has underlined ahead of the London Games, not to act to minimize the carbon footprint of the event 'is an option that carries significant reputational risk' (CSL 2009: 15). Indeed, Holden et al. (2008), in their discussion of British Columbia and the Vancouver Games, draw up a number of alternative scenarios of potential reputational enhancement and diminishment for the host territory, depending on the extent of successful implementation of the sustainability paradigm at the Games. But the centrality of sustainable development to the reputational and legitimational strategies of host coalitions is not restricted to global projections: Dansero et al. point out in their chapter here that sustainability is also integral to the securing of a local consensus for event staging, of convincing host populations to bear the (potential negative) impacts of the event. Various recent studies (Gursoy & Kendall 2006, Jie et al. 2010) have found that, perhaps unsurprisingly, the analysis of potential benefits and costs by the host community is closely related to the extent of local support for the mega-event. For Gursoy and Kendall, local community attachment, including the 'ecocentric attitude or degree of environmental sensitivity' (2006: 606), plays a significant role in the calculation of these costs and benefits, and is especially critical for residents' attitudes to staging of the Winter Games.

Second, as outlined at the end of the previous section, sustainable development is also central to the cultural transformation strategies of host coalitions. In environmental terms, the justification of staging mega-events – a highly carbon intensive operation – centres on their capacity not just to be remediated through footprint minimization or the dubious practice of carbon offsetting (eventually rejected as unsound by London 2012), but through their capacity to accelerate or produce policy and, especially, collective lasting social change. As critical moments, sports mega-events are thus legitimated through their avowed function as a 'spike' for long-term cultural change.

Mol and Zhang underline that mega-events can function as powerful 'sustainability attractors' for the diffusion of environmental technologies and technical norms; in their analysis, hosting the 2008 Olympics has accelerated environmental reform in Beijing, with the Games acting as dramatic time-space compressions of wider dynamics in environmental policy. Sports mega-events thus have the considerable capacity to function as global plat-forms for the development and dissemination of environmental best prac-tice and sustainable technologies, facilitating the creation and growth of new markets. One sense in which we should see mega-events therefore is through their capacity to act as powerful agents of technology transfer and technical norm diffusion, with a wide mimetic potential in both geographic and public policy sector terms, allied to a demonstration effect designed to encourage positive, individual lifestyle changes amongst civic populations. In their chapter, Caratti and Feragutto argue that attention to event man-agement procedures reveals the extent to which many of the direct environ-mental impact problems of mega-event staging can now be identified and solved; in Turin, for example, the adoption of EIA procedures for sensitive environments led to the relocation of the Pragelato ski-jump to a different site; at Aichi, the involvement of NGOs and citizens groups in planning for the 2005 World Expo resulted in its relocation in a less environmentally sensitive area. Moreover, they argue that the adoption of an environmental management system two years ahead of the Turin Games should be seen as a major achievement, enabling a comprehensive and integrated approach to tackling environmental impacts. In Paris, as Hayes argues, the Greens in City Hall saw the city's Olympic bid as an opportunity to achieve rapid cultural change across public procurement and regulatory decisions.

Yet three problems arise from these strategies. At various points in this book, contributors have referred to the literature on 'mega-projects' to work through the planning and coordination challenges of event staging. There is of course a key difference between a mega-event and a mega-project: the former's immoveable timetable for completion. Whilst mega-projects are typically characterized by their tendency for delay in delivery, mega-events cannot be allowed to be delivered late (this is, perhaps, a major line of frac-ture between our conceptions of a six-month World Expo as a mega-event, and a four-week sporting tournament televised live to a global audience). Indeed, the race to completion becomes both a dramatic narrative of pre-event time, often inscribed with (generalized, facile and frequently pejora-tive) narratives of national cultural identity, and a key determinant in event hosting decisions. This has numerous consequences for the type of event that is being staged, including the sustainability technology showcased by the mega-event. Inevitably it is relatively conservative: the desire to dem-onstrate world leader status has to be balanced against proven reliability; power generation and cooling systems have to work at the given time on the given day; there are no second chances available.

The second consequence of the timetable is that civil society actors, whether NGOs demonstrating scientific capacity, or locally organized move-ment organizations mobilizing around territorial use issues, may enjoy a sense of leverage over organizing committees, particularly in regimes with encoded commitments to the freedom of civil association. WWF have been particularly active in their involvement with hosting committees and corporate sponsors, from China to Canada to London. For CSOs, their main tactical advantages – their scientific know-how on the one hand, and (for the more radical) their ability to enforce costs and delays on to developers through popular mobilization, or to tarnish the reputational value of the hosting coalition through criticism on the other – are accentuated when set against non-negotiable, non-deferrable deadlines. The third consequence, however, is the reverse of the coin: the pragmatic problem-solving orientations of organizing committees will be balanced against the enforcement of security measures by states designed to preclude the materialization of the potential threat wielded by civil society actors. We can expect the degree and forms of such measures to vary as a function of regime organization, policy styles and cultural norms; but we can also expect to see them imposed and enforced in all host polities.

The second major area of worry is that, as the chapters in the second part of this book emphasize, the premise of change is based on standards, regulations and knowledge transfer, but rarely addresses, in any systematic way, one of the fundamental principles of sustainable development: the inclusion of civic publics in deliberative or participatory forms of decision-making. In an autocratic regime such as China, it is perhaps unsurprising that the technical aspects of sustainability were privileged over the democratic and redistributive ones. But the evidence also of the chapters by Hayes, Whitson and Dansero et al. point to top-down decisional processes, with environmental impact regulations ignored (Vancouver), even junior coalition partners excluded from key bid decisions and forced into a choice to support or oppose (Paris), and dedicated consultative structures being unable to produce much in the way of real consultation (Turin). Moreover, for all the advances they detect in environmental management at Turin, Caratti and Ferraguto also highlight that, once the Games were awarded, popular inclusivity in decision-making was limited, generic and peripheral, 'a symptom of the lack of a deeper involvement of the general body of citizens'. Dansero et al. concur: whereas Games models have sought to incorporate civil society groups in decision-making, this incorporation is limited at best, takes place at a stage too late for citizen input to influence fundamental decisions and leads to progressive marginalization.

Finally, there is, so far at least, little evidence that mega-event staging can produce long-term changes in social behaviour patterns or on public policy, beyond the real, non-negligible but nonetheless market-driven and incremental improvements in the construction, energy systems, event design and plastics industries in particular. Of course, as Karamichas points out with respect to Athens, there have been considerable improvements in the general infrastructure and transport system, as is typical of mega-event staging. But as he also argues in his discussion of the effects of the Athens and Sydney Games, there is little evidence from a series of key social and environmental indicators that Games hosting has produced structural social or political change in this regard. Moreover, given the crucial importance of effective event delivery, and the relatively short time frames for organization and construction, the experience of successive Games and tournaments tends to show that organizers give little thought as to how the event will bring about such change other than through a top-down demonstration effect. As we stressed in the introduction, neither the IOC nor FIFA seems willing to play a more interventionist role in standard definition or policy learning from iteration to iteration.

This is visible in the IOC's refusal, for instance, to respond positively to the Play Fair at the Olympics Campaign (PFOC) on labour rights abuses and the use of child labour in the IOC garment supply chains. The campaign seeks the insertion of a supply chain labour standards clause into the Olympic charter and of binding language into IOC sponsorship and licensing contracts, the establishment of an IOC mechanism for addressing labour abuses and effective oversight of National Organizing Committees (Miller 2005: 5–6, PlayFair2008 2007: 6). The IOC's position is that this is a matter for local hosting coalitions, and the Vancouver and London Organizing Committees both subsequently committed to include high ethical labour standards in official accreditation for clothing supply contracts (Clean Clothes 2007). But the campaign has been damning in its verdict on the IOC, repeatedly arguing that the IOC has 'categorically refused to take responsibility for workers' rights in the Olympics sponsorship and licensing programmes at global, national and Games level', instead stonewalling, spreading misinformation and sidestepping human rights issues (PlayFair 2008 (2007): 6–8).

The same pattern is visible for sustainability issues. As the chapters by Broudehoux and Polo in this volume testify, behavioural change is a key narrative of mega-event staging, but, as we have pointed out elsewhere (Hayes & Horne 2011), there is little or no evidence that the positivist demonstration effect of major event programming has successfully brought about collective behavioural change with respect to environmental sustainability practices, despite the affirmations of mega-event organizers. Moreover, even if we accept the dubious possibility of Games staging as a 'cultural spike' in the terms set out by the organizers, this pre-supposes the existence of both a domestic and global event governance regime with the capacity and will to ensure that it is implemented. Yet, at the global level, the IOC (and, indeed, FIFA) steadfastly refuses to play such a role, devolving responsibility for methodology and standards to OCOGs and national football associations. Public post-event compliance, reporting and monitoring is weak to non-existent, as Caratti and Ferraguto underline. Reliable and comparable data collection from previous iterations and similar events is a key problem (Collins et al. 2009). There are few mechanisms for diffusion of processes or practices between mega-event iterations, beyond the official IOC mechanism to share knowledge between host cities. True, Sochi will adopt the HECTOR mechanism from Turin, but this is clearly a step backwards from the carbon accounting methodology developed by London 2012, as we argue in the introduction. Even in the systemic terms of event hosting, there is an evident lack of strategic planning and thinking.

Future steps?

The studies presented in this volume have all contributed, in different ways and across different disciplines, to the identification and the analysis in sports mega-events of what we term here the *systemic contradictions* and *critical moments* that they produce as *disputed places*. It is the challenge of future research to deepen our understanding of these contradictions and the processes that sustain them, and the developing nature of the civic and social responses to them. We believe that four areas of how mega-events impact on civil societies need further examination:

1. the processes through which global NGOs (IOC, FIFA, etc.) develop their organizational capacity and reputational legitimacy as *social* actors, and the long-term post-event effects of this action;
2. the specific procedures and methodologies that hosting coalitions develop in order to create public support for or participation in event decision-making, and the putative development of a more actively participatory and inclusive event regime;
3. the mutation of and transmission of knowledge between mega-event iterations, particularly within the context of the developing social role of event governance regimes;
4. the organization of civic publics aiming to impact effectively on event logics, to provide a democratic counter-weight to mega-event governance – both at the 'point of touchdown' and between and across iterations.

As the contributions to this volume have emphasized, the top-down transformational projects and planning structures of mega-events give rise to impacts that are neither spatially nor socially neutral, and these impacts are visible across iterations in contrasting social and political contexts. Faced with this pattern, many of the contributions in this volume point to accruing but piecemeal evidence that civic actors mobilize either in opposition to mega-event hosting systemically, or organize constructively to minimize the social costs and accentuate the benefits of a given event iteration. In event management, this has perhaps brought more procedural and agenda-setting gains than it has immediate substantive changes.

One of the key questions for civic mobilizations against global events therefore concerns the capacity and opportunity of what McAdam et al. term 'scale shift', or the 'moderately complex process within which the relative salience of diffusion and brokerage varies, but passage through attribution of similarity and emulation regularly produces a transition from localization to large-scale coordination of action' (2001: 339). In many senses, the capacity of local civic actors to 'jump scale', and form effective coalitions at the transnational level – whether through the development of analyses and counter-hegemonic discourses between and across event iterations, or at the global level of contesting the premises and operation of NGOs such as the IOC and FIFA – appears crucial to changing the way in which such mega-events may operate in the future.

Much recent social movement literature has focused on the development of transnational advocacy networks and global movements (see inter alia Smith et al. (1997), Guidry et al. (2000), Smith & Johnston (2002), Della Porta & Tarrow (2005), Della Porta & Caiani (2009)). Keck and Sikkink (1998) underline that transnational advocacy networks have been most apparent and effective in the humanitarian, rights, environmental and latterly global justice movements. Renou's account of action against the Olympic torch is thus instructive; it is precisely in the area of human rights that international coordination and activism was evident, and in a (broadly) liberal democratic state that effective action was possible. True, the action discussed by Renou was largely symbolic, produced little ultimate effect on event staging, and stimulated a series of significant pro-China counter-mobilizations in most of its global city stops. Yet the pro-Tibet mobilizations severely disrupted the torch relay in cities in Europe, North America and Asia (London, Paris, San Francisco, New Delhi, Jakarta, Canberra and Nagano; see Gillon et al. 2010 for a detailed map), reinforcing a critical discourse of rights and developing awareness of China's occupation of Tibet, drawing attention to the disjuncture between the Olympic Movement's commitments and the concrete processes and conditions of event staging.

There is also some evidence that activists from Salt Lake, Vancouver and London have developed links to learn lessons and coordinate strategy (Boykoff 2011a). Yet the capacity to jump scale may be less relevant for the development of local civic responses to mega-event hosting. Defining the properties of mega-events as systemic (as do, for example, Chappelet & Kübler-Mabbott 2008) implies privileging the local conditions, problematics and processes of event organization. One implication of this is that the social and spatial inequalities generated by mega-event hosting (among other manifestations of neo-liberal urban development) need to be placed within the specific long-run local contexts in which they occur as well as within the repeated dynamics of event iteration. In other words, though mega-event staging produces particular developmental patterns, these patterns are not imposed from above but rather reveal and are consistent with ongoing social and spatial policies in the host city or region. Neo-liberal urban transformation or conceptions of environmental sustainability are not generated by the IOC or FIFA or any such organization; rather, they are consistent with the corporate goals of these organizations, and provide an opportunity for local host coalitions to accelerate their realization. As Boykoff notes of Vancouver, the Winter Games did not invent the city's sharp inequalities between rich and poor, or the 'spatial contradictions' between its reputation for being the most liveable but least affordable global city (Boykoff 2011b: 51–2). Rather, we can argue that event staging *emphasizes* contradictions and dynamics that are already manifest, in three ways: through the global legitimization brought by event conferment; through the global mediatization brought by event preparation and staging; and through the enhancement of existing socio-spatial inequalities via social displacement and policy alignment.

Civic responses consequently also imply the development of local strategies of action, and the mobilization of existing organizations, actors and networks. As Boykoff also notes of Vancouver, 'micro-struggle' successes by activists and civil libertarians – forcing City Hall to backtrack on a number of local regulations restricting civic freedoms – 'demonstrates the importance of organizing early and often around questionable measures' (2011b: 51). This is, of course, easier said than done. We will accordingly leave the last word in this volume to Tom Tresser (2009), one of the organizers of No Games Chicago, reflecting on the city's (ultimately unsuccessful) bid to host the 2016 Olympic Games:

> This seems to me to be the most under-reported and most corrosive aspect of the 2016 saga. Namely, the complete emasculation of Chicago's entire civic and academic infrastructure around compliance with 2016 dogma. No arms-length critical studies were done by any good government group. No cautions from groups who are supposed to be protecting the common good, protecting our parks, protecting the taxpayers. No calls to action from grassroots groups who usually can be counted on to defend neighborhoods against exploitation or neglectful politicians. Aside from one report from DePaul's Egan Center, which raised a number of important questions, there was no arms-length review or study of the project or scan of the vast Olympic research done by the groups who have staff and who should've been critical of the bid from the get go.

References

Akagül, D. (1995), 'Dynamismes et Pesanteurs Economiques', in D. Akagül, S. Vaner & B. Kaleagasi (eds), *La Turquie en Mouvement*, pp. 60–70. Brussels: Editions Complexe.

Alfred, G. T. (2005), *Wasáse: Indigenous Pathways of Action and Freedom*. Toronto: Broadview Press.

Anagnost, A. (1997), *National Past-times: Narratives, Representation and Power in Modern China*. Durham & London: Duke University Press.

Andersen, M. S. (2002), 'Ecological Modernization or Subversion? The Effect of Europeanization on Eastern Europe', *American Behavioral Scientist*, 45/9, pp. 1394–416.

Anderson, B. (1991), *Imagined Communities. Reflections on the Origins and Spread of Nationalism*, rev. edn. London: Verso.

Anderson, P. (2007), 'Jottings on the Conjuncture', *New Left Review*, 48, pp. 5–36.

Andranovich, G., Burbank, M. & C. Heying (2001), 'Olympic Cities: Lessons Learned from Mega-event Politics', *Journal of Urban Affairs*, 23/2, pp. 113–31.

Androulidakis, I. & I. Karakassis (2006), 'Evaluation of the EIA system Performance in Greece, Using Quality Indicators', *Environmental Impact Assessment Review*, 26, pp. 242–56.

Ang, I. (1996), *Living Room Wars: Rethinking Media Audiences for a Postmodern World*. London & New York: Routledge.

Anon. (2009), 'Tourism Slumps in 2008 Despite Olympics', *China Economic Review*, 9 January.

ANP (2010), *Australian National Parks*. http://www.australiannationalparks.com/, accessed 10 June 2010.

Appadurai, A. (1996), *Modernity at Large: Cultural Dimensions of Globalization*. Minnesota: University of Minnesota Press.

APUR (2007), 'Les Chiffres du Logement Social à Paris en 2006', *Note de 4 Pages*, 26, January. Paris: Atelier Parisien d'Urbanisme.

Arnaud, P. (1998), 'Le Sport, Vecteur des Représentations Nationales des Etats Européens', in P. Arnaud & J. Riordan (eds), *Sport et Relations Internationales (1900–1941)*, pp. 11–26. Paris: L'Harmattan.

Arnaud, P. & T. Terret (1993), *Le Rêve Blanc: Olympisme et Sports d'Hiver en France, Chamonix 1924, Grenoble 1968*. Bordeaux: Presses universitaires de Bordeaux.

ASSDA (2009), *Australian Election Study, 2007*. http://nesstar.assda.edu.au/webview/index.jsp?object=http://nesstar.assda.edu.au/obj/fCatalog/Catalog17, consulted 15 January 2010.

ATHOC (1996), *Athens 2004: Candidate City*. Athens: ATHOC.

Australian Bureau of Statistics (ABS) (2006a), 'What Do Australians Think About Protecting the Environment?', Paper prepared for the 2006 Australian State of the Environment Committee. Canberra: Department of Environment and Heritage.

Australian Bureau of Statistics (ABS) (2006b), *2006 Year Book Australia. A Comprehensive Source of Information about Australia*. Canberra: ABS.

Bagnasco, A. (1986), *Torino. Un Profilo Sociologico*. Turin: Einaudi.

Balkeley, H. (2001), 'Governing Climate Change: The Politics of Risk Society?', *Transactions of the Institute of British Geographers*, 26/4, pp. 430–47.

Barry, J. (2005), 'Ecological Modernisation', in J. S. Dryzek & D. Schlosberg (eds), *Debating the Earth: The Environmental Politics Reader*, 2nd edn. Oxford: Oxford University Press.

Bauman, Z. (2003), 'City of Fears, City of Hopes', CUCR (Centre for Urban and Community Research) Occasional Paper, http://www.goldsmiths.ac.uk/cucr/pdf/city.pdf. London: Goldsmiths College.

Baykan, B. G. (2007), 'From Limits to Growth to Degrowth within French Green Politics', *Environmental Politics*, 16/3, pp. 513–17.

Beck, U. (1992), *Risk Society: Towards a New Modernity*. London: Sage.

Beijing 2008 (2007a), 'Beijing Citizens Report Polluters and Gain Rewards', http://en.beijing2008.cn/47/82/article214008247.shtml, 11 January 2007.

Beijing 2008 (2007b), 'Beijing Olympics Sponsors Form Environment Group', http://en.beijing2008.cn/47/82/article214008247.shtml, 28 July 2007.

Bennett, T. (1991), 'The Shaping of Things to Come: Expo '88', *Cultural Studies*, 5/1, pp. 33–51.

Benneworth, P. & H. Dauncey (forthcoming, 2011), 'International Urban Festivals as a Catalyst for Governance Capacity Building', *Environment & Planning C: Government & Policy*, 29.

BEPB (2009), *Beijing Environmental Quality Bulletin*. Beijing: Beijing Environmental Protection Bureau (in Chinese).

BIE (1928 [2009]), *Convention Relating to International Exhibitions*. Paris: Bureau International des Expositions.

Bina, O. (2007), 'A Critical Review of the Dominant Lines of Argumentation on the Need for Strategic Environmental Assessment', *Environmental Impact Assessment Review*, 27/7, pp. 585–606.

Black, D. R. & J. Van der Westhuizen (2004), 'The Allure of Global Games for "Semi-Peripheral" Polities and Spaces: A Research Agenda', *Third World Quarterly*, 25/7, pp. 1195–1214.

Blanc, C. & J-M. Eysseric (1992), *Official Report of the XVI Olympic Winter Games of Albertville and Savoie*. Albertville: Organizing Committee.

Blanchard, B. (2008), 'Heard the One about You and Me?', *Reuters*, 13 August.

Bobbio, L. & E. Dansero (2007), *The TAV and the Valle di Susa. Competing Geographies*. Turin: Umberto Allemandi & Co.

Bobin, F. & Z. Wang (2005), *Pékin en mouvement. Des innovateurs dans la ville*. Paris: Editions Autrement.

BOCOG (2008), 'Beijing 2008', http://en.beijing2008.com, accessed 20 June 2008.

Böhm, S. & S. Dabhi (eds) (2009), *Upsetting the Offset: The Political Economy of Carbon Markets*. London: Mayfly Books.

Bolsmann, C. & K. Brewster (2009), 'Mexico 1968 and South Africa 2010: Development, Leadership and Legacies', *Sport in Society*, 12/10, pp. 1284–98.

Boltanski, L. & L. Thévenot (1991), *De la Justification. Les économies de la grandeur*. Paris: Gallimard.

Boltanski, L. & L. Thévenot (1999), 'The Sociology of Critical Capacity', *European Journal of Social Theory*, 2/3, pp. 359–77.

Bondonio, P. (2006), 'Torino, Its Olympic Valleys and the Legacy: A Perspective', in N. Müller, M. Messing & H. Preuss (eds), *From Chamonix to Turin. The Winter Games in the Scope of Olympic Research*, pp. 395–417. Kassel: Agon Sportverlag.

Bondonio, P. & N. Campaniello (2006), 'Torino 2006: What Kind of Olympic Games Were They? A Preliminary Account from an Organizational and Economic Perspective', *Olympika: The International Journal of Olympic Studies,*, XV, pp. 1–33.

264 *References*

Bondonio, P., Guala, C. & A. Mela (2008) 'Torino 2006 OWG: Any Legacies for IOC and Olympic Territories?', in R. K. Barney, M. K. Heine, K. B. Wamsley & G. H. MacDonald (eds), *Pathways: Critiques and Discourse in Olympic Research: Proceedings of the Ninth International Symposium for Olympic Research*, pp. 151–65. London, Ontario: International Centre for Olympic Studies.

Boniface, P. (2002), *La Terre est Ronde Comme un Ballon*. Paris: Seuil.

Bonnot, C. (ed.) (2008), 'Pékin 2008: La Face Cachée des JO', *Science & Vie* (special issue), August.

Booth, P. (2003), 'Promoting Radical Change: The *Loi Relative à la Solidarité et au Renouvellement Urbains* in France', *European Planning Studies*, 11/8, pp. 949–63.

Bora, T. (2000), 'Football and Its Audience: Staging Spontaneous Nationalism', in S. Yerasimos, G. Seufert & K. Vorhoff (eds), *Civil Society in the Grip of Nationalism*, pp. 375–402. Istanbul: Orient-Institut-IFEA.

Börzel, T. A. (2003), *Environmental Leaders and Laggards in Europe: Why There Is (Not) A 'Southern Problem'*. Aldershot: Ashgate.

Botetzagias, I. (2008), 'The Environmental Impact Assessment and Auditing Process and Greece: Evidence from the Prefectural Level', *Impact Assessment and Project Appraisal*, 26/2, pp. 115–25.

Bourdieu, P. (1990), *Other Words*. Cambridge: Polity.

Bowdin, G. A. J., Allen, J., O'Toole, W., Harris, R. & I. McDonnell (2006), *Events Management*, 2nd edn. Oxford: Butterworth-Heinemann.

Bowerman, G. (2009), 'Beijing Records Highest Global Hotel Occupancy Drop', *China Business News*, 21 January.

Boyd, D. (2006) 'Civil Disobedience, Eagleridge Bluffs, and Joni Mitchell', *North Shore Outlook*, 13 April.

Boykoff, J. (2011a), 'Space Matters: The 2010 Winter Olympics and Its Discontents', *Human Geography*, forthcoming.

Boykoff, J. (2011b), 'The Anti-Olympics', *New Left Review*, 67, pp. 41–59.

Brady, A-M. (2009), 'The Beijing Olympics as a Campaign of Mass Distraction', *The China Quarterly*, 197, pp. 1–24.

Briggs, R., McCarthy, H. & A. Zorbas (2004), *16 Days. The Role of the Olympic Truce in the Toolkit for Peace*. London: Demos.

British Columbia (2004), *Spirit of 2010 Tourism Strategy*. http://www.llbc.leg.bc.ca/public/PubDocs/bcdocs/369378/tourism.pdf.

Brohm, J-M. (1983), *Jeux Olympiques à Berlin*. Brussels: Complexe.

Brohm, J-M. (1997), *Les Shootés du Stade*. Paris: Méditerranée.

Brohm, J-M., Perelman, M. & P. Vassort (2005), 'Non à l'Imposture Olympique!', *Le Monde Diplomatique*, July, p. 3.

Bromberger, C. (1995), *Le Match de Football. Ethnologie d'une Passion Partisane à Marseille, Naples et Turin*. Paris: Editions de la Maison des sciences de l'homme.

Broudehoux, A-M. (2004), *The Making and Selling of Post-Mao Beijing*. London: Routledge.

Broudehoux, A-M. (2007), 'Spectacular Beijing: The Conspicuous Construction of an Olympic Metropolis', *Journal of Urban Affairs*, 29/4, pp. 383–99.

Brownell, S. (1995), *Training the Body for China: Sports in the Moral Order of the People's Republic*. Chicago: University of Chicago Press.

Brownell, S. (2008), 'Western Centrism in Olympic Studies and Its Consequence in the 2008 Beijing Olympics', in R. K. Barney, M. K. Heine, K. B. Wamsley & G. H. MacDonald (eds), *Pathways: Critiques and Discourse in Olympic Research. Ninth International Symposium for Olympic Research*, pp. 20–30. London, Ontario: International Centre for Olympic Studies.

Bryman, A. (2004), *The Disneyization of Society*. London: Sage.

Bulletin d'Histoire Politique (2003), 'Sport et Politique', 11/2, special issue.

Burbank M. J., Andranovich, G. D. & C. H. Heying (2000), 'Antigrowth Politics or Piecemeal Resistance? Citizen Opposition to Olympic-Related Economic Growth', *Urban Affairs Review*, 35/3, pp. 334–57.

Burbank, M. J., Andranovich, G. D. & C. H. Heying (2001), *Olympic Dreams: The Impact of Mega-Events on Local Politics*. Boulder: Lynne Rienner.

Burbank, M. J., Andranovich, G. D. & C. H. Heying (2002), 'Mega-Events, Urban Development and Public Policy', *The Review of Policy Research*, 19/3, pp. 179–202.

Bureau of Shanghai World Expo Coordination (2007), *Expo Shanghai Online Participants Manual*. Shanghai: Bureau of Shanghai World Expo Coordination. http://www.expo2010.cn/expo/expoenglish/ps/download/userobject1ai48689/00000000.pdf, accessed 12 September 2009.

Bureau of Shanghai World Expo Coordination (2009a), *Better City, Better Life – The Development of the Shanghai World Expo Theme*. Shanghai: Orient Publication Center (in Chinese).

Bureau of Shanghai World Expo Coordination/SEPB (2009b), *Environmental Report: Expo 2010 Shanghai China*. Shanghai: Orient Publication Centre. http://www.sepb.gov.cn/platform/UserFiles/File/2009-07-30-15-57-28-005+08002332992585646130556.pdf, accessed 10 October 2009.

Buttel, F. H. (2000a), 'Ecological Modernisation as Social Theory', *Geoforum*, 31/1, pp. 57–65.

Buttel, F. H. (2000b), 'Classical Theory and Contemporary Environmental Sociology: Some Reflections on the Antecedents and Prospects for Reflexive Modernization Theories in the Study of Environment and Society,' in G. Spaargaren, A. P. J. Mol & F. H. Buttel (eds), *Environment and Global Modernity*, pp. 17–39. London: Sage.

Buttel, F. H. (2003), 'Environmental Sociology and the Explanation of Environmental Reform', *Organization and Environment*, 16/3, pp. 306–44.

Callick, R. (2008), 'On Top of the World in Beijing – After the Games', *Weekend Australian*, 23 August.

Cantelon, H. & Letters, M. (2000), 'The Making of the IOC Environmental Policy as the Third Dimension of the Olympic Movement', *International Review for the Sociology of Sport*, 35/3, pp. 294–308.

Cao, J., Li, W., Tan, J., Song, W. et al. (2009), 'Association of Ambient Air Pollution with Hospital Outpatient and Emergency Room Visits in Shanghai, China', *Science of the Total Environment*, 407, pp. 5531–6.

Caratti, P. (2004), *ANSEA – Un approccio analitico alla Valutazione Ambientale Strategica*. FEEM Rapporto sullo sviluppo sostenibile.

Caratti, P. & D. Lanzetta (2006), *Sviluppo e tutela dell'ambiente attraverso I grandi eventi. Il caso delle Olimpiadi Torino 2006*. Bologna: Il Mulino.

Caratti, P., Dalkmann, H. & R. Jiliberto (eds) (2004), *Analysing Strategic Environmental Assessment: Towards Better Decision-making*. London: Edward Elgar.

Carter, N. T. & A. P. J. Mol (eds) (2007), *Environmental Governance in China*. London: Routledge.

Cashman, R. (2003), 'What Is "Olympic Legacy"?', in M. de Moragas, C. Kennett & N. Puig (eds), *The Legacy of the Olympic Games 1984–2000*, pp. 31–42. Lausanne: International Olympic Committee.

Cashman, R. (2006), *The Bitter-Sweet Awakening: The Legacy of the Sydney 2000 Olympic Games*. Petersham, NSW: Walla Walla Press.

Cashman, R. & S. Darcy (2008), *Benchmark Games: The Sydney 2000 Paralympic Games*. Petersham, NSW: Walla Walla Press.

Cashman, R. & A. Hughes (eds) (1999), *Staging the Olympics: The Event and Its Impact*. Sydney: University of New South Wales Press.

Chalkley, B. & S. Essex (1999a), 'Urban Development through Hosting International Events: A History of the Olympic Games', *Planning Perspectives*, 14, pp. 369–94.

Chalkley, B. & S. Essex (1999b), 'Sydney 2000: The "Green Games"?', *Geography*, 84/4, pp. 299–307.

Chappelet, J-L. (2002), 'From Lake Placid to Salt Lake City: The Challenging Growth of the Olympic Winter Games Since 1980', *European Journal of Sport Science*, 2, pp. 1–21.

Chappelet, J-L. & B. Kübler-Mabbott (2008), *The International Olympic Committee and the Olympic System: The Governance of World Sport*. London & New York: Routledge.

Chernushenko, D. (1994), *Greening Our Games – Running Sports Events and Facilities That Won't Cost the Earth*. Ottawa: Centurion.

Choi, Y. S. (2004), 'Football and the South Korean Imagination. South Korea and the 2002 World Cup Tournament', in W. Manzenreiter & J. Horne (eds), *Football Goes East: Business, Culture and the People's Game in China, Japan and South Korea*, pp. 133–47. London & New York: Routledge.

Christoff, P. (2005), 'Policy Autism or Double-Edged Dismissiveness? Australia's Climate Policy under the Howard Government', *Global Change, Peace and Security*, 17/1, pp. 29–44.

City of Cape Town (2009), *Green Goal Progress Report*. Cape Town: City of Cape Town.

Clean Clothes (2007), 'Beijing Olympics Here We Come', *Clean Clothes Newsletter*, 24, 1 October.

Clifton, S-J. (2009), *A Dangerous Obsession: The Evidence Against Carbon Trading and for Real Solutions to Avoid a Climate Crunch*. London: Friends of the Earth England, Wales and Northern Ireland.

Close, P., Askew, D. & X. Xin (2007), *The Beijing Olympiad. The Political Economy of a Sporting Mega-Event*. London & New York: Routledge.

CNDP (2006), *Extension du Tramway à Paris. Bilan établi par le Président de la Commission Nationale du Débat Public*. Paris: Commission Nationale du Débat Public.

CNOSF (2008), *Procédure de sélection nationale d'une ville requérante aux Jeux Olympiques et Paralympiques d'hiver de 2018*. Paris: Comité National Olympique et Sportif Français.

CNOSF (2009), *Rapport de la Commission d'Evaluation du CNOSF pour la Procédure de Sélection Nationale d'une Ville Requérante aux Jeux Olympiques et Paralympiques d'Hiver De 2018*. Paris: Comité National Olympique et Sportif Français.

Coakley, J. & P. Donnelly (2004), *Sports, Society*. Toronto: McGraw-Hill Ryerson.

Cochrane A., Peck J. & A. Tickell (1996), 'Manchester Plays Games: Exploring the Local Politics of Globalisation', *Urban Studies*, 33/8, pp. 1319–36.

Cohen, J. L. & A. Arato (1992), *Civil Society and Political Theory*. Cambridge, MA: MIT Press.

Cohen, M. J. (1998), 'Science and the Environment: Assessing Cultural Capacity for Ecological Modernization', *Public Understanding of Science*, 7/2, pp. 149–67.

Cohen, M. J. (2000), 'Ecological Modernisation, Environmental Knowledge and National Character: A Preliminary Analysis of the Netherlands', in A. P. J. Mol & D. A. Sonnenfeld (eds), *Ecological Modernisation Around the World. Perspectives and Critical Debates*, pp. 77–106. London & Portland: Frank Cass.

COHRE (2007), *Fair Play for Housing Rights: Mega-Events, Olympic Games and Housing Rights*. Geneva: The Centre on Housing Rights and Evictions.

COHRE (2008), *One World, Whose Dream? Housing Rights Violations and the Beijing Olympic Games*. Geneva: The Centre on Housing Rights and Evictions.

Collectif anti-jeux olympiques (2008), *Pékin 2008, Boycott!: l'Olympisme Totalitaire*. Paris: L'Harmattan.

Collins, A., Jones, C. & M. Munday (2009), 'Assessing the Environmental Impacts of Mega Sporting Events: Two Options?', *Tourism Management*, 30, pp. 828–37.

Comité anti-olympique de Grenoble (2009), 'Moins Vite, Moins Haut et Moins Fort', http://cao38.eu.org/-Comite-Anti-Olympique, accessed 12 December 2009.

Comune di Roma (1999), *Documento di Valutazione Ambientale Strategica per la Giornata Mondiale della Gioventù del 2000*. Rome: Comune di Roma.

Contassot, Y. (2007), *De l'Environnement au Développement Soutenable. Elements de Bilan Parisien 2001–2007*. http://www.yvescontassot.eu/share/files/paris%20environnement%202001-2008%20V1-1.pdf, posted 2 March 2007, accessed 20 July 2007.

Conti, S. & P. Giaccaria (2001), *Local Development and Competitiveness*. Kluwer: Dordrecht.

Cornelissen, S. (2004), 'It's Africa's Turn! The Narratives and Legitimations Surrounding the Moroccan and South African Bids for the 2006 and 2010 FIFA Finals', *Third World Quarterly*, 25/7, pp. 1293–309.

Cornelissen, S. (2010), 'Football's Tsars: Proprietorship, Corporatism and Politics in the 2010 FIFA World Cup', in P. Alegi & C. Bolsmann (eds), *South Africa and the Global Game: Football, Apartheid and Beyond*. London: Routledge.

Coulthard, G. (2007), 'Subjects of Empire: Indigenous Peoples and the "Politics of Recognition" in Canada', *Contemporary Political Theory*, 6/4, pp. 437–60.

Crivello, S. Dansero, E. & A. Mela (2006), 'Torino, the Valleys and the Olympic Legacy: Exploring the Scenarios', in N. Müller, M. Messing & H. Preuss (eds), *From Chamonix to Turin. The Winter Games in the Scope of Olympic Research*, pp. 377–94. Kassel: Agon Sportverlag.

CSEB (2006a), 'Cost Backgrounder', www.eagleridgebluffs.ca, 15 May.

CSEB (2006b), 'Provincial Government Forced to Again Suspend Logging of Overland Route Due to Violation of Commitments', www.eagleridgebluffs.ca, 16 March.

CSEB (2006c), http://www.eagleridgebluffs.ca/initiatives/injunction.php, 15 May.

CSL (2009), *Extinguishing Emissions? A Review of the Approach Taken to Carbon Measurement and Management Across the London 2012 Programme*. London: Commission for a Sustainable London 2012.

Cunneen, C. (2000), 'Public Order and the Sydney Olympics: Forget About the Right to Protest', *Indigenous Law Bulletin*, 5/1, pp. 26–27.

Curran, G. (2009), 'Ecological Modernisation and Climate Change in Australia,' *Environmental Politics*, 18/2, pp. 201–17.

Dailly, D., Kukawka, P., Préau, P., Servoin, F. & R. Vivian (1992), *Albertville '92: l'Empreinte Olympique*. Grenoble: Presses universitaires de Grenoble.

Dansero, E. & A. Mela (2007), 'Olympic Territorialization: The Case of Torino 2006', *Revue de Géographie Alpine – Journal of Alpine Research*, 95/3, pp. 16–26.

Dansero, E. & M. Puttilli (2010), 'Mega-events Tourism Legacies: The Case of Torino 2006 Winter Olympic Games. A Territorialization Approach', *Leisure Studies*, 29/3, pp. 321–41.

Dansero, E., De Leonardis, D. & A. Mela (2006), 'Torino 2006: Territorial and Environmental Transformations', in N. Müller, M. Messing & H. Preuss (eds), *From Chamonix to Turin. The Winter Games in the Scope of Olympic Research*, pp. 359–76. Kassel: Agon Sportverlag.

Dansero, E., Mela, A. & A. Segre (2003), 'Spatial and Environmental Transformations towards Torino 2006: Planning the Legacy of the Future', in M. de Moragas, C. Kennett

& N. Puig (eds), *The Legacy of the Olympic Games 1984–2000*, pp. 83–93. Lausanne: International Olympic Committee.

Dauncey, H. (1997), 'Choosing and Building the "Grand stade de France" – National Promotion through Sport and "incompétence technocratique"?', *French Politics and Society*, 15/4, pp. 32–40.

Dauncey, H. (1998), 'Building the Finals: Facilities and Infrastructure', *Culture, Sport, Society*, 1/2, pp. 98–120.

Dauncey, H. (2004), 'Les Jeux Olympiques de Londres 1948. "Figure imposée", ou "vitrine"?', in P. Milza et al. (eds), *Le Pouvoir des Anneaux: Les Jeux Olympiques à la Lumière de la Politique 1896–2004*, pp. 183–98. Paris: Vuibert.

Dauncey, H. & G. Hare (1999), *France and the 1998 World Cup. The National Impact of a World Sporting Event*. London: Frank Cass.

De Lange, P. (1998), *Games Cities Play: The Staging of the Greatest Socio-Economic Event in the World*. Pretoria: Sigma Press.

Defrance, J. (2000), 'La Politique De l'Apolitisme. Sur l'Autonomisation Du Champ Sportif', *Politix*, 50, pp. 13–27.

Del Corpo B. (2006), 'Torino 2006. Conflitti, sfide e valori in un Luogo Conteso', *Bollettino della Società Geografica Italiana*, XII, XI/3, pp. 643–71.

Del Corpo, B. & E. Dansero (2007), 'Torino 2006: Environment, Challenges and Conflicts in a Contended Place', Paper presented at the European Sociological Association Conference, Glasgow 3–6 September.

Della Porta, D. (ed.) (2006), *The Global Justice Movement. Cross-national and Transnational Perspectives*. New York: Paradigm.

Della Porta, D. & M. Caiani (2009), *Social Movements and Europeanization*. Oxford: Oxford University Press.

Della Porta, D. & M. Diani (2004), *Movimenti senza Protesta? L'ambientalismo in Italia*. Bologna: Il Mulino.

Della Porta, D. & L. Mosca (2003), *Globalizzazione e movimenti sociali*. Rome: Manifestolibri.

Della Porta, D. & S. Tarrow (eds) (2005), *Transnational Protest & Global Activism*. Boston: Rowman & Littlefield.

Della Seta, R. (2000), *La difesa dell'ambiente in Italia. Storia e cultura del movimento ecologista*. Milan: F. Angeli.

Densham, A., Czebiniak, R., Kessler, D. & R. Skar (2009), *Carbon Scam: Noel Kempff Climate Action Project and the Push for Sub-national Forest Offsets*. Amsterdam: Greenpeace.

Desai, A. & G. Vahed (2010), 'World Cup 2010: Africa's Turn or the Turn on Africa?', in P. Alegi & C. Bolsmann (eds), *South Africa and the Global Game: Football, Apartheid and Beyond*. London: Routledge.

Dimopoulos, P., Bergmeier, E. & P. Fischer (2006), 'Natura 2000 Habitat Types of Greece Evaluated in the Light of Distribution, Threat and Responsibility', *Proceedings of the Royal Irish Academy*, 106/3, pp. 175–87.

Dine, P. (2003), 'The End of an Idyll? Sport and Society in France, 1998–2002', *Modern & Contemporary France*, 11/1, pp. 33–43.

Dobson, A. (1995), *Green Political Thought*, 2nd edn. London & New York: Routledge.

Dong, M. Y. (2000), 'Defining Beijing: Urban Reconstruction and National Identity, 1928–1936', in J. W. Esherick (ed.), *Remaking the Chinese City: Modernity and National Identity, 1900–1950*, pp. 121–38. Honolulu: University Hawaii Press.

Dorronsoro, G. (ed.) (2005), *La Turquie Conteste: Mobilisations Sociales et Régime Sécuritaire*. Paris: CNRS.

Doyle, T. (2005), *Environmental Movements in Majority and Minority Worlds*. New Brunswick, NJ, & London: Rutgers University Press.

Doyle, T. (2010), 'Surviving the Gang Bang Theory of Nature: The Environmental Movement during the Howard Years', *Social Movement Studies*, 9/2, pp. 155–69.

Duara, P. (2001), 'The Discourse of Civilization and Pan-Asianism', *Journal of World History*, 12/1, pp. 99–130.

Dyreson, M. & M. Llewellyn (2010), 'Los Angeles Is *the* Olympic City: Legacies of the 1932 and 1984 Olympic Games', in J. A. Mangan & M. Dyreson (eds), *Olympic Legacies: Intended and Unintended, Political, Cultural, Economic and Educational*, pp. 108–35. London & New York: Routledge.

Eccleston, C. H. & R. B. Smythe (2002), 'Integrating Environmental Impact Assessment with Environmental Management Systems', *Environmental Quality Management*, 11/4, pp. 1–13.

EEA (2008), *Greenhouse Gas Emission Trends and Projections in Europe 2008: Tracking Progress towards Kyoto Targets*. EEA Report 5/2008. Copenhagen: European Environment Agency.

Ehrenberg, A. (1991), *Le Culte de la Performance*. Paris: Calman-Levy.

Eichberg, H. (2004), 'The Global, the Popular and the Inter-Popular: Olympic Sport between Market, State and Civil Society', in J. Bale & M. K. Christensen (eds), *Post-Olympism? Questioning Sport in the Twenty-first Century*, pp. 65–80. Oxford: Berg.

Elafros, Y. (2007), 'As Disaster Looms for the Global Environment, Greece Is still Ruining the Atmosphere with Bad Energy Sources', *Kathimerini*, 10 February.

Elias, N. (2000), *The Civilizing Process. Sociogenetic and Psychogenetic Investigations*, revised edn. Oxford: Blackwell.

Elias, N. & E. Dunning, (1987), *Sport Et Civilisation. La Violence Maîtrisée*. Paris: Fayard.

Elliot, M. & I. Thomas (2009), *Environmental Impact Assessment in Australia. Theory and Practice*, 5th edn. Sydney: The Federation Press.

Elliott, R. (2008) 'Blue Skys over Beijing or Statistical Manipulation?', *Globalisation and the Environment*, http://globalisation-and-the-environment.blogspot.com/2008/01/blue-skys-over-beijing.html, posted 10 January.

Essex, S. & B. Chalkley (1998), 'The Olympics as a Catalyst of Urban Renewal', *Leisure Studies*, 17/3, pp. 187–206.

Essex, S. & B. Chalkley (2002), 'L'Evoluzione degli Impatti Infrastrutturali delle Olimpiadi Invernali, 1924–2002', *Bollettino Della Società Geografica Italiana*, 12/7, pp. 831–51.

Essex, S. & B. Chalkley (2004), 'Mega-sporting Events in Urban and Regional Policy: A History of the Winter Olympics', *Planning Perspectives*, 19/2, pp. 201–4.

European Commission (2009), 'European Attitudes towards Climate Change', *Eurobarometer*, 72.1. Brussels: European Commission.

Fainstain, S. S. (2009) 'Mega-Projects in New York, London and Amsterdam', *International Journal of Urban and Regional Research*, 32/4, pp. 768–85.

Featherstone, D. (2005), 'Towards the Relational Construction of Militant Particularisms: Or Why the Geographies of Past Struggles Matter for Resistance to Neoliberal Globalisation', *Antipode*, 37, pp. 250–71.

FIFA (2007a), *FIFA World Cup™: TV Viewing Figures*. Zurich: Fédération Internationale de Football Association, http://www.fifa.com/mm/document/fifafacts/misc-tele/52/01/27/fs-401_05a_fwc-tv-stats.pdf.

FIFA (2007b), *Our Commitment*. Zurich: Fédération Internationale de Football Association.

FIFA (2011), *FIFA Financial Report 2010*. Zurich: Fédération Internationale de Football Association.

FIFA (n.d.), *Football For Hope. Football's Commitment to Social Development*. Zurich: Fédération Internationale de Football Association.

Fischer, T. B. (2002), 'Strategic Environmental Assessment Performance Criteria – The Same Requirements for Every Assessment?', *Journal of Environmental Assessment and Policy Management*, 4/1, pp. 83–99.

Fischer, T. B & K. Seaton (2002), 'Strategic Environmental Assessment: Effective Planning Instrument or Lost Concept?', *Planning Practice & Research*, 17/1, pp. 31–44.

Flyvberg, B., Bruzelius, N. & W. Rothengatter (2003), *Megaprojects and Risk: An Anatomy of Ambition*. Cambridge: Cambridge University Press.

Folke, C., Hahn, T., Olsson, P. & J. Norberg (2005), 'Adaptive Governance of Social-Ecological Systems', *Annual Review of Environment and Resources*, 30, pp. 441–73.

Foucault, M. (1975), *Surveiller et Punir: Naissance de la Prison*. Paris: Gallimard.

Foucault, M. (1976), *Histoire de la Sexualité, 1: La Volonté de Savoir*. Paris: Gallimard.

Francillon, C. (1991), *Chamonix '24, Grenoble '68, Albertville '92: le roman des Jeux*. Paris: Glénat.

Frappat, P. (1979), *Grenoble: Le Mythe blessé*. Paris: Alain Moreau.

Frey, M., Iraldo, F. & M. Melis (2007), 'The Impact of Wide Scale Sport Events on Local Development: An Assessment of the XXth Torino Olympics through the Sustainability Report', Paper presented at the Conference of the Regional Studies Association, Lisbon; available at http://papers.ssrn.com/sol3/papers.cfm?abstract_id=1117967.

Friedman, S. L. (2004), 'Embodying Civility: Civilizing Processes and Symbolic Citizenship in Southeastern China', *Journal of Asian Studies*, 63, pp. 687–718.

Frijns, J., Phuong P. T. & A. P. J. Mol. (2000), 'Ecological Modernisation Theory and Industrialising Economies: The Case of Viet Nam', in A. P. J. Mol & D. A. Sonnenfeld (eds), *Ecological Modernisation around the World. Perspectives and Critical Debates*, pp. 257–92. London: Frank Cass.

Gambino, R., Mondini, G. & A. Peano (2005) (eds), *Le Olimpiadi per il Territorio. Monitoraggio Territoriale del Programma Olimpico di Torino 2006*. Milan: Il Sole 24 Ore – Pirola Editore.

Gang, C. (2009), *Politics of China's Environmental Protection. Problems and Progress*. Singapore: World Scientific.

Garcia, B. (2007), 'Sydney 2000', in J. R. Gold & M. M. Gold (eds), *Olympic Cities. City Agendas, Planning, and the World's Games, 1896–2012*, pp. 235–64. London & New York: Routledge.

Garel, S., Nenner, C. & B. Maris (2005), 'Des JO à Paris? Non, merci...', *Le Monde*, 9 March.

Géopolitique, Revue de l'Institut international de géopolitique (1999), 'Sport et politique', 66, special issue.

Getz, D. (2008), 'Event Tourism: Definition, Evolution, and Research', *Tourism Management*, 29, pp. 403–28.

Getz, D. (2009), 'Policy for Sustainable and Responsible Festivals and Events: Institutionalization of a New Paradigm', *Journal of Policy Research in Tourism, Leisure & Events*, 1/1, pp. 61–78.

Giddens, A. (1989), *The Consequences of Modernity*. Cambridge: Polity.

Giddens, A. (1991), *Modernity and Self-Identity: Self and Society in the late Modern Age*. Cambridge: Polity.

Giddens, A. (2006), *Sociology*, 5th edn. Cambridge: Polity.

Gilbert, N. (1995), 'Emergence in Social Simulation', in N. Gilbert & R. Conte (eds), *Artificial Societies*, pp. 144–56. London: UCL Press.

Gillon, P., Grosjean, F. & L. Ravenel (2010), *Atlas du sport mondial. Business et spectacle: l'idéal sportif en jeu*. Paris: Autrement.

Gilstrap, D. L. (2005), 'Strange Attractors and Human Interaction: Leading Complex Organization through the Use of Metaphors', *Complexity: An International Journal of Complexity and Education*, 2/1, pp. 55–69.

Girginov, V. & L. Hills (2010), 'A Sustainable Sports Legacy: Creating a Link between the London Olympics and Sports Participation', in V. Girginov (ed.), *The Olympics. A Critical Reader*, pp. 430–47. London & New York: Routledge.

Giulianotti, R. (2005), *Sport. A Critical Sociology*. Cambridge: Polity.

Giulianotti, R. & R. Robertson. (eds) (2007), *Globalization and Sport*. Oxford: Blackwell.

Godard, O. (2008), 'Le Grenelle de l'Environnement Met-il la France sur la Voie du Développement Durable ?', *Regards sur l'Actualité*, 338, pp. 37–46.

Gold, J. R. & M. M. Gold (2008), 'Olympic Cities: Regeneration, City Rebranding and Changing Agendas', *Geographical Compass*, 2/1, pp. 300–18.

Gold, M. M. (2007), 'Athens 2004', in J. R. Gold & M. M. Gold (eds), *Olympic Cities. City Agendas, Planning, and the World's Games, 1896–2012*, pp. 265–85. London & New York: Routledge.

Gollain, F. (2006), 'Les Débats autour du Mouvement pour la Décroissance', *French Politics, Culture & Society*, 24/2, pp. 114–28.

Gössling, S., Broderick, J., Ceron, J-P., Dubois, G., Peeters, P., Strasdas, W. & J. P. Upham (2007), 'Voluntary Carbon Offsetting Schemes for Aviation: Efficiency, Credibility and Sustainable Tourism', *Journal of Sustainable Tourism*, 15/3, pp. 223–48.

Governa, F. Rossignolo, C. & S. Saccomani (2009), 'Torino. Urban Regeneration in a Post-Industrial City', *Journal of Urban Regeneration And Renewal*, 3/1, pp. 20–30.

Grabowski, C. (2006) 'Showdown on the Bluffs', *The Tyee*, 26 May.

Gratton, C., Shibli, S. & R. Coleman, (2006), 'The Economic Impact of Major Sports Events: A Review of Ten Events, the UK', in J. Horne & W. Manzenreiter (eds), *Sports Mega-Events: Social Scientific Analysis of a Global Phenomenon*, pp. 41–58. Oxford: Blackwell.

Green Games Watch 2000 (1999), *Environmental Compliance of Selected Olympic Venues*. Bondi Junction, New South Wales: Green Games Watch 2000.

Greene, S. J. (2003), 'Staged Cities: Mega-Events, Slum Clearance, and Global Capital', *Yale Human Rights & Development Law Journal*, 6, pp. 161–87.

Greenpeace (2000), *Greenpeace's Olympic Environmental Guidelines: A Guide to Sustainable Events*. http://www.greenpeace.org/raw/content/china/en/reports/guideline.pdf.

Greenpeace (2003), *Pinocchio 2004. Olympic Games – Athens 2004. Promises: Always Green, Always Forgotten. The Environmental Landscape a Year before the Olympics*. Athens: Greenpeace.

Greenpeace (2004a), *How Green the Games? A Greenpeace Assessment of the Environmental Performance of the Athens 2004 Olympics*. Athens: Greenpeace.

Greenpeace (2004b), *Olympic Games – Athens 2004. Promises: Always Green, Always Forgotten. An Assessment of the Environmental Dimension of the Games*. Athens: Greenpeace.

Groupe Ecologie et Solidarité (2008), *La Face Cachée de la Candidature de Grenoble aux Jeux Olympiques d'Hiver de 2018*. Grenoble: Hôtel de Ville.

Gruneau, R. (2002), 'Foreword', in M. D. Lowes, *Indy Dreams and Urban Nightmares. Speed Merchants, Spectacle, and the Struggle over Public Space in the World-Class City*, pp. ix–xii. Toronto: Toronto University Press.

Gu, H. (2007), 'Adaptive State Corporatist Solution of Environmental Conflicts: A New Paradigm of State-Society Relations in China and Japan', Paper presented at the 4th Annual ECPR General Conference, Pisa, September.

Guala, C. (2002), 'Per una tipologia dei mega eventi', *Bollettino Della Società Geografica Italiana*, 12/7, pp. 743–55.

Guala, C. (2007), 'Il monitoraggio dell'opinione pubblica', in P. Bondonio, E. Dansero, C. Guala, A. Mela & S. Scamuzzi (eds), *A Giochi Fatti*, pp. 105–122. Rome: Carocci.

Guala, C. (2009), 'Torino: i XX Giochi Olimpici e la stagione dei mega eventi', *Territorio*, 48, pp. 103–109.

Guala, C. & S. Crivello (2006), 'Mega-Events and Urban Regeneration. The Background and Numbers behind Turin 2006', in N. Müller, M. Messing & H. Preuss (eds), *From Chamonix to Turin. The Winter Games in the Scope of Olympic Research*, pp. 323–42. Kassel: Agon Sportverlag.

Guidry, J. A., Kennedy, M. D. & M. N. Zald (2000), 'Globalizations and Social Movements', in J. A. Guidry, M. D. Kennedy & M. N. Zald (eds), *Globalizations and Social Movements. Culture, Power and the Transnational Public Sphere*, pp. 1–32. Ann Arbor: University Michigan Press.

Gursoy, D. & K. W. Kendall (2006), 'Hosting Mega Events. Modeling Locals' Support', *Annals of Tourism Research*, 33/3, pp. 603–23.

Guttmann, A. (1994), *Games and Empires: Modern Sports and Cultural Imperialism*. New York: Columbia University Press.

Hajer, M. A. (1995), *The Politics of Environmental Discourse. Ecological Modernization and the Policy Process*. Oxford & New York: Oxford University Press.

Hajer, M. A. (1996), 'Ecological Modernisation as Cultural Politics', in S. Lash, B. Szerszynski & B. Wynne (eds), *Risk, Environment and Modernity: Towards a New Ecology*, pp. 246–68. London, Thousand Oaks, CA, & New Delhi: Sage.

Hall, C. M. (2001), 'Imaging, Tourism and Sports Event Fever: The Sydney Olympics and the Need for a Social Charter for Mega-Events', in C. Gratton & I. P. Henry (eds), *Sport in the City: The Role of Sport in Economic and Social Regeneration*, pp. 166–83. London: Routledge.

Hall, C. M. (2005), *Tourism: Rethinking the Social Science of Mobility*. Harlow: Prentice-Hall.

Hall, C. M. (2006), 'Urban Entrepreneurship, Corporate Interests and Sports Mega-Events: Thin Policies of Competitiveness Within the Hard Outcomes of Neoliberalism', in J. Horne & W. Manzenreiter (eds), *Sports Mega-Events: Social Scientific Analysis of a Global Phenomenon*, pp. 59–70. Oxford: Blackwell.

Hall, C. M. & J. Hodges (1996), 'The Party's Great, but What about the Hangover? The Housing and Social Impacts of Mega-Events with Special Reference to the 2000 Sydney Olympics', *Festival Management & Event Tourism*, 4, pp. 13–20.

Hamel, P. (2000), 'The Fragmentation of Social Movements and Social Justice. Beyond the Traditional Forms of Localism', in P. Hamel, H. Lustiger-Thaler & M. Mayer (eds), *Urban Movements in a Globalising World*, pp. 157–74. London: Routledge.

Hannigan, J. (1998), *Fantasy City*. London: Routledge.

Hannigan, J. (2002), 'The Global Entertainment Economy', in D. R. Cameron & J. G. Stein (eds), *Street Protests and Fantasy Parks: Globalization, Culture, and the State*, pp. 20–48. Columbia: UBC Press.

Hannigan, J. (2006), *Environmental Sociology*, 2nd edn. London & New York: Routledge.

Hare, G. (2003), *French Football: A Cultural History*. Oxford: Berg.

Hari, J. (2010), 'The Wrong Kind of Green', *The Nation*, 4 March.

Harrell, S. (1995), 'Introduction: Civilizing Projects and the Reaction to Them', in S. Harrell (ed.), *Cultural Encounters on China's Ethnic Frontiers*, pp. 3–36. Seattle: University Washington Press.

Harris, P. J, Harris, E., Thompson, S., Harrris-Roxas B. & L. Kemp (2009), 'Human Health and Wellbeing in Environmental Impact Assessment in New South Wales, Australia: Auditing Health Impacts within Environmental Assessment of Major Projects', *Environmental Impact Assessment Review*, 29/5, pp. 310–18.

Harvey, D. (2002), 'The Art of Rent: Globalization, Monopoly and Cultural Production', *Socialist Register*, pp. 93–110.

Harvey, D. (2005), *A Brief History of Neo-Liberalism*. Oxford: Oxford University Press.

Harvey, D. (2006), *Spaces of Global Capitalism. Towards a Theory of Uneven Geographical Development*. London & New York: Verso.

Hayes, G. (2002), *Environmental Protest and the State in France*. Basingstoke: Palgrave.

Hayes, G. & J. Horne (2011), 'Sustainable Development, Shock and Awe? London 2012 and Civil Society', *Sociology*, 45/5, forthcoming.

Held, D. (2004), *Global Covenant. The Social Democratic Alternative to the Washington Consensus*. Cambridge: Polity.

Held, D. & A. McGrew (2002), 'Introduction', in D. Held & A. McGrew (eds), *Governing Globalization: Power, Authority and Global Governance*, pp. 1–21. Cambridge: Polity.

Heper, M. (1991), 'The State and Interest Groups with Special Reference to Turkey', in M. Heper (ed.), *Strong State and Economic Interest Groups: The Post-1980 Turkish Experience*, pp. 2–22. Berlin & New York: De Gruyter.

Heynen, N. & P. Robbins (2005), 'The Neoliberalization of Nature: Governance, Privatization, Enclosure and Valuation', *Capital, Nature, Socialism*, 16/1, pp. 5–8.

Hill, C. R. (1994), 'The Politics of Manchester's Olympic Bid', *Parliamentary Affairs*, 47/3, pp. 338–54.

Hiller, H. H. (1998), 'Assessing the Impact of Mega-Events: A Linkage Model', *Current Issues in Tourism*, 1/1, pp. 47–57.

Hiller, H. H. (2000), 'Toward an Urban Sociology of Mega-Events', *Research in Urban Sociology*, 5, pp. 181–205.

Hoberman, J. (1995), 'Towards a Theory of Olympic Internationalism', *Journal of Sport History*, 22, pp. 1–37.

Hoberman, J. (2004), 'Sportive Nationalism and Globalization', in J. Bale & M. K. Christensen (eds), *Post-Olympism? Questioning Sport in the Twenty-first Century*, pp. 177–88. Oxford: Berg.

Hoberman, J. (2008), 'Think Again. The Olympics', *Foreign Policy*, 167, pp. 22–28.

Holden, M., MacKenzie, J. & R. Van Wynsberghe (2008), 'Vancouver's Promise of the World's First Sustainable Olympic Games', *Environment and Planning C: Government and Policy*, 26, pp. 882–905.

Holt, R. (1989), *Sport and the British*. Oxford: Clarendon Press.

Horne, J. (2007), 'The Four 'Knowns' of Sports Mega-Events', *Leisure Studies*, 26/1, pp. 81–96.

Horne, J. & Manzenreiter, W. (2006), 'An Introduction to the Sociology of Sports Mega-events', in J. Horne & W. Manzenreiter (eds), *Sports Mega-events. Social Scientific Analyses of a Global Phenomenon*, pp. 1–24. Oxford: Blackwell.

Houlihan, B. (1994), *Sport and International Politics*. London: Harvester Wheatsheaf.

Howe, P. D. (2008), *The Cultural Politics of the Paralympic Movement – Through an Anthropological Lens*. London: Routledge.

Hug, P.-A., Van Griethuysen, P. & C. Dubi (2002), *Olympic Games Management: from the Candidature to the final Evaluation, an Integrated Management Approach*, paper presented at the Symposium on The Legacy of the Olympic Games, Lausanne.

Hughes, G. (1999), 'Urban Revitalization: The Use of Festive Time Strategies', *Leisure Studies*, 18/2, pp. 119–35.

IDFE (2004), *Jeux Olympiques 2012. Pour Sauver la Candidature de Paris*. Paris: Ile de France Environnement.

INSEE (2007), *Chiffres clés. Enquêtes annuelles de recensement de 2004 à 2006: Paris*. Paris: Institut National de la Statistique et des Études Économiques. http://www.insee.fr/fr/recensement/nouv_recens/resultats/repartition/chiffres_cles/n2/n2_75056.pdf, posted July 2007, accessed 30 July 2007.

IOC (1992), *Manual for Cities Bidding for the Olympic Games*. Lausanne: International Olympic Committee.

IOC (1996), *Manual for Candidate Cities for the Games of the XXVIII Olympiad 2004*. Lausanne: International Olympic Committee.

IOC (1999), *Sport for Sustainable Development. Olympic Movement's Agenda 21*. Lausanne: Sport and Environment Commission of the International Olympic Committee.

IOC (2000), *Manual for Candidate Cities for the Games of the XXIX Olympiad 2008*. Lausanne: International Olympic Committee.

IOC (2003), *Olympic Charter*. Lausanne: International Olympic Committee.

IOC (2004), *2012 Candidature Procedure and Questionnaires. Games of the XXX Olympiad in 2012*. Lausanne: International Olympic Committee.

IOC (2005), *Report of the IOC Evaluation Commission for the Games of the XXX Olympiad in 2012*. Lausanne: International Olympic Committee.

IOC (2007), *Olympic Charter*. Lausanne: International Olympic Committee.

IOC (2010), *Olympic Marketing Fact File*, 2010 edn. Lausanne: International Olympic Committee.

Japan Association for the 2005 World Exposition (2006), *Environmental Report*. Aichi: Japan Association for the 2005 World Exposition.

Jennings, W. & Lodge, M. (2009), 'Governing Mega-events: Tools of Security Risk Management for the London 2012 Olympic Games and FIFA 2006 World Cup in Germany', unpublished paper presented to PSA annual conference, Manchester, April.

Jie, Y.,Yingkan, G. & H. Ping (2010), 'A Structural Equation Model of Residents' Support for Mega Event', *Chinese Journal of Population, Resources and Environment*, 8/1, pp. 71–9.

Juris, J. (2004), 'Networked Social Movements: Global Movements for Global Justice', in M. Castells (ed.), *The Network Society: A Cross-cultural Perspective*, pp. 341–62. Cheltenham: Edward Elgar.

Juris, J. (2008), *Network Futures: The Movements against Corporate Globalization*. Durham: Duke University Press.

Kahn, J. & J. Yardley (2007), 'As China Roars, Pollution Reaches Deadly Extremes', *New York Times*, 26 August.

Kananaskis G8 Summit Management Office (2002), *Strategic Environmental Assessment of the June 2002 G8 Summit in Kananaskis, Alberta*. Kananaskis: Kananaskis G8 Summit Management Office.

Karamichas, J. (2003), *Civil Society and the Environmental Problematic. A Preliminary Investigation of the Greek and Spanish Cases*, paper presented at the ECPR joint sessions of workshops, University of Edinburgh, 28 March – 2 April.

Karamichas, J. (2005), 'Risk versus National Pride: Conflicting Discourses over the Construction of a High Voltage Power Station in the Athens Metropolitan Area for Demands of the 2004 Olympics', *Human Ecology Review*, 12/2, pp. 133–42.

Karamichas, J. (2007), 'The Impact of the Summer 2007 Forest Fires in Greece: Recent Environmental Mobilizations, Cyber-Activism and Electoral Performance', *South European Society & Politics*, 12/4, pp. 521–34.

Karamichas, J.(2008), 'Red and Green Facets of Political Ecology. Accounting for Electoral Prospects in Greece', *Journal of Modern Greek Studies*, 26, pp. 311–36.

Karrer, F. (2002), 'La VAS per la Giornata Mondiale della Gioventù nell'ambito delle opere per il Giubileo del 2000', in A. Busca & G. Campeol (eds), *La Valutazione Ambientale Strategica e la Nuova Direttiva Comunitaria*, pp. 95–98. Rome: Palombi Fratelli.

Kasimati, E. (2003), 'Economic Aspects and the Summer Olympics: A Review of Related Research', *International Journal of Tourism Research*, 5, pp. 433–44.

Katzel, C. T. (2007), *Event Greening: Is This Concept Providing a Serious Platform for Sustainability Best Practice?* Master's Degree Thesis, Stellenbosch University, https://etd.sun.ac.za/jspui/handle/10019/431, retrieved 16 January 2010.

Kavouridis, K., Chaloulos, K., Leontidis, M. & C. Roubos (2005), 'The Exploitation of Lignite in Greece with Economic and Environmental Criteria. Current Situation-Prospects', paper presented at the meeting 'Lignite and Natural Gas in the Generation of Electricity in the Country', TEE Athens, 9–10 June.

Kazantzopoulos, G. (2002), 'The Environmental Strategy for the Olympic Projects', in Hellenic Environmental Law Society (ed.), *Olympic Games and the Environment – Conference Proceedings Athens 1-2/02/2001*, pp. 109-12. Athens & Komotini: Sakkoulas.

Kearins, K. & K. Pavlovich (2002), 'The Role of Stakeholders in Sydney's Green Games', *Corporate Social Responsibility and Environmental Management*, 9/3, pp. 157–69.

Keck, M. & K. Sikkink (1998), *Activists beyond Borders. Advocacy Networks in International Politics*. Ithaca & London: Cornell University Press.

Kempf, H. (2009), 'Mobilisation pour des Jeux Olympiques "Verts"', *Le Monde*, 21 January.

Keysar, E. (2005), 'Procedural Integration in Support of Environmental Policy Objectives: Implementing Sustainability', *Journal of Environmental Planning and Management*, 48/4, pp. 549–69.

Klein, N. (2007a), *The Shock Doctrine: The Rise of Disaster Capitalism*. London: Allen Lane/Penguin.

Klein, N. (2007b), *The Shock Doctrine Podcast*. http://itunes.apple.com/us/podcast/the-shock-doctrine-podcast/id266551506.

Klein, N. (2008), 'The Olympics: Unveiling Police State 2.0', *The Huffington Post*, http://www.huffingtonpost.com/naomi-klein/the-olympics-unveiling-po_b_117403.html, posted 7 August 2008.

Koutalakis, C. (2004), 'Environmental Compliance in Italy and Greece: The Role of Non-state Actors', *Environmental Politics*, 13/4, pp. 754–74.

Kovac, I. (2003), 'The Olympic Territory. A Way to an Ideal Olympic Scene', in M. de Moragas, C. Kennett & N. Puig (eds), *The Legacy of the Olympic Games 1984–2000*, pp. 110–117. Lausanne: International Olympic Committee.

Kozanoğlu, C. (1999), 'Beyond Edirne: Football and the National Identity Crisis in Turkey', in G. Armstrong & R. Giulianotti (eds), *Football Cultures and Identities*, pp. 117–25. London: Macmillan.

Krüger, A. (2004), 'Was the 1936 Olympics the First Postmodern Spectacle?', in J. Bale & M. K. Christensen (eds), *Post-Olympism? Questioning Sport in the Twenty-First Century*, pp. 33–50. Oxford: Berg.

Krüger, A. (2005), 'The Nazi Olympics of 1936', in K. Young & K. B. Wamsley (eds), *Global Olympics. Historical and Sociological Studies of the Modern Games*, pp. 43–58. London: Elsevier.

Kukawka, P., Préau, P. Servoin F. & R. Vivian (1991), *Albertville '92: Les Enjeux Olympiques*. Grenoble: PU de Grenoble.

Kurscheidt, M. (2000), 'Strategic Management and Cost-Benefit Analysis of Major Sport Events: The Use of Sensitivity Analyses Shown for the Case of the Soccer World Cup 2006 in Germany', Universität Paderborn, Working Paper No. 69, October.

Kwasniak, A. (2009), 'The Eviscerating of Federal Environmental Assessment in Canada', *Alberta Wilderness Association Journal*, 17/2, pp. 6–8.

La Spina, A. & G. Sciortino (1993), 'Common Agenda, Southern Rules: European Integration and Environmental Change in the Mediterranean States', in J. D. Liefferink, P. D. Lowe and A. P. J. Mol (eds), *European Integration & Environmental Policy*, pp. 99–113. London & New York: Belhaven.

Lake, D. A. (1999), 'Global Governance: A Relational Contracting Approach', in A. Prakash & J. A. Hart (eds), *Globalization and Governance*, pp. 31–53. London & New York: Routledge.

Lascoumes, P. (2008), 'Leviers d'Action et Obstacles à la Mise en Oeuvre d'une Politique de Développement Soutenable', *Regards sur l'Actualité*, 338, pp. 47–58.

Latham, A. (2003), 'Urbanity, Lifestyle and Making Sense of the New Urban Cultural Economy', *Urban Studies*, 40/9, pp. 1699–1724.

Latouche, S. (2003), 'L'Imposture du Développement Durable ou les Habits Neufs du Développement', *Mondes en Développement*, 121, pp. 23–30.

Latour, B. (1999), 'On recalling ANT', in J. Law & J. Hassard (eds), *Actor Network Theory and After*, pp. 15–25. Oxford: Blackwell.

Lebel, L., Anderies, J. M., Campbell, B., Folke, C., Hatfield-Dodds, S., Hughes T. P. & J. Wilson (2006), 'Governance and the Capacity to Manage Resilience in Regional Social-Ecological Systems', *Ecology and Society*, 11/1, Art. 19.

Lenskyj, H. J. (1996), 'When Winners are Losers: Toronto and Sydney Bids for the Summer Olympics', *Journal of Sport and Social Issues*, 20/4, pp. 392–410.

Lenskyj, H. J. (1998), 'Sport and Corporate Environmentalism: The Case of the Sydney 2000 Olympics', *International Review for the Sociology of Sport*, 33/4, pp. 341–54.

Lenskyj, H. J. (2000), *Inside the Olympic Industry: Power, Politics and Activism*. Albany, NY: State University of New York Press.

Lenskyj, H. J. (2002), *The Best Olympics Ever? Social Impacts of Sydney 2000*. Albany: State University of New York Press.

Lenskyj, H. J. (2008), *Olympic Industry Resistance: Challenging Olympic Power and Propaganda*. Albany: State University of New York Press.

Leonardsen, D. (2007), 'Planning of Mega Events: Experiences and Lessons', *Planning Theory & Practice*, 8/1, pp. 11–30.

Lesjø, J. H. (2000), 'Lillehammer 1994', *International Review for the Sociology of Sport*, 35/3, pp. 282–93.

Levitt, T. (1983), 'The Globalization of Markets', *Harvard Business Review*, May–June.

Lewicki, R. J, Gray, B. & M. Elliott (2003), *Making Sense of Intractable Environmental Conflicts: Frames and Cases*. Washington DC: Island Press.

Ley, D. & K. Olds (1988), 'Landscape as Spectacle: World's Fairs and the Culture of Heroic Consumption', *Society & Space*, 6, pp. 191–212.

Leys, S. (1997), *Analects of Confucius*. New York: Norton.

Liao, H. & A. Pitts (2008) 'A Brief Historical Review of Olympic Urbanization', in B. Majumdar & S. Collins (eds), *Olympism: The Global Vision. From Nationalism to Internationalism*, pp. 145–65. London: Routledge.

Lindblom, C. (1982), 'The Market as Prison', *The Journal of Politics*, 44/2, pp. 324–36.

LOCOG (2009), *Towards a One Planet 2012. London 2012 Sustainability Plan*, 2nd edn. London: London Organizing Committee of the Olympic Games and Paralympic Games.

Lowes, M. (2004) 'Neoliberal Power Politics and the Controversial Siting of the Australian Grand Prix Motorsport Event in an Urban Park', *Loisir et Société/Society & Leisure*, 27/1, pp. 69–88.

Lowes, M. D. (2002), *Indy Dreams and Urban Nightmares. Speed Merchants, Spectacle, and the Struggle over Public Space in the World-Class City*. Toronto: Toronto University Press.

Luscombe, D. (1998), 'Promises', in R. Cashman & A. Hughes (eds), *The Green Games: A Golden Opportunity*, pp. 14–17. Sydney: University of New South Wales, Centre for Olympic Studies.

MacDonald, J. P. (2005), 'Strategic Sustainable Development using the ISO 14001 Standard', *Journal of Cleaner Production*, 13, pp. 631–43.

MacDonald, S. (2003), 'The 2008 Beijing Olympics: China Pushes towards Modernization', http://www.azete.com/preview/39204, accessed 20 June 2009.

MacKenzie, J. D. (2006), *Moving Towards Sustainability in the Olympic Games' Planning Process*. Burnaby, Canada: Simon Fraser University Library.

Madrigal, R., Bee, C. & M. LaBarge (2005), 'Using the Olympics and FIFA World Cup to Enhance Global Brand Equity', in J. Amis & T. B. Cornwell (eds), *Global Sport Sponsorship*, pp. 179–190. Oxford: Berg.

Mair, R. (2007), 'Let's Learn from Eagleridge Bluffs Protest', *The Tyee*, 5 March.

Mairie de Paris (2005), *Recommandations Environnementales pour les Acteurs de la Construction et de l'Aménagement*. Paris: Direction de l'Urbanisme / Direction des Parcs, Jardins et Espaces Verts.

Mairie de Paris (2006), *Concertation pour l'Aménagement du Secteur Clichy-Batignolles. Les Principaux Elements de l'Exposition Publique*. Paris: Mairie de Paris / Mairie du Dix-septième Arrondissement.

Mairie de Paris (2007), *Plan Climat de Paris. Plan Parisien de Lutte contre le Dérèglement Climatique*. Paris: Mairie de Paris / Conseil de Paris.

Malfas, M., Theodoraki, E. & B. Houlihan (2004), 'Impacts of the Olympic Games as Mega-Events', *Municipal Engineer*, 157/ME3, pp. 209–20.

Mangan, J. A. (1988), *Pleasure, Profit, Proselytism: British Culture and Sport at Home and Abroad, 1700–1914*. London: Frank Cass.

Manzenreiter, W. (2006), 'Sports Spectacles, Uniformities and the Search for Identity in Late Modern Japan', in J. Horne & W. Manzenreiter (eds) (2006), *Sports Mega-Events: Social Scientific Analyses of a Global Phenomenon*, pp. 144–59. Oxford: Blackwell.

Marcou, J. (2002), 'Préface', in Chabal P. & A. Raulin (eds), *Les Chemins de la Turquie vers l'Europe*. Arras: Artois Presses Université.

Marivoet, S. (2006), *Euro 2004™: Um Evento Global Em Portugal*. Lisbon: Livros Horizonte.

Marks, J. (1999), 'The French National Team and National Identity: 'Cette France d'un "bleu métis"', in H. Dauncey & G. Hare (eds), *France and the World Cup: The National Impact of a World Sporting Event*, pp. 41–57. London: Frank Cass.

Martin, D-C. (2002), 'Les OPNI, l'Essence du pouvoir et le Pouvoir des Sens', in D-C. Martin (ed.), *Sur la Piste des OPNI (Objets politiques non identifiés)*, pp. 11–45. Paris, Karthala.

Marx, K. (1867/1973), *Capital, Vol 1*. London: Penguin.

Matheson, V. & R. Baade (2003), *Mega-Sporting Events, Developing Nations: Playing the Way to Prosperity?* Working Papers 0404, College of the Holy Cross, Department of Economics, http://ideas.repec.org/p/hcx/wpaper/0404.html.

Mawhinney, M. (2002), *Sustainable Development. Understanding the Green Debates*. Oxford: Blackwell.

May, V. (1995), 'Environmental Implications of the 1992 Winter Olympic Games', *Tourism Management*, 16/4, pp. 269–75.

McAdam, D., Tarrow, S. & C. Tilly (2001), *Dynamics of Contention*. Cambridge: Cambridge University Press.

McGeoch, R. (1999), 'The Green Games: The Legal Obligations That have Arisen from the "Green" Bid', *The University of New South Wales Law Journal*, 22/3, pp. 708–20.

McNeill, D. (2004), *New Europe: Imagined Spaces*. London: Hodder Arnold.

Melucci, A. (1989), *Nomads of the Present. Social Movements and Individual Needs in Contemporary Society*. London: Hutchinson-Rodins.

Meyer, M. (2008), *Last Days of Old Beijing. Life in the Vanishing Backstreets and a City Transformed*. New York: Walker.

Migdal, J. (1988), *Strong Societies and Weak States: State-Society Relations and State Capabilities in the Third World*. Princeton: Princeton University Press.

Miller, D. (2005), *A Play Fair Alliance Evaluation of the WFSGI Response to the Play Fair at the Olympics Campaign*. http://www.sweatsoap.org/documents/05-04-PFOC-eva-WFSGI.pdf.

Milza, P. (1984), 'Sport et Relations Internationales', *Relations Internationales*, 38, pp. 155–74.

Milza, P. (2002), 'Un Siècle de Jeux Olympiques', *Relations Internationales*, 111, pp. 299–310.

Mitchell, D. (1995), 'The End of Public Space? People's Park, Definitions of the Public, and Democracy', *Annals of the Association of American Geographers*, 85/1, pp. 108–33.

Mol, A. P. J. (1995), *The Refinement of Production: Ecological Modernization Theory and the Chemical Industry*. Utrecht, Netherlands: Van Arkel.

Mol, A. P. J. (1996), 'Ecological Modernisation and Institutional Reflexivity: Environmental Reform in the Late Modern Age', *Environmental Politics*, 5/2, pp. 302–23.

Mol, A. P. J. (2006), 'Environment and Modernity in Transitional China: Frontiers of Ecological Modernisation', *Development and Change*, 37/1, pp. 29–56.

Mol, A. P. J. (2008), *Environmental Reform in the Information Age. The Contours of Informational Governance*. Cambridge: Cambridge University Press.

Mol, A. P. J. (2009), 'Environmental Governance through Information: China and Vietnam', *Singapore Journal of Tropical Geography*, 30/1, pp. 114–29.

Mol, A. P. J. & G. Spaargaren (2000), 'Ecological Modernisation Theory in Debate: A Review', in A. P. J. Mol & D. A. Sonnenfeld (eds), *Ecological Modernisation around the World. Perspectives and Critical Debates*, pp. 17–49. London: Frank Cass.

Mol, A. P. J., Sonnenfeld, D. A. & G. Spaargaren (eds) (2009), *The Ecological Modernization Reader: Environmental Reform in Theory and Practice*. London: Routledge.

Monbiot, G. (2007), *Untroubled by Democracy*, http://www.monbiot.com/archives/2007/03/20/untroubled-by-democracy/, posted 20 March 2007.

Muhovic-Dorsner, K. (2005), 'Evaluating European Climate Change Policy: An Ecological Justice Approach', *Bulletin of Science, Technology & Society*, 25/3, pp. 238–46.

Muñoz, F. (2006), 'Olympic Urbanism and Olympic Villages: Planning Strategies in Olympic Host Cities, London 1908 to London 2012', in J. Horne & W. Manzenreiter (eds), *Sports Mega-Events. Social Scientific Analyses of a Global Phenomenon*, pp. 175–87. Oxford: Blackwell.

Nerlich, B. & N. Koteyko (2009), 'Compounds, Creativity and Complexity in Climate Change Communication: The Case of "Carbon Indulgences"', *Global Environmental Change*, 19/3, pp. 345–53.

New South Wales Government Department of Planning (1993), *SEPP No. 38 – Olympic Games and Related Development Proposals*, http://www.legislation.nsw.gov.au/full-html/repealed/epi+549+1993+cd+0+Y, accessed 13 May 2010.

Ngonyama, P. (2010), 'The 2010 World Cup in South Africa: Critical Voices from Below', in P. Alegi & C. Bolsmann (eds), *South Africa and the Global Game: Football, Apartheid and Beyond*. London: Routledge.

No2010 (2007a), *Why We Resist 2010*. http://www.no2010.com/node/18, posted 13 March 2007.

No2010 (2007b), 'No Olympics on Stolen Land', http://www.no2010.com/node/19.

O'Brien, C. (2008), 'Protest Zones in Parks Empty; Permit Process seen as a Trap', *The Washington Times*, 17 August.

O'Toole, W. J. (2002), 'Towards the Integration of Event Management Best Practice by the Project Management Process', http://www-personal.usyd.edu.au/~wotoole/conf_paper.htm, accessed 30 September 2009.

O'Toole, W. J. & P. Mikolaitis (2002), *Corporate Event Project Management*. New York: Wiley.

Öko-Institut (2003), *Green Goal – Environmental Objectives for the 2006 FIFA World Cup*. Berlin: Öko-Institut.

Öko-Institut (2007), *Green Goal Legacy Report*. Berlin: Öko-Institut.

Olds, K. (1989), 'Mass Evictions in Vancouver: The Human Toll of Expo '86', *Canadian Housing*, 6/1, pp. 49–53.

Olds, K. (1998), 'Urban Mega-Events, Evictions and Housing Rights: The Canadian Case', *Current Issues in Tourism*, 1/1, pp. 2–46.

Ollitrault, S. (2008), *Militer pour la Planète: Sociologie des Ecologistes*. Rennes: PU de Rennes.

Ong, R. (2004), 'New Beijing, Great Olympics: Beijing and Its Unfolding Olympic Legacy', *Stanford Journal of East Asian Affairs*, 4/2, pp. 35–49.

Owen, K. A. (2001), *The Local Impacts of the Sydney 2000 Olympic Games: Processes and Politics of Venue Preparation*. Sydney: University of New South Wales, Centre for Olympic Studies.

Özal, T. (1988), *La Turquie en Europe*. Paris: Plon.

Papadakis, E. (2002), 'Environmental Capacity Building in Australia', in H. Weidner & M. Jänicke (eds), *Capacity Building in National Environmental Policy*, pp. 19–44. London: Springer.

Papadopoulos, A. G. & C. Liarikos (2007), 'Dissecting Changing Rural Development Networks: The Case of Greece', *Environment and Planning C: Government and Policy*, 25, pp. 292–313.

Papageorgiou, K. & I. N. Vogiatzakis (2006), 'Nature Protection in Greece: An Appraisal of the Factors Shaping Integrative Conservation and Policy Effectiveness', *Environmental Science & Policy*, 9/5, pp. 476–86.

Partidário, M. R. & J. Arts (2005), 'Exploring the Concept of Strategic Environmental Assessment Follow-up', *Impact Assessment and Project Appraisal*, 23/3, pp. 246–57.

Pelloux, P. (2005), 'Clichy-Batignolles, le renouveau d'une emprise sous-utilisée', *Paris Projet*, 36–37, pp. 110–-20. Paris: Atelier Parisien d'Urbanisme.

Pietsch, J. & I. McAllister (2010), ' "A Diabolical Challenge": Public Opinion and Climate Change Policy in Australia', *Environmental Politics*, 19/2, pp. 217–36.

Pinson, G. (2002), 'Political Government and Governance: Strategic Planning and the Reshaping of Political Capacity in Turin', *International Journal of Urban and Regional Research*, 26/3, pp. 477–93.

Placé, J. V. (2004), *Jeux Olympiques 2012 – Intervention Verte au Conseil Régional*, http://idf.lesverts.fr/article.php3?id_article=100, posted 2 November 2004, accessed 13 June 2007.

PlayFair2008 (2007), *No Medal for the Olympics on Labour Rights*. http://www.playfair2008.org/docs/playfair_2008-report.pdf.

Politix (2000), 'Sport et Politique', 50, special issue.

Polo, J-F. (2003), ' "A Côté du Tour": Ambushing the Tour for Political and Social Causes', *International Journal of the History of Sport*, 2, pp. 246–66.

Polo, J-F. (2005), 'Avrupa Fatihi. Les Enjeux Européens du Sport en Turquie', in C. Guionnet & L. Arnaud (eds), *Les Frontières du Politique*, pp. 209–32. Rennes: PU de Rennes.

Polo, J-F. & C. Visier (2006), 'Les Séismes de 1999 en Turquie, un Laboratoire de la Société Civile', in J-F. Pérouse & T. Coanus (eds), *Pouvoirs Urbains et Sociétés Urbaines Face aux Risques, sur le Pourtour de la Méditerranée*, pp. 79–98. Paris: Anthropos-Economica.

Pope, J., Annandale, D. & A. Morrison-Saunders (2004), 'Conceptualising Sustainability Assessment', *Environmental Impact Assessment Review*, 24, pp. 595–616.

Pound, R. W. (2003), *Olympic Games Study Commission: Report to the 115th IOC Session, Prague, July 2003*. Lausanne: International Olympic Committee.

Preuss, H. (1998), 'Problematizing Arguments of the Opponents of Olympic Games', in R. K. Barney, K. G. Wamsley, S. G. Martyn & G. H. MacDonald (eds), *Global and Cultural Critique: Problematizing the Olympic Games*, pp. 197–218. London & Ontario: International Centre for Olympic Studies.

Preuss, H. (2004), *The Economics of Staging the Olympics. A Comparison of the Games 1972–2008*. Cheltenham: Edward Elgar.

Pridham, G. & M. Cini (1994), 'Environmental Standards in the European Union: Is There a Southern Problem?', in M. Faure, J. Vervaele & A. Weale (eds), *Environmental Standards in the EU in an Interdisciplinary Framework*, pp. 251–77. Antwerp: Maklu.

Psaropoulos, J. (2008), 'Greece's Suspension from Kyoto', *Athens News*, 28 June.

Quenault, B. (2004), 'Le Développement Durable comme Pierre d'Achoppement des Relations Nord/Sud au Sein des Négociations Commerciales Multilatérales à l'Organisation Mondiale du Commerce', *Mondes en Développement*, 127, pp. 11–27.

Reasons, C. (1984), 'It's Just a Game? The 1988 Winter Olympics', in C. Reasons (ed.), *Stampede City: Power and Politics in the West*, pp. 123–45. Toronto: Between the Lines.

Relations Internationales (2002), 'Olympisme et Relations internationales", 111–12, double special issue.

Riou, A. (2004) 'Candidature de Paris pour l'Organisation des Jeux Olympiques et Paralympiques en 2012', Speech to the Conseil de Paris, 19 October 2004, http://www.lesvertsparis.org/article.php3?id_article=768, posted 21 October 2004.

Ritchie, J. R. B. (1984), 'Assessing the Impact of Hallmark Events: Conceptual and Research Issues', *Journal of Travel Research*, 23/1, pp. 2–11.

Ritzer, G. (1993), *The McDonaldization of Society*. Newbury Park, CA: Pine Forge.

Robèrt, K-H., Schmidt-Bleek, B., Aloisi de Larderel, J., Basile, G., Jansen, J. L., Kuehr, R., Price Thomas, P., Suzuki, M., Hawken, P. & M. Wackernagel (2002), 'Strategic

Sustainable Development – Selection, Design and Synergies of Applied Tools', *Journal of Cleaner Production*, 10, pp. 197–214.

Roche, M. (1994), 'Mega-Events and Urban Policy', *Annals of Tourism Research*, 21/1, pp. 1–19.

Roche, M. (2000), *Mega-Events and Modernity: Olympics and Expos in the Growth of Global Culture*. London & New York: Routledge.

Roche, M. (2003), 'Mega-Events, Time and Modernity: On Time Structures in Global Society', *Time and Society*, 12/1, pp. 99–126.

Roche, M. (2006), 'Nationalism, Mega-Events and International Culture', in G. Delanty & K. Kumar (eds), *Handbook of Nations and Nationalism Studies*, pp. 260–72. London: Sage.

Rolnik, R. (2009), *Report of the UN's Special Rapporteur on Adequate Housing as a Component of the Right to an Adequate Standard of Living, and on the Right to Non-Discrimination in This Context*. New York: United Nations General Assembly, A/HRC/13/20.

Rootes, C. (2008), 'The First Climate Change Elections? The Australian General Election of 24 November 2007', *Environmental Politics*, 17/3, pp. 473–80.

Rosenau, J. N. (1990), *Turbulence in World Politics: A Theory of Change and Continuity*. Princeton: Princeton University Press.

Routledge, P. (2003), 'Convergence Space: Process Geographies of Grassroots Globalisation Networks', *Transactions of the Institute of British Geographers*, 28, pp. 333–49.

Ruggiero, V. & N. Montagna (eds) (2008), *Social Movements: A Reader*. London & New York: Routledge.

Rutheiser, C. (1996), *Imagineering Atlanta. The Politics of Place in the City of Dreams*. London & New York: Verso.

Rydell, R. W. (1993), *World of Fairs. The Century-of-Progress Expositions*. Chicago: University of Chicago Press.

Saccomani, S. (1994), 'The Preliminary Project of Turin's Master Plan', in B. Dimitriou & M. J. Thomas (eds), *When the Factories Close*, Oxford Brooks University, School of Planning, WP No. 147, Oxford, pp. 6–25.

Sadd, D. & I. Jones (2009), 'Long-term Legacy Implications for Olympic Games', in R. Raj & J. Musgrave (eds), *Event Management and Sustainability*, pp. 90–8. Wallingford: CABI Publishing.

Sandvoss, C. (2009), *A Game of Two Halves. Football, Television and Globalization*. London: Routledge.

Sassen, S. (1994), *Cities in a World Economy*. Thousand Oaks, CA: Pine Forge Press.

Sassen, S. (2007), *A Sociology of Globalization*. New York: W. W. Norton & Co.

Sauzay, L. (1998), *Louis Pradel, Maire de Lyon: Voyage au Coeur du Pouvoir Municipal*. Lyon: Editions Lyonnaises d'Art et d'Histoire.

Scamuzzi, S. (2006), 'Winter Olympic Games in Turin: The Rising Weight of Public Opinion', in N. Müller, M. Messing & H. Preuss (eds), *From Chamonix to Turin. The Winter Games in the Scope of Olympic Research*, pp. 343–58. Kassel: Agon Sportverlag.

Schimmel, K. S. (2006), 'Deep Play: Sports Mega-events and Urban Social Conditions in the USA', in J. Horne & W. Manzenreiter (eds), *Sports Mega-Events: Social Scientific Analyses of a Global Phenomenon*, pp. 160–74. Oxford: Blackwell.

Searle, G. (2002), 'Uncertain Legacy: Sydney's Olympic Stadiums', *European Planning Studies*, 10/ 7, pp. 845–60.

Searle, G. & M. Bounds (1999), 'State Powers, State Land and Competition for Global Entertainment: The Case of Sydney', *International Journal of Regional Research*, 23/1, pp. 165–72.

Segre, A. (2002), 'L'Ambiente delle Olimpiadi di Torino 2006', *Bollettino Della Società Geografica Italiana*, 7/4, pp. 895–912.

SEPB (2000), *Environmental Impact Assessment of the Master Plan of Expo 2010*. Shanghai: Shanghai Environmental Protection Bureau (in Chinese).

Shaw, C. (2008), *Five Ring Circus: Myths and Realities of the Olympic Games*. Vancouver: New Society.

Shi, M. (1993), *Beijing Transforms: Urban Infrastructure, Public Works, and Social Change in the Chinese Capital, 1900–1928*. Ph.D. dissertation, Columbia University.

Shi, Q. & T. Gong (2008), 'Life-cycle Environmental Friendly Construction of a Large Scale Project: A Case Study of the Shanghai World Expo 2010', *Journal of Sustainable Development*, 1/3, pp. 17–20.

Short, J. R. (2004), *Global Metropolitan. Globalising Cities in a Capitalist World*. London: Routledge.

Shoval, N. (2002), 'A New Phase in the Competition for the Olympic Gold: The London and New York Bids for the 2012 Games', *Journal of Urban Affairs*, 24/5, pp. 583–99.

Smith, J. & H. Johnston (eds) (2002), *Globalization and Resistance. Transnational Dimensions of Social Movements*. Boston: Rowman & Littlefield.

Smith, J., Chatfield, C. & R. Pagnucco (eds) (1997), *Transnational Social Movements and Global Politics: Solidarity beyond the State*. Syracuse, NY: Syracuse University Press.

Smith, K. (2007), *The Carbon Neutral Myth. Offset Indulgences for Your Sins*. Amsterdam: Transnational Institute/Carbon Trade Watch.

SOCOG (1993), *Sydney Olympics 2000 Bid*. Sydney: Sydney Organizing Committee for the Olympic Games.

Sonnenfeld, D. A. (1998), 'From Brown to Green? Late Industrialization, Social Conflict, and Adoption of Environmental Technologies in Thailand's Pulp Industry', *Organization and Environment*, 11/1, pp. 59–87.

Spaargaren, G. & A. P. J. Mol (1992), 'Sociology, Environment and Modernity. Ecological Modernization as a Theory of Social Change', *Society and Natural Resources*, 5/5, 1992, pp. 323–45.

Spaargaren, G., Mol, A. P. J. & F. H. Buttel (eds) (2006), *Governing Environmental Flows. Global Challenges for Social Theory*. Cambridge, Mass.: MIT Press.

Spilling, O. R. (1996), 'Mega-Event as Strategy for Regional Development: The Case of the 1994 Lillehammer Winter Olympics', *Entrepreneurship & Regional Development*, 8/4, pp. 321–44.

Spilling, O. R. (1998), 'Beyond Intermezzo? On the Long-term Industrial Impacts of Mega-Events: The Case of Lillehammer 1994', *Festival Management & Event Tourism*, 5, pp. 101–22.

Streets, D. G., Fu, J. S., Jang, C. J. et al. (2007), 'Air Quality during the 2008 Beijing Olympic Games', *Atmospheric Environment*, 41, pp. 480–92.

Sugden, J. & A. Tomlinson (1998), *FIFA and the Contest for World Football*. Cambridge: Polity.

Swart, K. & U. Bob (2004), 'The Seductive Discourse of Development: The Cape Town 2004 Olympic Bid', *Third World Quarterly*, 25/7, pp. 1311–24.

Szarka, J. (2004), 'Sustainable Development Strategies in France: Institutional Settings, Policy Style and Political Discourse', *European Environment*, 14, pp. 16–29.

Tajima, A. (2004), ' "Amoral Universalism": Mediating and Staging Global and Local in the 1998 Nagano Olympic Winter Games', *Critical Studies in Media Communication*, 21/3, pp. 241–60.

Terret, T. (2004) 'Lyon, the City which Never Hosted the Olympic Games', in I. Okubo (ed.), *Sport and Local Identity. Historical Study of Integration and Differentiation*, pp. 238–44. Sankt Augustin: Academia Verlag.

Terret, T. (2010), 'The Albertville Winter Olympics: Unexpected Legacies – Failed Expectations for Regional Economic Development', in J. A. Mangan & M. Dyreson (eds), *Olympic Legacies, Intended and Unintended, Political, Cultural, Economic and Educational*, pp. 20–38. London: Routledge.

Thériault, J. Y. (1987), 'Mouvements Sociaux et Nouvelle Culture Politique', *Politique*, 12, pp. 5–36.

Thrift, N. (2004), 'Intensities of Feeling: Towards a Spatial Politics of Affect', *Geografiska Annaler Series B, Human Geography*, 86/1, pp. 57–78.

Thrift, N. (2005), ' "But Malice Aforethought": Cities and the Natural History of Hatred', *Transactions of the Institute of British Geographers*, 30/2, pp. 133–50.

Tian, Q. W. & P. Brimblecombe (2008), 'Managing Air in Olympic Cities', *American Journal of Environmental Sciences*, 4/5, pp. 439–44.

Tomlinson, J. (1999), *Globalization and Culture*. Cambridge: Polity.

TOROC (1998a), *Bid File*. Turin: Organizing Committee for the XX Olympic Winter Games Torino 2006, http://www.torino2006.it/ENG/OlympicGames/spirito_olimpico/approfondimenti_sostenibilita.html.

TOROC (1998b), *Greencard*. Turin: Organizing Committee for the XX Olympic Winter Games Torino 2006, http://www.torino2006.it/ENG/OlympicGames/spirito_olimpico/approfondimenti_sostenibilita.html.

TOROC (2002), *Carta di Intenti in tema di responsabilità sociale*. Turin: Organizing Committee for the XX Olympic Winter Games Torino 2006 http://www.torino2006.it/ENG/OlympicGames/spirito_olimpico/approfondimenti_sostenibilita.html.

Tresser, T. (2009), 'What Happened in Copenhagen (When We Went to Influence the IOC – Again)', *The Huffington Post*, 12 October, http://www.huffingtonpost.com/tom-tresser/what-happened-in-copenhag_b_316371.html.

Tumblety, J. (2008), 'The Soccer World Cup of 1938: Politics, Spectacles, and *la Culture Physique* in Interwar France', *French Historical Studies*, 31/1, pp. 77–116.

Uhlin, A. (2009), *Democratic Legitimacy of Transnational Actors: A Framework for Analysis*. Paper prepared for the ECPR Joint Sessions, Workshop 29: Civil Society, Democracy and Global Governance, Lisbon 14–19 April.

UNEP (2007), *Beijing 2008 Olympic Games – an Environmental Overview*. Nairobi: United Nations Environmental Programme.

UNEP (2009a), *Independent Environmental Assessment Beijing 2008 Olympic Games*, Nairobi: United Nations Environmental Programme.

UNEP (2009b), *UNEP Environmental Assessment Expo 2010 Shanghai, China*. Nairobi: United Nations Environmental Programme.

UNSD (2009), *Greenhouse Gas Emissions. CO_2 Emissions in 2006*, http://unstats.un.org/unsd/ENVIRONMENT/air_co2_emissions.htm, accessed 15 January 2010.

Urry, J. (2000), *Sociology beyond Society*. London: Routledge.

Urry, J. (2003), *Global Complexity*. Cambridge: Polity.

Van der Heijden, H-A. (1999), 'Environmental Movements, Ecological Modernisation and political Opportunity Structures', in C. Rootes (ed.), *Environmental Movements. Local, National and Global*, pp. 199–221. London & Portland, OR: Frank Cass.

VANOC (2003), *Vancouver 2010 Bid Book*, vol. 3. Vancouver: Vancouver Organizing Committee for the 2010 Olympic and Paralympic Winter Games.

VANOC (2009), *The Carbon Management Program for Vancouver 2010: Fact Sheet*. Vancouver: Vancouver Organizing Committee for the 2010 Olympic and

Paralympic Winter Games, http://www.vancouver2010.com/dl/00/19/23/vanoc-carbon-management-fact-sheet_60d-jJ.pdf.

Vanolo, A. (2004), 'The External Images of Helsinki and Turin: Representing High Technology and Industry Vocation', Research and Training Network Urban Europe, Working Paper No. 8, http://www.urban-urope.net/working/08_2004_Vanolo.pdf.

Ville d'Annecy (2009), *Annecy 2018 au Coeur de l'Esprit Olympique*. Dossier de presse, 23 June.

Villepreux, O. (2008), *Feue la Flamme: Pour en Finir avec les JO*. Paris: Gallimard.

Visier, C. (2009), 'La Turquie: Instrument de Politisation, Objet de Politisation', *European Journal of Turkish Studies* [Online], 9, http://ejts.revues.org/index3709.html, accessed 6 FFebruary 2010.

Vogel, E. F. (1971), 'Preserving Order in the Cities', in J. W. Lewis (ed.), *The City in Communist China*, pp. 75–93. Stanford, CA: Stanford University Press.

Waitt, G. R. (2001). 'The Olympic Spirit and Civic Boosterism: The Sydney 2000 Olympics', *Tourism Geographies*, 3, pp. 249–78.

Waitt, G. R. (2003), 'Social Impacts of the Sydney Olympics', *Annals of Tourism Research*, 30/1, pp. 194–215.

Walsh, B. (2009), 'Why Global Warming May be Fuelling Australia's Fires', *Time*, 9 February.

WCED (1987), *Our Common Future*. Geneva: World Commission on Environment and Development.

Weidner, H. (2002), 'Capacity Building for Ecological Modernization. Lessons from Cross-National Research', *American Behavioral Scientist*, 45/9, pp. 1340–68.

Westerbeek, H. & A. Smith (2003), *Sport Business, The Global Marketplace*. London: Palgrave.

Whitelegg, D. (2000), 'Going for Gold: Atlanta's Bid for Fame', *International Journal of Urban and Regional Research*, 24, pp. 801–17.

Whitson, D. (2004), 'Bringing the World to Canada: 'The Periphery of the Centre"', *Third World Quarterly*, 25/7, pp. 1215–32.

Whitson, D. & J. Horne (2006), 'Underestimated Costs and Overestimated Benefits? Comparing the Outcomes of Sports Mega-Events in Canada and Japan', *The Sociological Review*, 54/2, pp. 73–89.

Williams, B. A. & A. R. Matheny (1995), *Democracy, Dialogue and Environmental Disputes: The Contested Languages of Social Regulation*. New Haven: Yale University Press.

Wilson, H. (1996), 'What Is an Olympic City? Visions of Sydney 2000', *Media, Culture & Society*, 18/3, pp. 603–18.

World Bank (1997), *Clear Water, Blue Skies. China's Environment in the New Century*. Washington D.C.: World Bank.

WWF-Greece (2004), *Environmental Assessment of the Athens 2004 Olympic Games*. Athens: WWF.

Xie, L. (2009), *Environmental Activism in China*. London & New York: Routledge.

Yardley, J. (2007), 'Mr. Tool. And Beijing Speaks to the World', *New York Times*, 17 April.

Yarsüvat, D. & P-H. Bolle (eds) (2004), 'La Violence dans le Sport et le Fanatisme', Actes du séminaire, les Journées de Galatasaray, (Sporda Şiddet ve Fanatizm Semineri Notlan), Istanbul, 20–21 December 2002. Istanbul: Editions de l'Université de Galatasaray.

Yeoman, I., Robertson, M., Ali-Knight, J., Drummond, S. & U. McMahon-Beattie (eds) (2003), *Festival and Events Management: An International Arts and Culture Perspective*. Oxford: Butterworth Heinemann.

Yerasimos, S., Seufert, G. & K. Vorhoff (eds) (2000), *Civil Society in the Grip of Nationalism*. Istanbul: Orient-Institut-IFEA.

York, G. (2008), 'Olympic Preparations Curb Capital's Freewheeling Spirit; Tourism Declines as Beijing Cracks Down on Security Ahead of 'Killjoy Games" ', *The Globe and Mail*, 27 June.

York, R. & E. A. Rosa (2003) 'Key Challenges to Ecological Modernization Theory. Institutional Efficacy, Case Study Evidence, Units of Analysis, and the Pace of Eco-Efficiency', *Organization & Environment*, 16/3, pp. 273–88.

Yurdsever, A. N. (2003), '19 Mayis Nasil Bayram Oldu', *Toplumsal Tarih*, 113, pp. 34–7.

Zbicz, D. C. (2009), *Asia's Future: Critical Thinking for a Changing Environment*. Washington, DC: USAID/ The Woodrow Wilson International Center for Scholars.

Zhang, A. (2008), *The Environment, Beijing and the 2008 Olympic Games*. Beijing: Greenpeace.

Zhi, W. & Y. Liu (2008), 'Post-Event Cities', *Architectural Design*, 78/5, pp. 60–3.

Žižek, S. (2005), 'The Empty Wheelbarrow', *The Guardian*, 19 February.

Index